N. W. B. Clarke was formerly with the Building Research Station in England as a Senior Principal Scientific Officer.

BURIED PIPELINES

Plate 1. A CRITICAL TEST – 60 in. concrete pipe with a traditional unreinforced *in-situ* concrete surround in the testing machine.

The surround became so highly stressed, or even cracked at its thinnest sections by shrinkage of the site concrete, before any load was applied, as to make the estimation of the load carrying capacity of the composite structure very uncertain or even impossible. This discovery led to the unavoidable conclusion that this method of strengthening a large pipe is technically unsound in principle.

BURIED PIPELINES

A MANUAL OF STRUCTURAL DESIGN AND INSTALLATION

by

N. W. B. CLARKE, M.ENG., M.I.C.E., A.M.I. WATER E.

Latterly a Senior Principal Scientific Officer of the Building Research Station, Garston, Watford, Herts.

FOREWORD by

SIR FREDERICK LEA, C.B., C.B.E., D.SC., F.R.I.C.

Former Director of Building Research

MACLAREN AND SONS   LONDON

85334 010 2

BURIED PIPELINES
BY N. W. B. CLARKE

FIRST PUBLISHED 1968
BY MACLAREN AND SONS LTD

THE MACLAREN GROUP OF COMPANIES

7 GRAPE STREET
LONDON WC2

MADE IN GREAT BRITAIN
AT THE PITMAN PRESS
BATH

Foreword

Progress with a programme of investigation of the structural behaviour of pipes and underground pipelines, which was started some seventeen years ago at the Building Research Station at the request of the Ministry of Housing and Local Government, has been reported from time to time in a number of published papers and articles.

The author has now retired from the Building Research Station, but the interest aroused in the profession and the industry generally has been such as to encourage the hope that this condensation of the fruits of the investigation will be welcomed not only by engineers and designers, but also by pipemakers and pipe-laying contractors.

F. M. LEA
(*Former Director of Building Research*)

Author's Preface

The growing interest in a rational method of design and construction of buried pipelines, and the fact that the information on the subject is scattered widely, has encouraged the author to collect, correct, and expand the work of the past seventeen years into an inclusive manual which will supersede and to some extent amplify all previous Building Research Station publications on the subject and which, it is hoped, will be of convenient service to all branches of the industry. Close liaison and co-operation between designers, constructors and pipemakers is essential to the efficiency and economy of any pipeline, since good design and poor construction can be as dangerous and expensive as poor design and good construction.

The estimation of the various crushing loads which may be imposed on a conduit is essentially an exercise in soil mechanics as was first realised by Professor Marston of Iowa in about 1913. Soil specialists are not entirely satisfied with the theory which he and his two famous colleagues Professors Spangler and Schlick developed, but admit that they cannot at present suggest any better practical guidance. The next move lies very largely with them.

These loads, and the ability of pipes to carry them, can both vary widely, depending on the method of installation of the conduit and the crushing strength of the pipes respectively. It is not surprising, therefore, that the survival of a conduit installed by the older traditional method, in which neither the crushing strength of the conduit nor the load imposed upon it were even estimated, is a matter of chance (Plate I). All the more so as the many other loads and forces which might be exerted on the conduit (and which do not appear in the Marston theory) were neither properly understood nor provided for.

Large steel pipes are being increasingly used as culverts under motor roads, and as protection to other conduits under railways. They have long been used as pressure pipes for many purposes. The introduction of a method of design, based on a new concept of the pipe-soil system, and its effect on the critical buckling stress in the pipes, should therefore need no apology. The suggested method of estimating the loads on each of several parallel rigid or flexible pipes under an embankment is also new.

The mathematical derivations of the values of the various load coefficients are given for the benefit of those readers who may be interested in fundamentals, and advantage has been taken of the availability of a computer to check all the embankment load coefficients and to prove that for the incomplete positive or negative conditions they are linear with respect to the ratio H/B_c or H/B_d, and to derive their linear equations for use in extrapolating their values for any value of H/B_c or H/B_d.

The suggestions for the rational design of flexible joints are tentative as regards the minimum draw to be provided. The remaining geometry depends on the length of the initial axial gap between spigot and socket, which is still controversial, and which depends, to some extent, on topography and soil characteristics.

Corrosion effects were felt to be too wide and specialised to be discussed here. The scope of the subject, which is essentially chemical, is indicated by Table Q and the brief bibliography on 'Corrosion and Protection', p. 215, but the engineer faced with such problems would be well advised to consult a specialist chemist.

The worked examples given in Appendix A are included to illustrate the use of the various formulae and design charts in detail. The working can be considerably shortened when the reader is fully familiar with the method and especially so with the help of N.B.S. Special Report No. 37[13] for pipes other than those laid under embankments and valley fills. The latter are dealt with in Chapter 8.

The Author would express his grateful acknowledgements—

to the pioneer work in this subject, extending over the past 50 years or more, by the late Dean Anson Marston and the late Professor W. J. Schlick, and to the great and continuing contribution by Professor M. G. Spangler, all of the Engineering Experiment Station of Iowa State University, U.S.A., whose work has been freely drawn upon in the compilation of this manual; also to Professor Spangler for help with a number of difficulties and permission to reproduce a number of his diagrams and charts.

He would also express his indebtedness—

to Professors D. L. Holl of the Iowa State University, N. M. Newmark of the University of Illinois and R. E. Fadum of Harvard University, for their work affecting surcharge loading and to them and Messrs John Wiley and Sons for permission to reproduce Table C and Chart C5;

to Professor Heger of the Massachusetts Institute of Technology for his assistance with the combined loading of reinforced concrete pressure pipes;

to Professor Meyerhoff of the University of Nova Scotia for permission to include the results of the work on steel pipes recently carried out by himself and his colleagues;

to Dr. Stutterheim for Plate 4;

to the A.S.C.E. and the W.P.C.F. for their many valued publications and permission to reproduce the material of Charts C9 and C10;

to the A.C.P.A. for Chart C1 and the material of Fig. 4.5;

to the U.S. Highway Research Board for many valuable papers;

to the A.W.W.A. for their publications on many associated topics and especially on the design of prestressed concrete, cast iron, and steel pressure pipes;

to the successive chief engineers of the Ministry of Housing and Local Government for stimulation and encouragement in the investigation and patience with the results;

to British Railways for help with railway loading;

to various Local Government engineers for the anonymous photographs of failures and for their advice, encouragement and support;

to the Road Research Laboratory and the British Rubber Producers Research Association for valuable technical advice and help;

to many consultants for their continuing interest in the subject;

to the Institutions of Civil Engineers and Public Health Engineers for their help in publishing and discussing former papers;

to the British makers of all kinds of pipes for much valuable practical advice, guidance and support;

and to the Controller of Her Majesty's Stationery Office for permission to reproduce Crown Copyright material.

Appreciative thanks are offered to the former Chairman of the Building Research Board of the Dept. of Scientific and Industrial Research, Sir Herbert Manzoni, to the former Director of Building Research, Sir Frederick Lea, and to senior colleagues at the Building Research Station for their constant encouragement and help throughout the investigation. The help of Drs L. F. Cooling and W. H. Ward soils matters and of junior colleagues at all times with many details is also gratefully acknowledged and special thanks are extended to Mr E. W. Watson and Mrs E. M. Sobolev for their computer and statistical work, to Mr D. C. Teychenné for Table U, to Mrs E. J. Cibula for long and patient editorial assistance, to Messrs J. Rogers and D. F. Gifford for filmwork, to Mr S. A. Jones for help with drawings and diagrams and to Mrs Ranscombe, Mrs Allsop and the ever helpful typists.

Finally he gladly acknowledges his indebtedness to, and great appreciation of, his immediate colleagues on the research team, Messrs O. C. Young and J. H. Smith, for their invaluable help in the experimental work, with the film and for fruitful discussions throughout the investigation, also for their essential work on the evaluation of wheel loads and the Charts C6, 7 and 8, and the former for his work on bedding theory. Without their enthusiastic and sustained support over the years this Manual could not have been produced.

Garston, Watford, Herts.
July 1967

Contents

SECTION E SECONDARY LOADS AND FORCES—FLEXIBLE JOINTS

List of Plates

List of Figures in the Text

Fig. No.

List of Design Charts

BURIED PIPELINES

List of Tables

List of Worked Examples in Appendix A

Notation

In the following notes the word 'conduit' includes both circular and non-circular sections. The word 'pipe' is used specifically for circular sections.

SYMBOLS

		Unit
A	Area of steel per inch of length of a pipe wall.	in^2
a	Width of annular gap between spigot and socket in a flexible joint.	in.
B	Breadth of a rectangle.	ft
B_c	Overall diameter or width of a conduit.	ft
B_{cA}, B_{cB}	Overall diameter or width of individual pipes, *A*, *B*, etc.	ft
B_d	Width of a trench at the top of a conduit, i.e. effective width.	ft
B_{dA}, B_{dB}	Effective width of a trench for individual pipes, *A*, *B*, etc.	ft
B_t	Effective width of a tunnel or heading.	ft
b	Width of contact area of rubber 'O' ring with surface of spigot or socket in a flexible joint.	in.
C_{BS}	Crushing test conversion factor for rigid pipes.	—
C_c	Load coefficient for fill load, positive projection and 'wide' trench conditions.	—
C_c'	Load coefficient for fill load, plus uniform surcharge load in conditions as for C_c.	—
C_{cs}	Load coefficient for fill load in positive projection and 'wide' trench conditions, when submerged.	
C_d	Load coefficient for fill load in 'narrow' trench conditions.	—
C_{ds}	Load coefficient for fill load in 'narrow' trench conditions, when submerged.	—
C_n, C_n'	Load coefficients, for fill load, negative projection and induced trench conditions.	—
C_{ns}	Load coefficient for fill load, negative projection and induced trench conditions, when submerged.	—
C_t	Load coefficient for a single concentrated surcharge, uncorrected.	—

C_{us}	Load coefficient for uniformly distributed surcharge of large extent, for 'narrow' trench conditions.	—
C_{sus}	Load coefficient for a static uniform surcharge of limited extent, in all conditions.	—
C_{sust}	Load coefficient for a transient surcharge of limited extent, in all conditions.	—
c	Radial compression of a rubber 'O' ring in a flexible joint.	in.
D	Internal diameter of a pipe.	ft or in.
d	Uncompressed diameter of a rubber 'O' ring in a flexible joint.	in.
d_c	Vertical deflection of a flexible pipe.	in.
E	Young's Modulus.	lb/in^2
e	Base of natural logarithms.	—
E_s	Modulus of deformation of a soil = 1·5 Rk.	lb/in^2
F_{lag}	Deflection lag factor for flexible pipes.	—
F_i	Impact factor for live loads.	—
F_L	Live load correction factor.	—
F_m	Trench bedding factor.	—
F_p	Bedding factor for positive projection, induced trench, and very wide trench conditions.	—
F_s	Safety factor for rigid non-pressure pipes against crushing.	—
F_{se}	Safety factor for rigid pressure pipes against crushing.	—
F_{si}	Safety factor for rigid pressure pipes against bursting.	—
f_a	Permissible compressive hoop stress in a thin steel pipe wall.	lb/in^2
f_c	Critical hoop stress in a steel pipe wall.	lb/in^2
f_s	Tensile hoop stress in a steel pressure pipe.	lb/in^2
f_y	Yield stress for steel in a pipe wall.	lb/in^2
H	Depth of cover (not submerged) above the top of a conduit.	ft
H'	(i) $= H + U_s/\gamma_s$. Equivalent depth of cover including uniform surcharge of large extent in 'wide' trench and positive or negative projection conditions.	ft

H'	(ii) $= H - p'Bd$, depth of cover above natural ground level, negative projection.	ft
H_e	Height of plane of equal settlement above top of conduit.	ft
H_e'	Height of plane of equal settlement above natural ground level, negative projection.	ft
H_s	Depth of water above the top of a submerged conduit.	ft
h	(i) Depth of an element of fill below the surface.	ft
	(ii) Hardness of a rubber sealing ring in a flexible joint.	B.S. degrees
I	Moment of inertia of pipe wall per inch of length.	in^4
I_σ	Influence value for concentrated surcharge loads or for uniform surcharge loads of small extent (see Chart C5).	—
K	Rankine's coefficient of lateral earth pressure. $= (\sqrt{\mu^2 + 1} - \mu)/(\sqrt{\mu^2 + 1} + \mu)$.	—
K_c	Constant of soil reaction for a clay soil.	$lb/in^2/in.$
K_s	Constant of soil reaction for a sandy soil.	$lb/in^2/in.$
K_θ	Bedding constant for flexible pipes.	—
k	Coefficient of soil reaction for a flexible pipe.	$lb/in^2/in.$
L	Length of an individual pipe, or of a rectangle.	ft
L_n	Effective length of an individual flexibly jointed pipe.	ft
m	(i) Exposure ratio used in computing F_p (see Chart C10).	—
m	(ii) Poisson's ratio.	—
m	(iii) $= B/H$ (see Chart C5) used for determination of I_σ.	—
n	(i) $= L/H$ (see Chart C5) used for determination of I_σ.	—
n	(ii) = void ratio for a coarse granular soil $= \dfrac{\text{Vol. of voids}}{\text{Total Vol.}}$	—
N, N'	Factors used in computing F_p.	—
P, P_1, P_2	Magnitude of individual concentrated surcharges.	lb
p	Projection ratio, positive projection and wide trench condition.	—

Symbol	Definition	Units
p'	Negative projection ratio, negative projection and induced trench conditions.	—
p_c	Critical pressure (collapse pressure) of a thin steel pipe.	lb/in²
p_e	External fluid pressure acting on a pipe wall.	lb/in²
p_f	Friction head in a pumping main.	lb/in²
p_g	Effective external ground water or submergence pressure.	lb/in²
p_h	Maximum static internal water pressure in a pressure pipe.	lb/in²
p_i	Effective internal fluid pressure in a pressure pipe line.	lb/in²
p_o	The unit pressure produced at the top of a flexible pipe by external loading other than fluid pressure.	lb/in²
p_{oc}	The internal pressure required to produce zero compressive ring stress in a prestressed concrete pressure pipe wall.	lb/in²
p_{peak}	Maximum radial pressure imposed on a pipe socket in a flexible joint.	lb/in²
p_s	Maximum surge pressure in a pressure pipe line.	lb/in²
p_t	Maximum internal test pressure in a pipe line.	lb/in²
p_{ult}	Bursting pressure of a rigid pipe when carrying no external load.	lb/in²
p_v	Velocity head in a pressure pipeline.	lb/in²
p_{vac}	The negative pressure induced in a pressure pipeline by a partial vacuum.	lb/in²
p_w	Working internal pressure in a pressure pipeline.	lb/in²
q	Ratio of horizontal load/vertical load on a conduit, used in computing F_p.	—
R	External radius of a smooth steel pipe, or mean radius of a corrugated steel pipe.	in.
R_n, R_m	Min. radius of curvature of a flexibly jointed pipeline.	ft
r_{sd}	Settlement ratio for positive or negative projection or wide trench conditions.	—
S	Max. permissible angle of slew in a flexible joint.	degrees
S_d	Settlement of the fill in the height $p'B_d$, negative projection.	in.

S_f	Settlement of the pipe invert, positive or negative projection.	in.
S_g	Settlement of the natural ground, positive or negative projection.	in.
S_m	Settlement of the fill in the height pB_c, positive projection.	in.
t	Pipe wall thickness.	in.
t_i	Wall thickness required to resist internal pressure in a flexible pressure pipe.	in.
t_e	Wall thickness required to resist external pressure in a flexible pipe.	in.
U_s	Intensity of uniform surcharge of large extent.	lb/ft²
U_{sus}	Intensity of static uniform surcharge of small extent.	lb/ft²
U_{sust}	Intensity of transient uniform surcharge of small extent.	lb/ft²
V	The load on an element of fill at depth h in the interior prism, positive or negative projection condition.	lb/lin. ft
V'	The load on an element of fill at depth h in the exterior prism, positive or negative projection condition.	lb/lin. ft
W_c	Magnitude of load caused by fill in 'narrow' trench, negative projection, or induced trench conditions.	lb/lin. ft
W_c'	Magnitude of load caused by fill in positive projection and 'wide' trench conditions.	lb/lin. ft
W_c''	Magnitude of load caused by positive or negative projection or 'wide' trench fill plus uniform surcharge of large extent.	lb/lin. ft
W_{cs}	Magnitude of load caused by a concentrated surcharge, uncorrected.	lb/lin. ft
W_{csu}	Magnitude of load caused by concentrated surcharge corrected for non-uniform distribution and worst position of the wheels.	lb/lin. ft
W_e	Magnitude of total effective external design load for non-pressure pipes.	lb/lin. ft
W_e'	Magnitude of total effective external design load for pressure pipes as affected by internal pressure.	lb/lin. ft

W_f	Magnitude of load which a bedded rigid pipe can carry when buried $= W_T C_{BS} F_m$ (field strength).	lb/lin. ft
W_L	Magnitude of load caused by 2 loads of 7,000 lb each at 3 ft centres, excluding impact effects.	lb/lin. ft
W_o	Magnitude of load caused by B.S. 153 Type HB wheel loading, excluding impact effects.	lb/lin. ft
W_p	Magnitude of load caused by self weight of pipe.	lb/lin. ft
W_R	Initial radial load on pipe socket imposed by ring compression in a flexible joint.	lb/in.
W_{sus}	Magnitude of load caused by a static uniform surcharge of U_s lb/ft² of limited extent.	lb/lin. ft
W_{sust}	Magnitude of load caused by a transient uniform surcharge of U_s lb/ft² of limited extent.	lb/lin. ft
W_T	Safe crushing test strength of a rigid pipe.	lb/lin. ft
W_{us}	Magnitude of load caused by a uniform surcharge of U_s lb/ft² and of large extent in narrow trench, negative projection, or induced trench conditions.	lb/lin. ft
W_{us}'	Magnitude of load caused by a uniform surcharge of U_s lb/ft² of large extent in positive projection or 'wide' trench conditions.	lb/lin. ft
W_w	Magnitude of load caused by water contained in a pipe.	lb/lin. ft
X	Horizontal distance between parallel pipes.	ft
x, x'	(i) Factors used in computing F_p	—
x, x_1, x_2	(ii) Height of top of pipe above or below natural ground surface, see Fig. 4.7.	ft
Y	(i) Depth of top of lower pipe below trench bottom of upper pipe (see Fig. 4.13).	ft
	(ii) Minimum thickness of granular bed for a rigid pipe. (See Charts C9 and C10.)	in.
z	Initial width of horizontal gap between spigot and socket in a flexible joint.	in.
Δ_x	Deflection of the horizontal diameter of a flexible pipe.	in.
Δ_y	Deflection of the vertical diameter of a flexible pipe.	in.
γ_d	Drained density of coarse grained soils $= \gamma_s - n\gamma_w$.	lb/ft³
γ_s	Saturated density of soil.	lb/ft³
γ_w	Density of water $= 62{\cdot}4$ lb/ft³.	lb/ft³

Σ	Sign of summation.	—
ϕ	Angle of friction of a soil sliding on soil.	degrees
ϕ'	Angle of friction of fill sliding on undisturbed soil.	degrees
σ	Intensity of vertical stress.	lb/in^2 or lb/ft^2
θ	Bedding angle for flexible pipes.	degrees
μ	Coefficient of internal friction of the fill material $= \tan \phi$.	—
μ'	Coefficient of sliding friction of fill on undisturbed soil $= \tan \phi'$.	—
μ_r	Coefficient of static friction of rubber on the surface of a pipe.	

Abbreviations

ACPA	American Concrete Pipe Association.
ASCE	American Society of Civil Engineers.
ASTM	American Society for Testing Materials.
AWWA	American Waterworks Association.
H.R.L.	Hydraulics Research Laboratory. (GB)
I.S.O.	International Standards Organisation.
M.W.B.	Metropolitan Water Board. (London)
NRPRA	Natural Rubber Producers' Research Association. (GB)
WPCF	Water Pollution Control Federation. (USA)

Pipelayers' Song

Here's a health to the pipe, may it stay one piece,
To cracking an end, to strength increase!
Come, let's help it bear its load,
Flex its joints, 'neath bumpless road.
And they who would this help deny,
In a sticky clay trench,
Down in mucky water,
Down, down, down, down,
Stuck in muddy water,
Let them lie!

To the air of '*Down among the dead men*'
(John Dyer, 1700–58)

Section A

Introduction
Basic Principles of Design
Definitions

Chapter 1
Introduction

1.1 Scope of the Manual

The purpose of this Manual is to present in convenient and readily accessible form the most widely used methods of rational structural design of underground pipelines and the complementary methods of their construction. It collects the essential theoretical and practical matter from a vast literature and presents all the essential design formulae, graphs and other data for both *rigid* and *flexible* conduits, when laid under various conditions in trenches, headings or by thrust boring, or under embankments or valley fills, to enable the safe load carrying capacity of a conduit and the loads it must carry to be estimated and equated with sufficient accuracy for most practical purposes.

The detailed methods of computation of both loads and pipe strength are given and illustrated by worked examples. Complementary methods of construction are described in some detail.

1.2 Historical Notes

A programme of work on underground pipelines was begun some years ago at the Building Research Station, to investigate their structural behaviour and, if possible, to rationalise their design. This involved a preliminary survey of the methods commonly adopted in the design and construction of sewer pipelines, a review of the extensive literature, investigation of many failures in service, and numerous laboratory tests. As a result, it has been possible to draw some general conclusions about pipeline design and to make the specific recommendations given in the following chapters.

It was found that failures occurred, not infrequently in British practice, in new sewer pipelines during, or shortly after, backfilling, and sometimes even before the application of any load, and that their causes were not fully understood. These pipelines, however large, were not 'designed' in the structural sense of relating strength to imposed load, but were laid in accordance with current regulations, codes of practice, or traditional methods. The effects of thermal and moisture changes, and of ground movements and differential settlements, were largely ignored, except in mining areas, and little attention was given to essential site operations, such as the quality and placing of site concrete and the proper compaction of the backfill. It was also found that in the relevant British Standards for sewer pipes (as distinct from pressure pipes) little or no attention had been given to requirements for crushing strength, and the relevant codes of practice gave insufficient attention to pipe design and laying practice.

More rational methods of design and construction had been developed, mainly in the United States of America. They were concerned with the estimation of the primary vertical loading imposed on, and the load-carrying capacity of, the pipes under various bedding conditions. They differentiated between trench and embankment loading and between the behaviour of 'rigid' (e.g. cast iron, concrete, clayware) and 'flexible' (e.g. steel, pitch fibre, plastics) pipes. However, they paid little attention to the secondary loads and forces imposed on the pipes by thermal and moisture changes, ground movements and differential settlement which are responsible for most of the fractures occurring in the smaller sizes of rigid pipes.

'Rigid' pipes are weak in tension and crack under very small deformation. Investigations of failures revealed that the major causes of cracking were:

1. Inadequate load-carrying capacity of the pipes even when provided with *in-situ* concrete haunches or surrounds (Plates 1, 2, 3).
2. Uneven or inadequate bedding (Plates 12, 13).
3. Ill-conceived construction methods which adversely affect either the load on, or the load-carrying capacity of, the pipes, or both (Plate 11).
4. The use of rigid jointing materials, and the consequent lack of axial flexibility and extensibility in the pipeline (Plate 6).
5. Differential thermal deformations or moisture movements in the body of the pipes (Plates 9, 10).
6. Differential settlement of the pipes and the structures to which they are connected (manholes, buildings, etc.).

'Flexible' pipes do not crack under slow deformation even if this is large, but fail by buckling or flattening if they are inadequately supported laterally. Field experience in Great Britain with this type of sewer pipe is limited so far and is confined largely to pitch fibre and uPVC (unplasticated poly vinyl chloride), in which failure has been caused mainly by errors or carelessness in construction. Steel pressure pipes have been used for many years.

Both types of pipe, and their jointing or bedding materials, are subject to chemical attack and, however well pipelines have been designed and constructed, they may fail in time from this cause. Thus sulphates, in the soil or groundwater, may attack concrete pipes externally, whilst, under adverse conditions, such pipes may be severely attacked internally above the waterline by sulphuric acid derived by bacterial action from sludge and slime in the sewers or, more generally, by aggressive industrial effluents (Plate 4).

As a preliminary to establishing the loads imposed on pipes by fill and surface surcharges, a careful study was made of the work of Marston, Spangler and Shlick of the University of Iowa, U.S.A., extending over the previous 40 years.[16,17,19,21,22] It was established that their methods of computing these loads, whilst still empirical to some extent, and not completely satisfying on theoretical grounds, were practically sound and that, pending a more fundamental soil-mechanics approach, they could be accepted. A theoretical and experimental check of the effects of various types of bedding confirmed the Iowa findings.

Crushing tests on concrete and clayware pipes also confirmed Marston's view that theory could not be relied upon to predict the crushing test strength of rigid pipes, still less that of the composite construction when such pipes are surrounded with site concrete. This cast suspicion on the latter practice as commonly used in the United Kingdom, particularly as the crushing strength of such structures could not be tested after laying. Attention was therefore turned to alternative methods of strengthening concrete pipes. These included thickening the wall, reinforcing with steel or glass fibre, or prestressing with steel. All of these alternatives could be tested before laying and all gave more consistent crushing strengths in laboratory tests than the composite construction. Moreover, they all depended on factory control rather than site control, for their inherent strength.[5]

The relation of jointing methods to fractures was closely studied and the beneficial effect of flexible jointing was established.[81],[81a]

Meanwhile, the behaviour of plain concrete pipes surrounded with plain *in situ* concrete was studied experimentally. It was found that in the larger sizes of pipe, consistent crushing strengths could not be obtained even with rigorous control of the site concrete. This variation was eventually traced to the effects of thermal and drying shrinkage in the site concrete. High tensile strains, sometimes approaching the ultimate, were found at the pipe springings before the pipes were loaded[31] (Plate 1). This discovery threw light on a number of failures involving both crushing and transverse fractures which had been observed in pipelines of this type, and finally condemned the method as unsound in principle for large pipes. During the course of this experimental work, Young found that the strengthening effect of concrete cradles could be greatly increased by reinforcing them and increasing their depth and peripheral extent.[39]

This work further emphasised the advisability of strengthening pipes in the factory rather than in the field and led to a request, supported by the manufacturers, for the revision of the British Standards for both concrete and glazed vitrified clayware pipes to include specified minimum crushing test strengths increasing with diameter and in two or more strength classes. This has now been achieved, and 'extra strength' rigid pipes of clayware, concrete and asbestos-cement are now available, thus enabling rational designs to be made without difficulty. See Tables H and J.

With the active support of the Ministry of Housing and Local Government, attention was directed to the major problems of revision of the relevant codes and ministerial requirements for sewer pipe design. A greatly simplified presentation of the method of design was devised which would enable rational designs to be prepared by small local authorities without the rather tedious computations involved in the method proper, though possibly with some sacrifice of economy.[10] This was a 'stop-gap' and more precise and economical methods of load estimation together with comprehensive load charts were subsequently prepared.[13] Designers are recommended to use the new charts[13] and this Manual wherever possible.

In conjunction with the Road Research Laboratory, the Ministry of Transport and the British Transport Commission, attention was given to estimating the loads imposed on pipelines under roads and railways by vehicles and locomotives.* For the heaviest type of road transport in England and Wales it was found to be necessary to consider the effects of eight-wheel loads acting simultaneously. This computation proved to be exceedingly tedious and an attempt was made to do it once and for all, and to present the results in the form of design graphs.[11] Fortunately, it was found that this road loading was also adequate for 1961 railway loading. The complicating factor is that under some conditions, these live loads are not uniformly distributed across the width of the pipe. Analysis showed that Spangler's method of dealing with this effect could be too conservative under some conditions; a method of computing equivalent uniform loads was therefore devised and applied in the preparation of the design graphs referred to above[11] and included here as Charts C6, 7, and 8.

The load-carrying capacity of a pipeline depends largely on the manner in which it is constructed. As already mentioned, many failures in service have been due to inadequate or ill-conceived construction methods. It is essential that construction should complement design. To illustrate the essential principles of the new construction technique for inspectors, foremen and pipelayers, a film entitled *Modern methods of Pipelaying* showing the appropriate site operations has been prepared and is available on

* But see Table D for the latest recommendations contained in Ref. 10*a*.

loan from the Film Library of the Ministry of Public Buildings and Works or by purchase from the Building Research Station.

The use of flexible joints for all the commonly used kinds of rigid pipes has increased rapidly in recent years and more and more pipe manufacturers are producing joints suitable for use with the new methods of pipeline construction, either as variants of types already in use, or as fresh designs. A survey of such joints has led to suggestions for a more rational approach to their design and for the revision and extension of the appropriate British Standard.[81a, 81b]

Finally, a study has been made of the loads imposed on pipelines under embankments[14] (see Chapter 4) and of the most recent proposals for the design of thin walled steel (flexible) conduits [43] [44] (see Chapter 6).

Chapter 2
Basic Principles of Design and Definitions

2.1 Basic Principles of Design

2.1 (*a*) A buried pipeline, together with its soil environment, is a structure which is subject to the same laws of mechanics as any other structure. This means:

1. That the load-carrying capacity of a conduit must be sufficient for the maximum load to be imposed upon it with an adequate factor of safety against fracture or excessive deformation.
2. That the design of the wall thickness of '*rigid*' non pressure and pressure pipes is based on the resistance of the pipe wall to the circumferential bending moments induced in it by the external loads; together with the circumferential tensile stress caused by the internal pressure if any.
3. That the design of '*flexible*' non pressure and pressure pipes is based on the yield strength of the material of the pipe wall in compression and on its resistance to circumferential buckling under the restraining forces exerted by the surrounding soil which restrict its vertical and horizontal deflection: and in addition, for pressure pipes, on the yield strength of the wall material in tension.
4. That the safety factor for rigid pipes is applied to their cracking or crushing test strength, whilst that for flexible pipes is applied to their vertical or horizontal deflection, and the critical (buckling) pressure or yield strength.
5. That the pipeline must be, or must be so constructed as to be, sufficiently flexible to conform to all the soil, thermal or moisture movements to which it may be subjected without loss of efficiency for its purpose, i.e. the conveyance of fluids without leakage either outwards or inwards.

2.1 (*b*) All the materials used in pipes, joints and bedding must be, or must be so protected as to be, sufficiently durable in the external and internal environments to which the pipeline may be subjected to ensure that structural stability and watertightness will not be impaired throughout the expected, and usually long, life of the pipeline.

2.1 (c) Construction methods must be such as to ensure that the assumptions made in design are realised as nearly as possible in practice.

2.1 (d) Of the optional methods of installation, the least expensive under any specific local conditions should usually be selected.

2.1 (e) It is usually preferable, either by design or construction, to eliminate a source of stress wherever possible rather than to make provision for its resistance.

2.1 (f) In estimating fill loads, the great variability of soil conditions must be considered and those giving the greatest load (i.e. the greatest soil density and the least favourable friction) in any given length of the pipeline must be selected and provided for.

2.1 (g) In view of the variability in soil characteristics, pipe strength and site workmanship, great precision in design is not to be expected and factors of safety should be selected with these variables in mind. Too much reliance should not be placed on theoretical values, unsupported by knowledge of the site and by good engineering judgement.

2.2 Definitions

2.2 (a) Trench Conditions

'NARROW' TRENCH AND 'NARROW' SUB-TRENCH. In trench and negative projection conditions, a trench or sub-trench in which the effective trench width (B_d) is equal to or less than the transition width for the given cover depth (H) and value of B_c and in which $W_c = C_d \gamma_s B_d^2$ (see Fig. 3.1).

'WIDE' TRENCH AND 'WIDE' SUB-TRENCH. A trench or sub-trench in which the effective trench width (B_d) is equal to or greater than the transition width for the given value of B_c and cover depth (H), and is less than $\{B_c + 2(H + mB_c)/\mu\}$ (and for which trench bedding factors F_m must be used) and in which

$$W_c' = C_c \gamma_s B_c^2 \text{ (see Fig. 4.8)}$$

'VERY WIDE' TRENCH AND VERY WIDE SUB-TRENCH. A trench or sub-trench in which the effective width (B_d) is equal to or greater than $\{B_c + 2(H + mB_c)/\mu\}$ (and in which embankment bedding factors (F_p) may be used) (rare), but $W_c' = C_c \gamma_s B_c^2$.

V TRENCH. A trench with strongly sloping self-supporting sides, with or without a sub-trench, increasingly used in open country in order to dispense with timbering (see Fig. 3.1).

EFFECTIVE TRENCH WIDTH (B_d). The overall width of a trench or sub-trench between the undisturbed soil in the trench sides, measured at (i) the level of the top of the conduit in trench conditions; (ii) at ground level in negative projection conditions; (iii) at the top of the initial fill in induced trench conditions (see Fig. 4.2).

TRANSITION WIDTH. (i) The effective trench width (B_d) at which for a given cover depth (H) the fill load (W_c) computed by the 'narrow' trench formula is equal to the fill load (W_c') computed for the given value of B_c by the 'wide' trench formula. At this width, the fill load is a maximum for the given cover depth and $W_c' = C_c \gamma_s B_c^2$ irrespective of greater trench width (see Figs. 4.9 (*a*) and (*b*)). (ii) The effective sub-trench width (B_d) at which for a given cover depth (H) and value of B_c the fill load (W_c) computed by the negative projection formula is equal to the fill load (W_c') computed by the positive projection formula.

TRANSITION DEPTH. The cover depth (H) for a given effective trench width (B_d) and value of B_c at which the 'narrow' trench fill load (W_c) equals the 'wide' trench fill load (W_c'). Thus a parallel-sided trench of uniform width may be 'wide' down to the transition depth and 'narrow' below that depth.

2.2 (*b*) *Tunnel, Heading and Thrust Bore Conditions* (see Fig. 3.1)

TUNNEL. An excavation made wholly below ground level and at any depth, usually of relatively large cross section, either to avoid breaking open the surface, or to maintain a specified gradient through rising ground, or to avoid underground obstructions, which, when suitably lined, constitutes a conduit.

HEADING. An excavation into which to lay a pipeline made wholly below ground level usually at a relatively shallow depth and of small cross section, used to avoid breaking open the surface or when more economical at the required depth than a trench, or to avoid existing congested underground works.

THRUST BORE. An excavation wholly below ground which is made by jacking pipes forward from a pit and excavating from inside the pipe as it advances, either by hand for 48-in. or larger pipes, or by means of an earth auger for smaller pipes.

SHAFT OR PIT. A vertical excavation giving access to a tunnel, or heading, or thrust bore, for men and materials, and providing an exit for spoil.

2.2 (*c*) *Embankment Conditions*

POSITIVE PROJECTION. The condition in which a conduit projects partially or completely above the original ground level under an embankment or valley fill, or partially or completely above the foundation level in a wide or very wide sub-trench (see Figs. 3.1 and 5.3).

NEGATIVE PROJECTION. The condition in which a conduit is laid in a sub-trench completely below the natural ground level, or below the bottom of a wide excavation, or below the top of the initial fill in the induced trench condition, under an embankment or valley fill (see Figs. 3.1 and 5.2).

NEUTRAL CONDITION. The negative projection condition in which a conduit is laid with its top at the same level as the natural ground surface, under an embankment or valley fill, or at the same level as the bottom of a wide excavation (see Fig. 5.2); or the positive projection condition in which the settlement ratio is zero (see Figs. 3.1 and 4.5).

PLANE OF EQUAL SETTLEMENT. The horizontal plane in an embankment or valley fill above which the settlement of the fill is the same over the width of the conduit as on either side of the conduit (see Figs. 4.4, 4.5 and 4.10).

COMPLETE POSITIVE PROJECTION CONDITION. The condition of a conduit laid with positive projection under an embankment or valley fill, the height of which is such that no plane of equal settlement lies within it. (Theoretically the plane of equal settlement is at or above the top of the embankment) (see Figs. 4.5 and 4.6 (*a*)).

INCOMPLETE POSITIVE PROJECTION CONDITION. The condition of a conduit laid with positive projection under an embankment or valley fill, the height of which is such that the plane of equal settlement lies within it and the settlement ratio is positive (see Figs. 4.5 and 4.6 (*b*)).

COMPLETE TRENCH (DITCH) CONDITION. The condition of a conduit laid with positive or negative projection under an embankment or valley fill, the height of which is such that no plane of equal settlement lies within it (see Figs. 4.5 and 4.11 (*a*)).

INCOMPLETE TRENCH (DITCH) CONDITION. The condition of a conduit laid with positive or negative projection under an embankment, the height of which is such that the plane of equal settlement lies within it and the settlement ratio is negative (see Figs. 4.5 and 4.11 (*b*)).

INDUCED TRENCH CONDITION. ('Imperfect trench' condition in Marston's terminology). The condition of a conduit laid with positive projection under an embankment when the fill is so placed as to convert the positive 'projection' condition into the 'negative projection' condition, either complete or incomplete, thereby materially reducing the fill load on the pipe (see Figs. 3.1 and 4.12).

PROJECTION RATIO

(i) *For positive projection conditions* is the ratio (p) of the height of the top of the conduit above the original ground level to the overall width of the conduit (B_c) (see Fig. 4.7).

(ii) *For negative projection and induced trench conditions* is the ratio (p') of the depth of the top of the conduit below the natural ground level; or below the top of the initial fill in Induced Trench Conditions, to the width of the sub-trench (B_d) (see Fig. 4.7).

SETTLEMENT RATIO (r_{sd}). A theoretically rational but not practically pre-determinable factor affecting the load imposed on conduits under embankments by the fill. Its value may be positive or negative for the positive projection condition but is always negative for the negative projection condition and, in both, is assessed from experimental load measurements and practical experience. Its sign for positive projection conditions depends on whether the top of the conduit settles more (−) or less (+) than the fill alongside the conduit, i.e. on the nature of the deformation of the critical plane (see Table B and Figs. 4.4 and 4.10).

CRITICAL PLANE

(i) *In positive projection conditions*, the horizontal plane originally passing through the top of the conduit after laying the conduit and before loading it by fill above that level (see Fig. 4.4).

(ii) *In negative projection conditions*, the original plane of the natural ground after laying the conduit and before loading it by fill above that level (see Fig. 4.10).

N.B. In either condition this plane may settle and bend as filling proceeds.

2.2 (d) Loads

FILL LOAD. The load imposed on a conduit either in trench, heading, or thrust bore, or under an embankment or valley fill, by the earth fill overlying the conduit.

UNIFORMLY DISTRIBUTED (OR UNIFORM) SURCHARGE OF LARGE EXTENT. A uniformly distributed flexible static load of constant intensity (U_3) and of unlimited extent applied to the surface of the ground or the top of an embankment over or in the vicinity of an underground conduit (e.g. piles of materials, or soil, or merchandise).

UNIFORMLY DISTRIBUTED (OR UNIFORM) SURCHARGE OF LIMITED EXTENT. A uniformly distributed static load of constant intensity U_{sus}, or a temporary or transient load of intensity U_{sust}, either being of known position relative to the conduit and of limited dimensions, the greater of which is more than half the cover depth, applied at or below the surface of the fill but above the top of the conduit (e.g. column or wall footings, tracked vehicles) (see Figs. 4.29 and 4.30).

CONCENTRATED SURCHARGE. A static or transient, or temporary, load of any magnitude (P) applied over a small area of contact, the greatest dimension of which is less than half the cover depth, applied at or below the surface of the ground, or the top of an embankment or valley fill, over or in the vicinity of an underground conduit (e.g. a vehicle wheel load which is transient when moving, temporary when stationary) (see Fig. 4.26).

COVER DEPTH (OR COVER). The distance (H) from the top of the conduit to the ground surface in a trench, heading, or thrust bore, or to the top of an embankment or fill, at any point in the length of the conduit (see Fig. 3.1).

SATURATED SOIL DENSITY. The weight per cubic foot of the fill material (γ_s) when fully saturated with water, but not submerged.

DRAINED DENSITY. The weight per cubic foot of a coarse grained, free draining, fill material (γ_d) which is wet but not submerged.

VOID RATIO. The ratio 'n' of the volume of voids in a coarse grained material to the total volume of the material.

SUBMERGED SOIL DENSITY. The weight per cubic foot of the fill material when submerged. It is equal to the saturated density less the density of water ($\gamma_s - \gamma_w$), or to the drained density less the weight of water displaced by the solid particles ($\gamma_d - (1 - n)\gamma_w$).

COEFFICIENT OF SOIL FRICTION. Is the ratio of the shear force per unit area required to maintain steady sliding along a shear plane, to the pressure per unit area acting normal to that plane.

(i) For undisturbed soil, or in the fill itself, $\mu = \tan \phi$.
or (ii) For fill sliding on undisturbed soil $\mu' = \tan \phi'$.
μ' may be equal to, but cannot exceed, μ.

RANKINE'S COEFFICIENT OF EARTH PRESSURE (K). The ratio of the lateral to the vertical pressure in a soil

$$K = (\sqrt{\mu^2 + 1} - \mu)/(\sqrt{\mu^2 + 1} + \mu)$$

ROAD. Any road in England and Wales, other than a private carriageway or access road, including verges and foot ways, and vehicle approaches across them.

ACCESS ROAD. A private road used by light vehicles only, e.g. farm tracks.

PRIVATE CARRIAGEWAY. A private drive used only by light vehicles (e.g. access for passenger cars, delivery vans, coaches, coal lorries and furniture vans).

FIELDS AND GARDENS. Land other than roads and carriageways which is used only by light agricultural type vehicles and including woodland, moorland, etc.

VEHICLE APPROACH. A roadway across a footpath or verge giving access for vehicles to premises alongside a road.

VERGE. A strip of land, usually unpaved, either alongside a road or between traffic lanes, which is not normally used by vehicles but which may be occasionally so used in emergency.

IMPACT FACTOR. The ratio (F_i) of the dynamic to the static load imposed by a transient surcharge on an underground conduit.

LIVE LOAD CORRECTION FACTOR. A factor (F_L) by which the non-uniformly distributed load over the width of a rigid conduit, caused by a concentrated surcharge, (W_{cs}) must be multiplied to convert it into an equivalent uniformly distributed load per linear ft (W_{csu}).

EQUIVALENT UNIFORMLY DISTRIBUTED (LIVE) LOAD (W_{csu}). The load per linear ft of conduit required to induce the same maximum circumferential bending moment in a rigid conduit wall as that caused by the non-uniformly distributed load caused by a concentrated surcharge, with the same class of bedding.

2.2 (*e*) *Pipes and Pipelines*

RIGID PIPE. A pipe which has a large inherent resistance to crushing (large stiffness) and which fails by brittle fracture under a very small deformation (e.g. clayware, concrete, cast or spun iron, asbestos cement).

RIGID PIPELINE. A line constructed of rigid pipes with rigid joints.

FLEXIBLE PIPELINE. A line constructed of rigid pipes with flexible joints, or of flexible pipes with rigid or flexible joints.

FLEXIBLE PIPE. A pipe which has little inherent resistance to crushing (small stiffness), permits very large deformations under sustained loading without cracking, and fails by buckling or flattening. Its load carrying capacity may depend largely on side support from the surrounding soil (e.g. pitch fibre, some plastics, steel and ductile iron pipes).

PRESSURE PIPE. A rigid or flexible pipe which is designed to withstand simultaneous external load and external pressure either alone or in combination with internal pressure.

NON-PRESSURE PIPE. A rigid or flexible pipe which is designed to withstand simultaneous external load and external pressure only.

PRESTRESSED CONCRETE CYLINDER-TYPE PIPE. A prestressed concrete pipe incorporating a thin steel cylinder between the prestressed core and the cover concrete. The cylinder acts as a waterproof membrane.

PRESTRESSED CONCRETE NON-CYLINDER-TYPE PIPE. A prestressed concrete pipe in which the cover concrete is monolithic or in direct contact with the prestressed core but with no waterproof cylindrical membrane.

'SAFE 2 OR 3-EDGE TEST STRENGTH' (W_T). The statistical minimum value of the ultimate or 0·01 in. crack load per linear foot of a pipe in a laboratory or works crushing test of a rigid pipe with 2 or 3-edge B.S. or A.S.T.M. test bearings. This value should be as specified by the appropriate British (or A.S.T.M.) Standard (or guaranteed by the pipemaker in the absence of such a standard).

For unreinforced rigid pipes (e.g. plain concrete, asbestos-cement, clayware, cast iron) it is the ultimate test load.

For steel reinforced concrete pipes, it is the test load required to produce a crack in the pipe nowhere wider than $\frac{1}{100}$ in. over a length of not less than 12 in., or 80% of the ultimate test load, *whichever is the less.*

For prestressed concrete cylinder pipes, which are not yet standardised, it is 90% of the test load required to produce a crack nowhere wider than $\frac{1}{1000}$ in. (Maker's value.)

For prestressed concrete non-cylinder type pipes it is not yet determined. (Refer to Makers.)

LABORATORY STRENGTH ($W_T C_{BS}$). The *theoretical* 2-edge test load required to produce the same circumferential bending moments in the pipe wall as the actual 2 or 3-edge test load W_T. It is obtained by multiplying the value of W_T by an appropriate bearing correction factor C_{BS} which is never greater than unity.

BEARING CORRECTION FACTOR (C_{BS}). The ratio of the laboratory strength of a rigid pipe to its actual 2 or 3-edge test strength, W_T

$$\text{i.e. } C_{BS} = \text{laboratory strength}/W_T$$

It is used to reduce actual test loads to equivalent idealised 2-edge test loads and thereby to eliminate unavoidable errors caused by the use of various practicable bearings in crushing tests. For practical 2-edge test bearings its value depends on the type of bearing, the O.D. of the pipe and the magnitude of the applied test load. For B.S. 3-edge test bearings its value is 1·0 (see Fig. 5.1).

THEORETICAL 2-EDGE BEARING. A theoretical test condition in which a rigid pipe is loaded and supported in the testing machine by knife edge bearings placed at the extremities of the vertical diameter.

THREE-EDGE BEARING. A practical test condition in which a rigid pipe is loaded in a testing machine through a narrow strip of rubber or timber placed at the top whilst supported by two similar strips placed at equal specified intervals from the pipe centre line under the invert (see Fig. 5.1).

B.S. 556 2-EDGE TEST BEARING. A practical test condition in which a rigid pipe is loaded in a testing machine through a rubber strip 6 in. wide and 1 in. thick placed at the crown of the pipe whilst supported by a similar rubber strip placed beneath the invert (see Fig. 5.1).

FIELD STRENGTH. The maximum load carrying capacity of a rigid pipe ($W_T\ C_{BS}\ F_m$) or ($W_T\ C_{BS}\ F_p$) as laid underground and as influenced by the class of bedding provided.

BEDDING FACTOR (F_m or F_p). The ratio of the ultimate crushing test strength of a rigid pipe, when uniformly loaded over its external diameter and supported by a specific class of bedding, to its Laboratory Strength (see Charts C9 and C10).

BEDDING CLASS. A specific construction method of distributing the reaction from the foundation over the lower periphery of an underground pipe and thereby reducing the maximum circumferential bending moment in a rigid pipe wall (see Charts C9 and C10).

FACTOR OF SAFETY F_s. A factor, not less than unity, by which the computed value of the Total Effective External Design Load (We) or the effective internal pressure (p_i) is multiplied to provide for uncertainties in the loading conditions, in rigid pipe strength and in the quality of workmanship in the site operations. It is not applicable to flexible pipes, (see Table G).

BEDDING CONSTANT (K_θ). A factor affecting the deflection of a flexible pipe without side support, which depends on the bedding angle θ (see Table L).

BEDDING ANGLE θ. The angle subtended at the centre of the pipe by the arc of contact of a flexible pipe with its bed (see Table L).

CRITICAL (OR COLLAPSE) PRESSURE (P_c). The external fluid pressure under which a thin flexible pipe becomes unstable and starts to buckle or yield plastically.

Section B

The Loads Imposed on Buried Conduits

Chapter 3
Classification and nature of the loads

3.1 Classification

The loads and forces which are, or may be, imposed on a buried pipeline are conveniently classified in three categories, namely:

(i) The externally applied, vertically acting, gravitational loads caused by the fill and surface surcharges.

(ii) Fluid pressures applied either externally or internally or both.

(iii) The largely fortuitous and frequently indeterminable forces exerted either axially or transversely in any direction, by thermal and moisture changes, in the conduit or in the soil, by soil movements, by tree roots, by the relative settlement of any structure to which the conduit is connected, or by irregularities in the foundation or bedding of the conduit.

Of these categories, (i) and (ii) are known as 'primary loads' and (iii) as 'secondary loads and forces'. The latter are discussed in Chapter 9.

3.2 Primary External Loads

The primary external loads of categories (i) and (ii) comprise:

(*a*) *Permanent loads* such as the fill load, including the effect of the water contained in the conduit and, for flexible pipes, the weight of the pipe, together with the worst probable simultaneous combination of:

(*b*) *Temporary loads* such as uniformly distributed surface surcharges of large and indefinite extent, e.g. additional temporary fill, stock piles of materials, cargo, or any similar load, which may probably be placed over, and extend both along and across, the line of the conduit anywhere.

(*c*) *Transient loads* such as the concentrated surcharges which may be applied at any point on the surface over or in the vicinity of the conduit, e.g. by vehicle wheel loads on roads, railways, industrial yards or airfields.

(*d*) *Permanent, temporary, or transient* loads such as uniformly distributed surcharges of small extent and known position, applied at or below the surface, but above the top of the conduit, and in its vicinity. These surcharges may be static, e.g. column or wall footings, pylon bases, crane or piledriver supporting pads, or transient and dynamic, e.g. tracked construction equipment, or tanks.

(*e*) *Permanent or temporary fluid pressure* applied to the external periphery of the conduit by immersion in water or by any possible submergence of the ground surface, e.g. at a river crossing, or in tidal areas, or by accidental flooding, or by partial vacuum in pressure pipes conveying liquids.

3.3 Primary Internal Loads

The primary internal fluid pressure loads of category (ii) may be temporary or transient and comprise the maximum surge or test pressure to which a pressure pipeline may be subjected in service. The working pressure may be regarded as normal and permanent as affecting sustained stresses in the conduit.

3.4 Assumptions Regarding the Worst Simultaneous Combination of Primary Loads

It is usually assumed:

(i) that the fill and water loads are always present,

(ii) that static uniformly distributed surcharges or large extent will not occur simultaneously with mobile concentrated surcharges,

(iii) that in pressure pipes the maximum internal test or surge pressure will not occur simultaneously with maximum mobile surface surcharge loading.

3.5 Factors affecting the Various Component Loads

3.5 (a) The Fill Load

This load is static and permanent and is usually assumed to be uniformly distributed over the width of the conduit and along the length under consideration. Its value depends on the saturated density (or the drained density of coarse granular materials) and the frictional characteristics of the fill, the depth of cover over the top of the conduit, and the effective width of the trench or the overall width of the conduit. The effective width depends on the method of installation which, in turn, affects the direction of the frictional forces acting on the shear planes within the fill and on the type of conduit (i.e. whether rigid or flexible). It is reduced by *permanent* submergence.

The various methods of installation covered by the theory are classified as follows, (Fig. 3.1), (see Chapter 4 for details).

(i) *In trench*

1. The 'narrow' trench condition which includes a wide main or 'V' trench with a narrow subtrench.

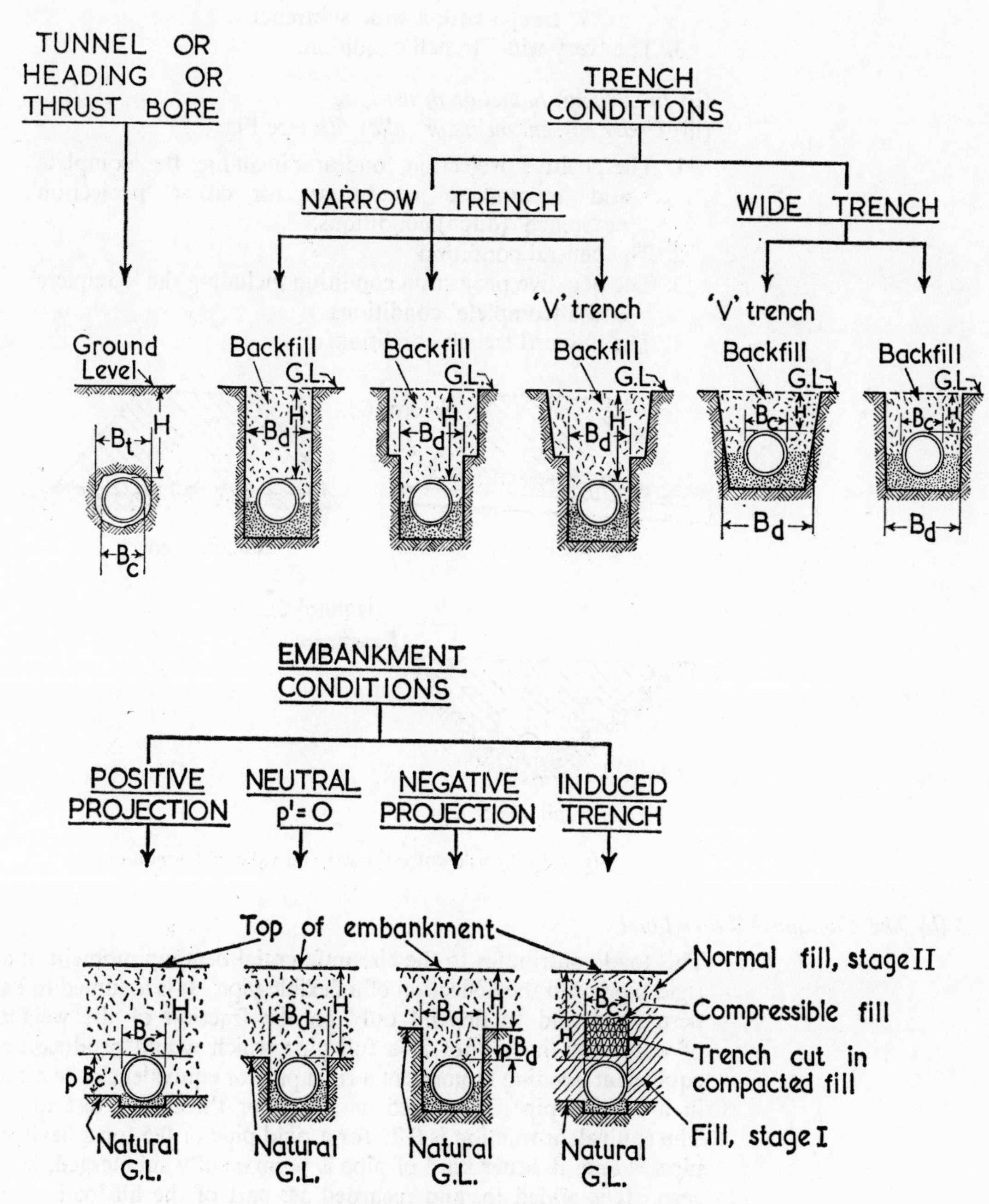

Fig. 3.1 Classification of fill loading conditions imposed by various construction methods

After Spangler—by permission

2. The 'wide' trench condition which includes a wide main or 'V' trench with a wide subtrench.
3. The 'very wide' trench condition.

(ii) *In heading, tunnel or thrust bore*
(iii) *Under embankments or valley fills* (see Fig. 3.2)

1. The positive projection condition including the 'complete' and 'incomplete' conditions, for either 'projection' or 'trench' (ditch) conditions.
2. The neutral condition.
3. The negative projection condition including the 'complete' and 'incomplete' conditions.
4. The induced trench condition.

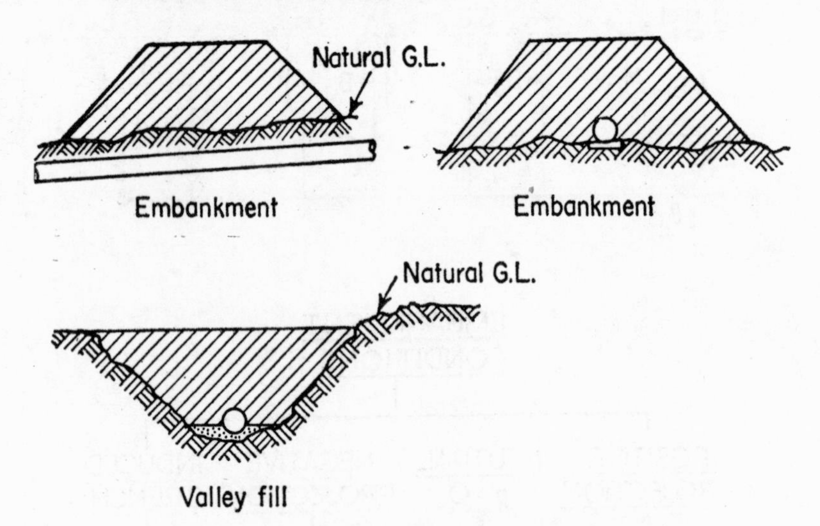

Fig. 3.2 Typical embankment and valley fill conditions

.5 (b) The Contained Water Load

This load contributes to the circumferential bending moment in a rigid pipe or to the deflection of a flexible pipe. It is assumed to be permanent and to depend only on the fraction of the weight of water per linear foot in a full pipe which would produce an equivalent bending moment in a rigid pipe or equivalent deflection in a flexible pipe, if applied as a two or three-edge test load. The equivalent fraction is 0·75 for a rigid pipe or 0·5 for a flexible pipe[33] and, if either kind of pipe is permanently submerged, it is zero. It is added to, and regarded as, part of the fill load. No ruling can be given for non-circular conduits, other than to add the total weight of water to the fill load.

If for either kind of pipe this load is less than say 5% of the total load, it may be ignored.

3.5 (c) The Self-weight of the Pipe

This is permanent but can usually be neglected in rigid pipes since its effect on the circumferential bending moments is included in the two or three-edge test load. In flexible pipes its effect on pipe deflection is insignificant in thin walled pipes. In thick walled flexible pipes, the equivalent deflection in two or three-edge tests is produced by half the weight per unit length which should be added to the total imposed load if it exceeds say 5% of the latter.

3.5 (d) The Uniformly Distributed Surcharge Loads of Large Extent

The loads imposed by these surcharges, of large and unlimited extent in plan overlying the conduit, are static, and usually temporary. They depend upon the intensity of the surcharge, the width of the trench, or of the conduit, the method of installation of the conduit, the cover depth, and the frictional characteristics of the fill. The surcharge is assumed to be frictionless in narrow trench, or complete negative projection, or induced trench conditions; and to have the same frictional effect as an equivalent height of fill in wide trench or positive projection conditions, or in incomplete negative projection or induced trench conditions. Its intensity is assumed to be uniform along and across the conduit.

For conduits in England and Wales, in the absence of a known value, an arbitrary minimum value of 500 lb/ft^2 is recommended for design purposes.*

On adjacent parallel pipes, either in the same trench or under an embankment, these loads are assumed to vary in sympathy with the corresponding fill loads. The possibility of submergence of the surcharge is ignored in design.

3.5 (e) The Loads Imposed by Concentrated Surcharges

These loads are usually dynamic and transient, and caused by vehicle wheels. They depend on the magnitude of the wheel loads and their positions relative to the conduit, on the depth of cover over the top of the conduit, and on its overall width, and on impact effects caused by irregularities in the road surface. Because of their transient nature the soil fill in any method of installation is assumed to be equivalent to the theoretical semi-infinite elastic solid on which the Boussinesq stress theory is based. In this theory the imposed loads do not depend upon soil friction and, because of the limited duration of the surcharge and its small area of surface contact, it does not contribute to the active soil pressure imposed on projecting conduits under embankment or very wide trench conditions. The imposed loads are assumed to be independent of the nature of the fill, or of the conduit, or of the presence of adjacent parallel conduits, or of

* But see latest recommendations in Ref. 10*a*, para. 235 (d) and Table D.

partial or complete submergence of the fill, or of the method of installation. If the largest dimension of the area of contact with the surface does not exceed half the cover depth, the surcharge may be regarded as equivalent to a theoretical point load, otherwise it is treated as a uniform surcharge of small extent (see Article 3.5 (*f*)).

Since the intensity of the imposed load varies both across the width of the conduit and along its length, an equivalent uniformly distributed load which produces the same maximum ring bending moment in a rigid conduit must be substituted for the theoretical load.

In England and Wales the maximum road loading to be considered in design at present consists of one 8-wheel group of B.S. 153 type HB loading, but this maximum is under current discussion by the responsible Ministries and may be changed.*

3.5 (f) The Loads Imposed by Uniformly Distributed Surcharges of Limited Extent

These loads are of a defined and limited area in plan applied at or below the surface but above the top of the conduit. They may be static and permanent, or dynamic and transient, provided that the largest dimension of the area of contact exceeds half the cover depth. They depend on the intensity of the surcharge but otherwise on the same factors as concentrated surcharge loads (see Article 3.5 (*e*) above). For surcharges which are eccentrically placed with reference to the axis of the conduit, the Boussinesq theory used only yields the vertical component of the load, which may be less than the actual inclined load.

The maximum load imposed by tracked vehicles, either during construction or subsequently, will usually occur when the centre of gravity of the vehicle is vertically above the axis of the conduit, unless the conduit is narrow and shallow, when a single track passing over it may cause the greater load. In either case impact effects must be considered. The magnitude and position of static surcharges must be anticipated if they are to be incorporated in the design.

3.5 (g) External Fluid Pressures

These pressures may be permanent or temporary and may fluctuate with the ground or free water level. They depend only on the height of the water surface above the axis of the conduit and produce an approximately uniform compressive stress in the walls of circular pipes. In rigid pipe conduit design they are usually ignored since they do not contribute to the circumferential bending moments other than by reducing the density of the portion of the fill which is permanently submerged. No such

* See Table D for the latest recommendations.

reduction of fill load is permissible in design when there is any possibility of subsequent lowering of the water level. In underwater rigid or flexible non-pressure pipes they may be the major external load and so control the pipe wall thickness.

In either pressure or non-pressure flexible pipes they are of importance since they contribute to the critical (buckling) pressure when the pipes are not under internal pressure.

In any pressure pipes which are permanently submerged the internal pressure may be partly or wholly neutralised by the external pressure and, in all conduits, the water load is neutralised by permanent submergence.

Partial vacuum which may occur in siphons or in water pipes or rising mains due to a burst or the blockage or inadequacy of an air valve is treated as an added temporary external fluid pressure.

3.5 (h) Internal Fluid Pressures

These pressures in gravity water mains and in water or sewage pumping mains may be subjected to large variations caused by surges[9,41] (Chapter 7) due to faulty operation of valves or pump power failures. The pipelines may also be subjected to test pressures which are higher than operating pressures and, from time to time, the pressure may be reduced to zero during maintenance operations, or in the event of a burst. In gravity mains the maximum static head, exclusive of surge, occurs when the velocity of flow is zero. In pumping mains the maximum pressure will be the sum of the maximum static head, the friction head at maximum flow, and any other losses of head in the system; or the maximum surge pressure due to sudden stoppage of the pumps and closing of non return valves in the absence of effective surge suppression.

Chapter 4
The theory and mathematical derivation of the load formulae

4.1 Narrow Trench Fill Loads (Fig. 4.1)

4.1 (a) Derivation of the Formulae

This theory, originally proposed by Marston and Talbot in 1913,[16 and subsequently confirmed experimentally, was based on Jannsen's earlier theory of the loads on grain silo floors. It postulate that the load on the conduit consists of the weight of the prism o earth fill contained within the trench above the top of the conduit less the sum of the frictional or shear forces acting on the trencl sides as the fill settles. The cohesion (if any) between the trench fil and the soil in the trench sides is ignored because of its variabl and uncertain value depending on moisture conditions. Th frictional values therefore depend upon the coefficient of slidin friction between the fill and the soil in the trench sides μ' and o the ratio K of the active horizontal pressure to the vertical pressur in the soil, (i.e. the shear force between the fill and each trencl wall is $K\mu'$ times the vertical pressure $H\gamma_s$).

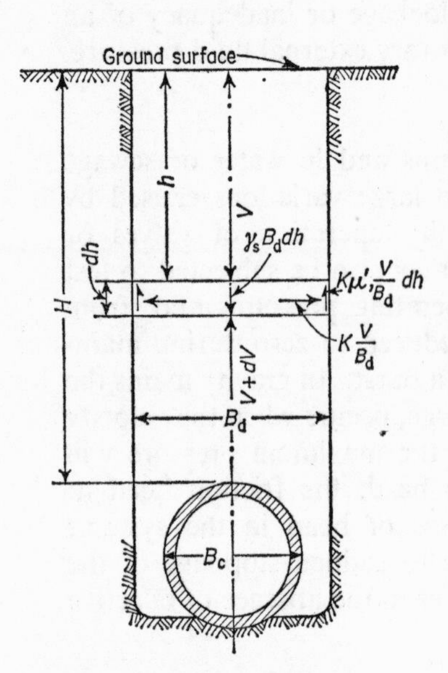

Fig. 4.1 Force diagram for a conduit in the 'narrow' trench condition

After Spangler—by permission

Thus, considering any element of fill of the width of the trench unit length, and depth dh (Fig. 4.1) and equating the forces actin on it, if V is the effective total load imposed on the upper surfac and μ' is the coefficient of sliding friction of the fill on the trencl wall,

$$V + dV = V + \gamma_s B_d dh - 2K\mu' \frac{V}{B_d} dh$$

The solution of this differential equation is:

$$V = \gamma_s B_d^2 \left(\frac{1 - e^{-2K\mu' h/B_d}}{2K\mu'} \right)$$

and when $h = H$ the total load on the pipe is:

Equation 4.1

$$W_c = \gamma_s B_d^2 \left(\frac{1 - e^{-2K\mu' H/B_d}}{2K\mu'} \right)$$

For convenience the expression in brackets is known as th load coefficient C_d and the formula is written:

Formula 4.2

$$W_c = C_d \gamma_s B_d^2$$

Marston took the value of K to be Rankine's value for active pressure, i.e.

$$K = \frac{\sqrt{\mu'^2 + 1} - \mu'}{\sqrt{\mu'^2 + 1} + \mu'} = \frac{1 - \sin \phi'}{1 + \sin \phi'}$$

where μ' ($= \tan \phi'$) is the coefficient of sliding friction of the fill material on the soil in the trench sides.

Note that when H/B_d becomes large, C_d approaches a limiting value of $1/(2K\mu')$ and that the limit is reached at shallower depths for high values of $K\mu'$ than for the lower values. Note also that μ' may be equal to or less than μ (for the undisturbed soil) but it cannot be greater; since if it were, shear would occur in the undisturbed soil. (Some more recent investigators[27] consider that Marston's assumption of active soil pressure is possibly open to question and that K would tend towards a static value of 0·5 in cohesive soils whilst in highly compacted granular materials it could be even higher, so giving a lower value of the fill load. The long duration values may differ and produce an increase in load however, as remarked by Marston.)

Marston's values of the load coefficient C_d for various values of $K\mu'$ and $\frac{H}{B_d}$ are plotted in Design Chart C1.

It was found that for a rigid conduit installed without special care in the compaction of the fill alongside it, the whole of the load W_c is carried by the conduit. If however the conduit is a flexible pipe and the side fill is well compacted, so that it is as stiff as the pipe, the distribution of the load across the trench width becomes virtually uniform and the load on the pipe is then only its proportion of the total load,[7]

Formula 4.3

$$\text{i.e. } W_c = C_d \gamma_s B_d^2 \left(\frac{B_c}{B_d}\right) = C_d \gamma_s B_c B_d$$

It is reasonable to assume that this formula (4.3) would be applicable to rigid conduits if the side fill were similarly adequately compacted.

4.1 (*b*) *Effective Density of the Fill Material*

Since the moisture content of the fill material will usually vary with weather conditions, its maximum unsubmerged density will occur when its voids are filled with water. This is known as the saturated density (γ_s) which is normally used in design. If any part of the fill is submerged, however, the effective density will be ($\gamma_s - \gamma_w$). Coarse gravels and similar materials, which are free draining, do not retain the void water and their maximum density occurs when the material is wet but drained. This is known as the drained density (γ_d). The submerged density is then:

$$\gamma_d - \gamma_w(1 - n)$$

where $n =$ the volume of voids per unit volume of soil. Note that

$$\gamma_d = \gamma_s - n\gamma_w \qquad \gamma_s = \gamma_d + n\gamma_w$$

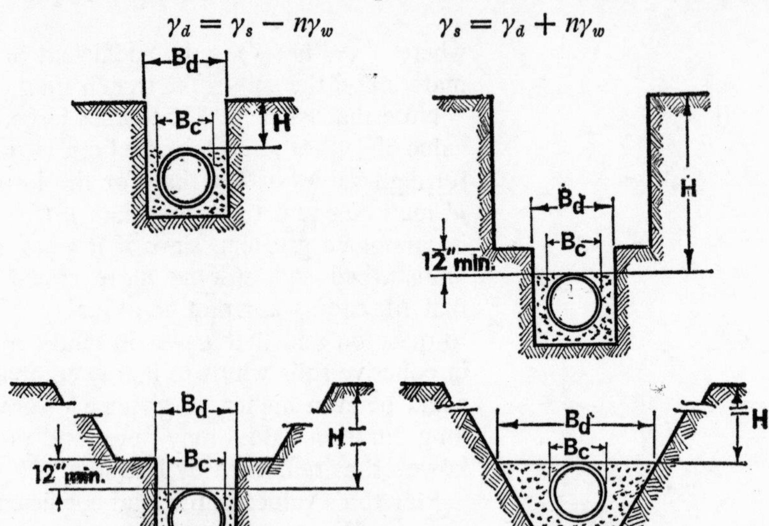

Fig. 4.2 (*a*) Effective trench width B_d and suggested minimum cover in sub-trenches

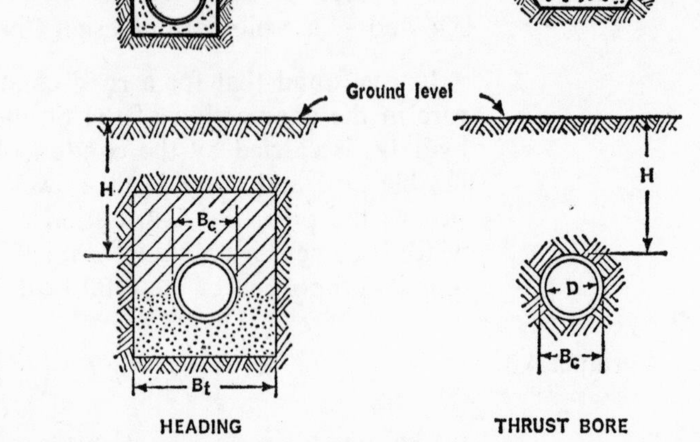

Fig. 4.2 (*b*) Effective widths of heading and thrust bore

Crown Copyright—by permission

4.1 (*c*) *Effective Trench Width* (B_d) (Fig. 4.2)

Schlick[19] has shown that the value of the trench width which governs the load in any type of trench is the width of the trench at the top of the pipe. Note however that in a 'V' trench or wide trench with a sub-trench, the frictional values depend on $K\mu$ and not on $K\mu'$, since the shear planes are now in the fill.

4.1 (*d*) *Effective Overall Conduit Width* (B_c) (Fig. 4.3)

Essentially this is the external diameter of a pipe, or the maximum overall width of any rigid encasement which is placed above the pipe axis, or the maximum overall width of a non-circular conduit.

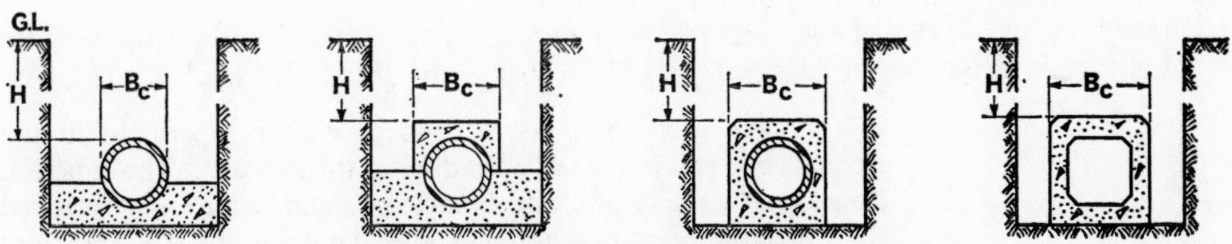

Fig. 4.3 Effective values of overall width of conduit B_c

Crown Copyright—by permission

4.2 Narrow Trench Loads Imposed by Uniform Surcharges of Large Extent

Since the frictional characteristics of the surcharge may be variable or non-existent it is usual to design for the worst condition of no relieving shear force in the surcharge. If the intensity of the surcharge is U_s lb/ft² there will then be a load over the width of the trench at the surface of B_dU_s.

Then in the differential equation given in Article 4.1 (*a*) above, when $h = 0$, $V = B_dU_s$ and the solution of the equation is

$$V = \gamma_s B_d^2 \left(\frac{1 - e^{-2K\mu' h/Bd}}{2K\mu'}\right) + B_dU_s e^{-2K\mu' h/Bd}$$

and when $h = H$, the total load on the conduit is

$$\gamma_s B_d^2 \left(\frac{1 - e^{-2K\mu' H/Bd}}{2K\mu'}\right) + B_dU_s e^{-2K\mu' H/Bd}$$

but the first term in this expression is the fill load $C_d\gamma_s B_d^2$ and the load imposed by the surcharge is therefore

Equation 4.4
$$W_{us} = B_dU_s e^{-2K\mu' H/Bd}$$

or putting

$$e^{-2K\mu' H/Bd} = \text{a load coefficient } C_{us}$$

Equation 4.5
$$W_{us} = C_{us}B_dU_s$$

This load will decrease as $\frac{H}{B_d}$ increases, tending to zero when H/B_d is very large.

Marston's values of C_{us} for various values of $K\mu'$ and $\frac{H}{B_d}$ are plotted in the Design Chart C2.

Note that the value of $K\mu'$ (or $K\mu$) is the same as that used for the corresponding fill load; also that the proportion of this load transmitted to the conduit is the same as that of the fill load. Thus for a flexible pipe

Equation 4.6
$$W_{us} = C_{us}B_dU_s\left(\frac{B_c}{B_d}\right) = C_{us}B_cU_s$$

4.3 Heading or Thrust Bore Overburden (=fill) Load and Loads Imposed by Uniformly Distributed Surcharges of Large Extent (Figs. 3.1 and 4.2 (*b*))

Theoretically, cohesion as well as friction is operative in the undisturbed soil overlying a heading or thrustbore. The cohesion however is usually likely to be variable and it may be destroyed by any subsequent excavation above the conduit. It is therefore considered safe but conservative to regard the load imposed upon a conduit in a heading of width B_t as being the equivalent of that in a trench of the same width if the heading is not well packed. This assumes that shear planes will occur at each wall of the heading:

Then

Formula 4.7
$$W_c = C_d \gamma_s B_t^2$$

and

Formula 4.8
$$W_{us} = C_{us} B_t U_s$$

If the heading is well packed so that the sides are at least as stiff as the undisturbed soil, or if the conduit is installed by thrust boring, the shear planes may be assumed to occur at the limits of the overall width of the conduit B_c and the loads then become:

Formula 4.9
$$W_c = C_d \gamma_s B_c^2$$

Formula 4.10
$$W_{us} = C_{us} B_c U_s$$

Note that in both conditions the values of $K\mu$ and γ_s *for the undisturbed soil* are applicable. Also that excessive compaction of the packing over the top of the conduit may increase the load on it by mobilising resistant friction and cohesion in the overlying soil. It is theoretically possible in a heading to eliminate circumferential bending moments in a rigid pipe by so controlling the compaction of the side packing as to equalise the lateral and vertical loads. This device should not be relied upon where subsequent excavation could possibly relieve the lateral load on a rigid pipe designed to resist direct compressive stress alone, or on a thin walled flexible pipe where the lateral pressure is depended upon to limit deflection.

4.4 Positive Projection Fill Load Under an Embankment or Valley Fill (Fig. 4.4)

4.4 (*a*) *Theory*

In 1922 Marston[7] postulated that if a conduit of any shape or material is laid with its invert at or just below ground level, and its top above ground level, and it is then loaded with fill as in Fig. 3.1 to a uniform height H, it projects positively into the fill and is said to be in the *positive projection condition.*

Then, assuming for simplicity that the invert of the rigid conduit settles equally with the natural ground under the influence of the fill, the prism of fill overlying the overall width of the conduit (known as the interior prism) will be shorter than the prisms of fill on either side of it (known as the exterior prisms) and will therefore settle less than they do (Fig. 4.4). This relative movement will produce shear forces on the imaginary vertical planes bounding the interior prism and therefore a downward drag on them.

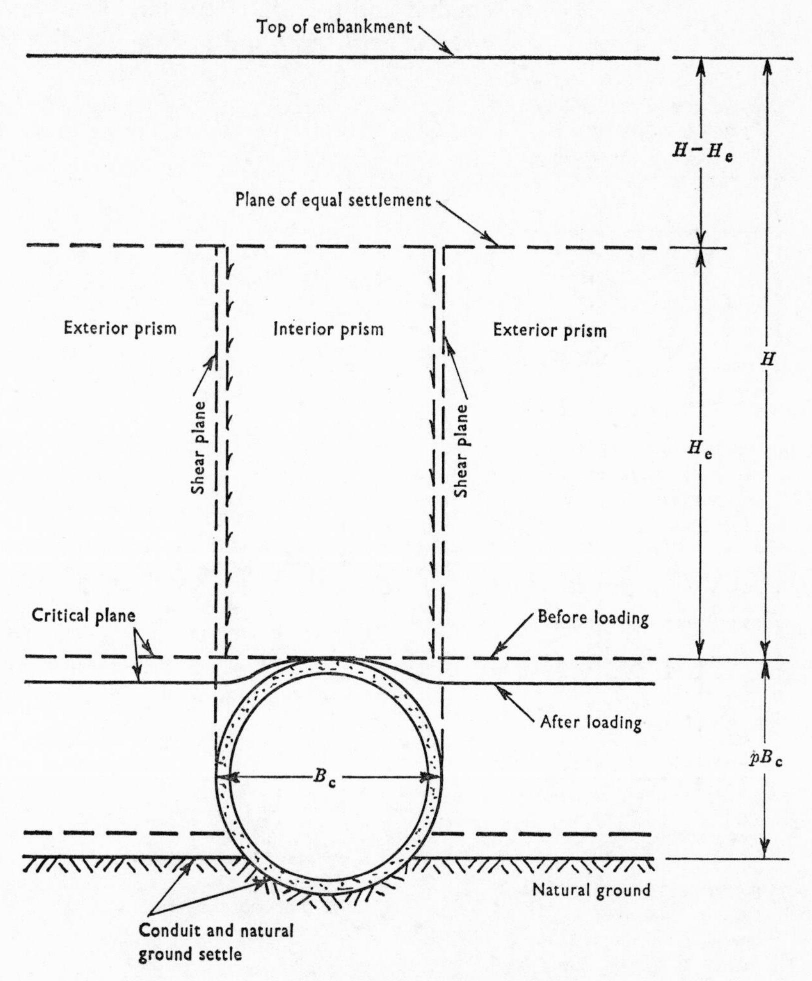

Fig. 4.4 Elements of the Positive Projection condition
After Spangler—by permission

The load on the conduit will therefore be greater than the weight of the interior prism by the frictional forces exerted upon it. These frictional forces (again ignoring cohesion) are assumed to be the summation of the active lateral pressure on the two vertical planes multiplied by the coefficient of internal (sliding) friction of the fill (μ).

In a comparatively low embankment the shear stresses extend up to the surface of the fill and the conduit is then said to be in the '*complete projection condition*' (Fig. 4.5 (*a*)).

If the height of the fill is sufficiently increased, however, it is found that the shear stresses do not extend to the surface but cease at some horizontal plane in the fill between the top of the conduit and the top of the fill. This plane is known as the *plane of equal settlement* and its height above the top of the conduit is

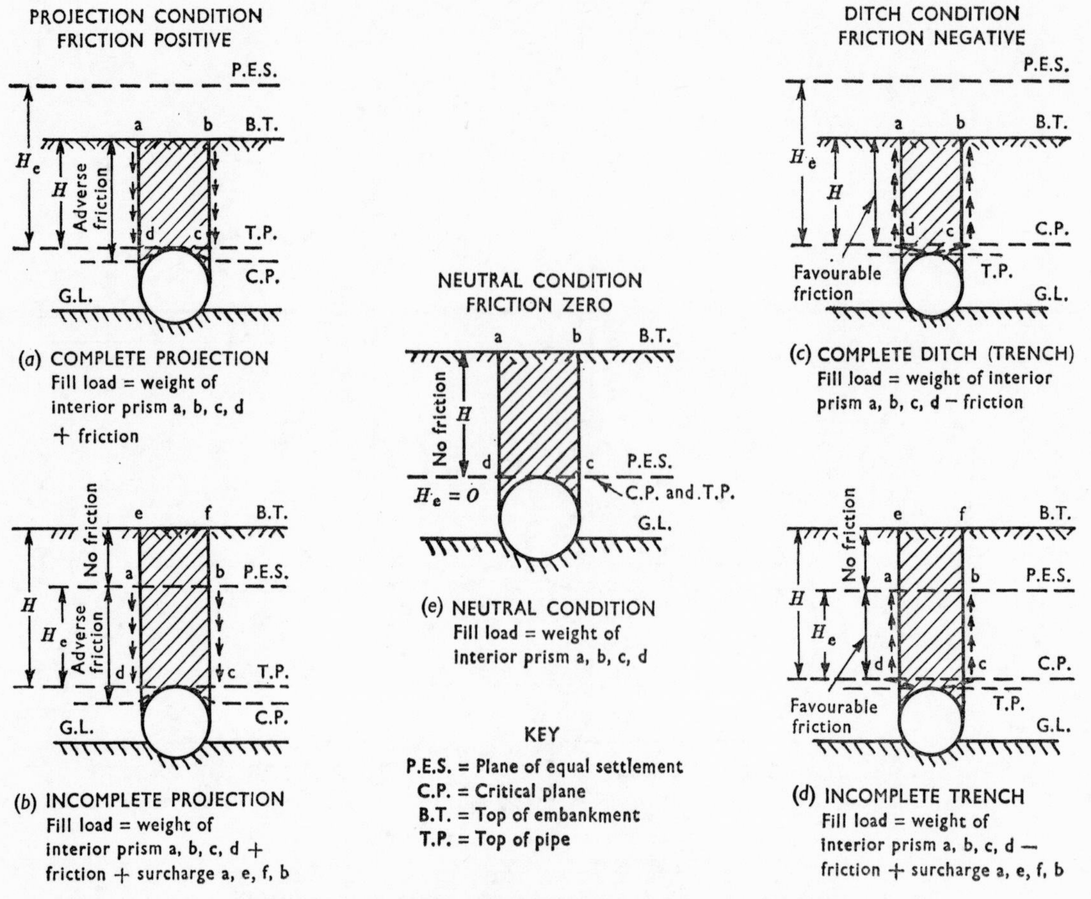

Fig. 4.5 The five possible loading conditions for Positive Projecting conduits

After Spangler—by permission

the *height of equal settlement* (He). When this occurs the conduit is said to be in the *incomplete projection condition* and no frictional forces are exerted above the plane of equal settlement (Fig. 4.5 (*b*)).

The horizontal plane through the top of the conduit after installation, and before the fill is placed, is known as the '*critical plane*'. In the projection conditions so far described it will be observed that, after loading, this plane is deformed downward relative to the top of the conduit (Fig. 4.5 (*a*) and (*b*)). If, however, the top of the conduit settles more than the fill alongside it, either because it is flexible and suffers a relatively large reduction in vertical diameter, or because it is founded on very yielding soil, or both, the frictional forces are reversed and the load on the conduit is then less than the weight of the interior prism of fill, i.e. somewhat similar to a trench load. Hence, if the shear stresses extend up to the surface of the fill, the conduit is said to be in the '*complete ditch* (*trench*) *condition*' (Fig. 4.5 (*c*)).

However, if the fill is sufficiently high a plane of equal settlement can occur within it (but no such plane is possible in a trench since the undisturbed soil in the trench sides is assumed not to settle after the conduit is installed) and the conduit is then said to be in the *incomplete ditch* (*trench*) *condition*. In this condition the 'critical plane' will be deformed *upward* relative to the top of the conduit (Fig. 4.5 (*d*)).

There is also an intermediate condition, in which the top of the conduit settles equally with the fill alongside it. This is known as the '*neutral condition*'. The plane of equal settlement is now coincident with the critical plane and the latter is not deformed at all relative to the top of the conduit. In this condition no shear stresses occur in the fill and the load is simply the weight of the interior prism of fill of width B_c and height H, i.e. $W_c = H\gamma_s B_c$ (Fig. 4.5 (*e*)). This load corresponds to the over burden load on the conduit.

The Projection and Settlement Ratios

The deformation of the 'critical' plane at the top of the conduit relative to that in the fill alongside it is an indicator of the frictional conditions present in the fill. It may be positive or negative and its magnitude modifies the height of the plane of equal settlement. Its value will evidently depend, *inter alia*, on the extent to which the top of the conduit projects above the surface of the natural ground, and the ratio of this projection to the width of the conduit is known as the *projection ratio* (p), i.e. the actual projection is pB_c (Figs. 4.4 and 4.7). Other quantities affecting the relative deformation are:

1. The settlement (compression) of the soil in the projection height (pB_c) caused by the overlying fill (Sm)
2. The total settlement of the pipe invert (Sf)
3. The settlement of the natural ground surface (Sg)
4. The vertical deflection (shortening) of the conduit (dc)

Referring to Fig. 4.6 (*b*) the relative deformation of the critical plane is then:

$$(S_m + S_g) - (S_f + d_c)$$

i.e. [settlement of the critical plane in the fill] minus [settlement of the top of the conduit].

This difference divided by *Sm* is a measure of the relative deformation, and is known as the '*settlement ratio*'. This theoretically valid ratio is conveniently used in the mathematical derivation of the height of equal settlement. It is positive in the incomplete projection condition and negative in the incomplete ditch condition and, in combination with the 'projection' ratio, modifies the load imposed on the conduit, as shown below. It cannot be predetermined, however, except by experienced estimation, based on its empirical value derived from existing conduits[23] (see Table B).

4.4 (*b*) *The Positive Projection Formulae and their Derivation*

4.4 (*b*) (i) *The Complete Positive Projection Condition Fill Load* ($H_e > H$)

Consider the equilibrium of a horizontal element of fill in the interior prism under the vertical forces postulated in this theory, (Fig. 4.6 (*a*))

$$V + dV = V + \gamma_s B_c dh \pm 2K\mu \frac{V}{B_c} dh$$

The solution of this differential equation is:

$$V = \gamma_s B_c^2 \left(\frac{e^{\pm 2K\mu h/B_c} - 1}{\pm 2K\mu} \right)$$

At the top of the conduit $V = W'_c$ and $h = H$, whence:

Equation 4.11

$$W'_c = \gamma_s B_c^2 \left(\frac{e^{\pm 2K\mu H/B_c} - 1}{\pm 2K\mu} \right)$$

or designating the expression in brackets as the load coefficient C_c

Formula 4.12

$$W'_c = C_c \gamma_s B_c^2$$

In equation (4.11) the upper signs are used for the *complete projection* condition and the lower signs for the *complete ditch* (*trench*) condition.

4.4 (*b*) (ii) *The Incomplete Positive Projection Condition Fill Load* ($H_e < H$)

Again considering the equilibrium of an element of fill in the interior prism, the same equation as for complete projection holds, but 'h' is now the depth of the element below the plane of equal settlement (Fig. 4.6 (b))

i.e. $$V + dV = V + \gamma_s B_c dh \pm 2K\mu \frac{V}{B_c} dh$$

but when $h = 0$, $V = (H - H_e)\gamma_s B_c$ and the solution of the equation is:

Equation 4.13 $$V = \gamma_s B_c^2 \left\{ \frac{e^{\pm 2K\mu h/B_c} - 1}{\pm 2K\mu} + \left(\frac{H}{B_c} - \frac{H_e}{B_c} \right) e^{\pm 2K\mu h/B_c} \right\}$$

and when $h = H_e$, $V = W'_c$

and

Formula 4.14 $$W'_c = \gamma_s B_c^2 \left\{ \frac{e^{\pm 2K\mu H_e/B_c} - 1}{\pm 2K\mu} + \left(\frac{H}{B_c} - \frac{H_e}{B_c} \right) e^{\pm 2K\mu H_e/B_c} \right\}$$

and again designating the expression in brackets as C_c

Formula 4.14a $$W'_c = C_c \gamma_s B_c^2$$

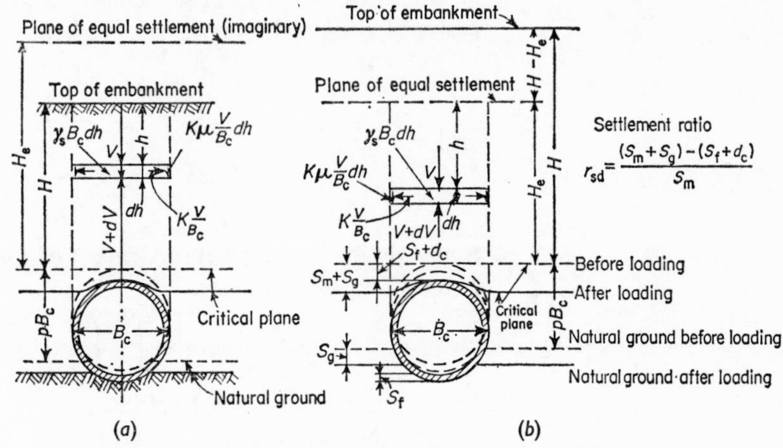

Fig. 4.6 Forces acting on a conduit in the Positive Projection conditions
(a) in the complete projection condition
(b) in the incomplete projection condition
After Spangler—by permission

As before the upper signs are used for the *incomplete projection* condition and the lower signs for the *incomplete ditch* condition.

Evaluation of H_e (Spangler's method[23]). The value of C_c now depends on the value of H_e however, and this value is obtained by equating the total settlements of the interior and exterior prisms of fill below the plane of equal settlement, and for this equation the load of an exterior prism must be known. Adopting Spangler's assumptions of the effective width of the exterior prisms (i.e. B_c for each) and that the total height of fill (H) affects the settlement, the sum of the widths of the interior prism and the two exterior prisms is $3B_c$ and the total load at the critical plane is $3H\gamma_s B_c$. But the load of the interior prism at the critical plane is, by definition, $C_c\gamma_s B_c^2$. Therefore the sum of the loads on the two exterior prisms is

$$3H\gamma_s B_c - C_c\gamma_s B_c^2$$

and their combined width is $2B_c$.

Then at any depth 'h' below the plane of equal settlement the sum of the loads on the outer prisms is:

$$2V' = 3(H - H_e + h)\gamma_s B_c - V$$

where V has the same value as in equation (4.13).

The stress in the fill of the exterior prisms at depth h is then $\frac{V'}{B_c}$ and the strain is $\frac{V'}{B_c E}$ where E is the mean elastic modulus of the soil.

Then if λ' is the total settlement of an *exterior* prism below the plane of equal settlement, the settlement of the element is

$$d\lambda' = \frac{V'dh}{B_c E}$$

and

$$\lambda' = \int_0^{H_e} \left\{ \frac{3(H - H_e + h)\gamma_s B_c - V}{2B_c E} \right\} dh$$

Similarly if λ is the total settlement of the *interior* prism below the plane of equal settlement

$$\lambda = \int_0^{H_e} \left(\frac{V}{B_c E} \right) dh$$

Then equating the total settlements of the interior and exterior prisms between the plane of equal settlement and ground level:

$$\lambda + S_f + d_c = \lambda' + S_m + S_g \text{ (see Article 4.4 (a))}$$

and for convenience putting

$$\frac{(S_m + S_g) - (S_f + d_c)}{S_m} = r_{sd}$$

the settlement ratio,

Equation 4.15

$$\lambda - \lambda' = r_{sd} S_m$$

The value of Sm. The vertical stress in an *exterior* prism at the critical plane is

$$\frac{3H\gamma_s B_c - C_c\gamma_s B_c^2}{2B_c}$$

and the settlement below the critical plane in the height pB_c is

$$S_m = \left(\frac{3H\gamma_s B_c - C_c\gamma_s B_c^2}{2B_c} \right) \frac{pB_c}{E}$$

Then substituting the values of λ, λ' and Sm in equation (4.15) and dividing by $3\gamma_s B_c^2/2E$,

$$\frac{e^{\pm 2K\mu H_e/B_c} - 1}{\pm 2K\mu}\left[\frac{1}{2K_u} \pm \left(\frac{H}{B_c} - \frac{H_e}{B_c}\right) \pm \frac{r_{sd}p}{3}\right]$$

$$\pm \tfrac{1}{2}\left(\frac{H_e}{B_c}\right)^2 \pm \frac{r_{sd}p}{3}\left(\frac{H}{B_c} - \frac{H_e}{B_c}\right) e^{\pm 2K\mu H_e/B_c}$$

Equation 4.16

$$-\frac{1}{2K\mu}\frac{H_e}{B_c} \mp \frac{H}{B_c}\cdot\frac{H_e}{B_c} = \pm\, r_{sd}pH/B_c.$$

As before the *upper* signs are used for the *incomplete projection* condition where r_{sd} is positive and the *lower* signs for the *incomplete trench* condition where r_{sd} is negative. This equation is solved by trial (or computer) to obtain the value of He and thence of the load coefficient C_c for given values of $\frac{H}{B_c}$, $r_{sd}p$ and $K\mu$ Formula (4.14)).

The values of C_c were computed by Spangler[8,23] for various values of $r_{sd}p$ and H/B_c with $K\mu$ at a constant value of 0·19, (i.e. maximum adverse friction) for projection conditions, and 0·13 (i.e. minimum favourable friction) for ditch conditions, Marston having previously shown that variations in $K\mu$ made little difference in the value of C_c.

Spangler's values of C_c have been recently checked at the Building Research Station by computer and shown to be linear for both the incomplete conditions. They have been replotted in the Design Chart C3 to which the equations to the linear portions of the curves have been added to facilitate extrapolation.

The validity of the theory and of the existence of the plane of equal settlement were subsequently established experimentally by Spangler and Shlick[7] and, still later, Spangler investigated the value of the settlement ratio of a number of existing conduits[23] and made the recommendations given in Table B. These values have not yet been checked elsewhere, so far as the author is aware, but they and the projection formulae have been widely used in the U.S.A. and other countries.[1,2]

4.5 Positive Projection Loads Imposed by a Uniform Surcharge of Large Extent

In view of the variable nature of possible surcharge materials, their frictional characteristics may vary considerably. For the sake of simplicity, and in the interests of safety, they are therefore assumed to have the same effect on the load imposed on the conduit as an equivalent added height of fill having the same density and frictional characteristics as the fill in the embankment. Thus if the surcharge is U_s lb/ft², the equivalent additional height of

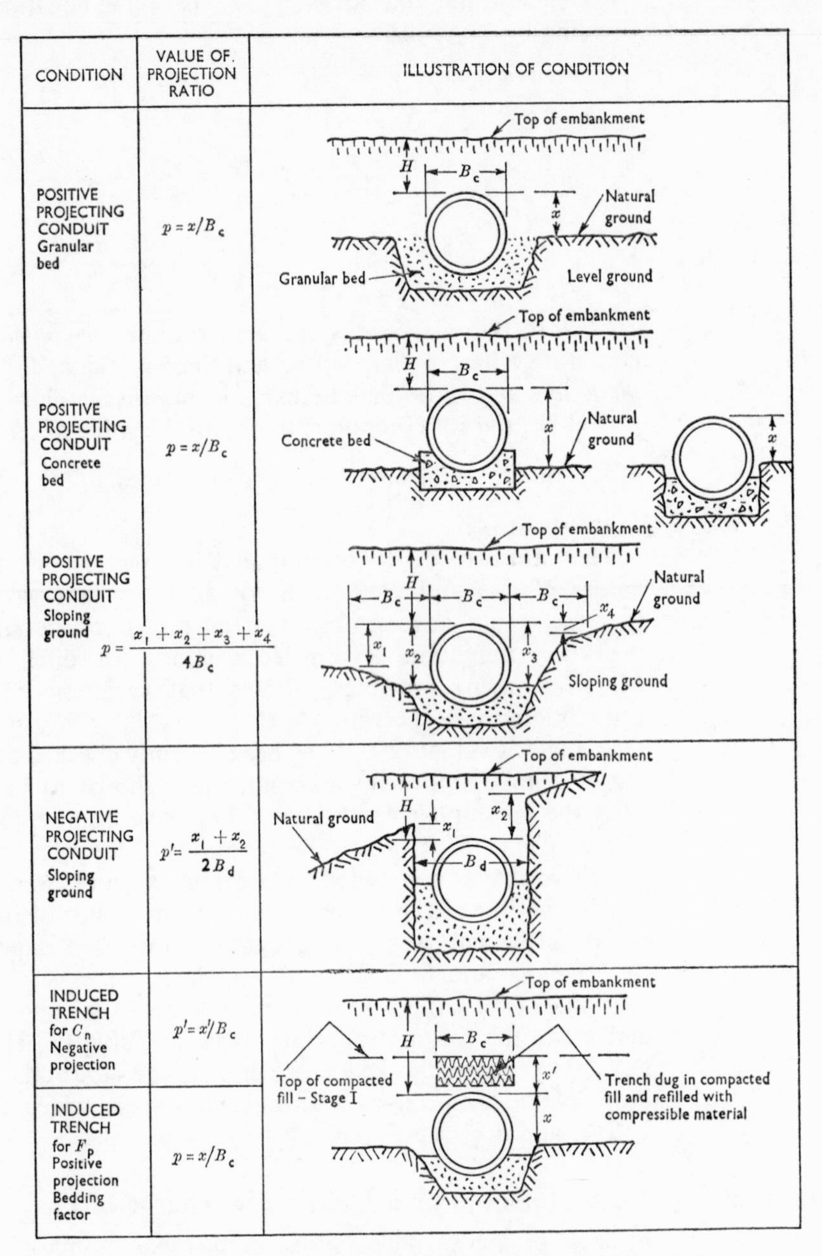

CONDITION	VALUE OF PROJECTION RATIO	ILLUSTRATION OF CONDITION
POSITIVE PROJECTING CONDUIT Granular bed	$p = x/B_c$	
POSITIVE PROJECTING CONDUIT Concrete bed	$p = x/B_c$	
POSITIVE PROJECTING CONDUIT Sloping ground	$p = \dfrac{x_1 + x_2 + x_3 + x_4}{4B_c}$	
NEGATIVE PROJECTING CONDUIT Sloping ground	$p' = \dfrac{x_1 + x_2}{2B_d}$	
INDUCED TRENCH for C_n Negative projection	$p' = x'/B_c$	
INDUCED TRENCH for F_p Positive projection Bedding factor	$p = x/B_c$	

Fig. 4.7 Values of the projection ratio for Positive and Negative Projection conditions

After Am. Conc. Pipe Assn.[2]*—by permission*

fill will be $\dfrac{U_s}{\gamma_s}$ ft. The effective equivalent total cover over the conduit is then $H' = H + U_s/\gamma_s$ and the combined fill and surcharge load will be $C'_c\gamma_s B_c^2$ where C'_c depends upon $\dfrac{H'}{B_c}$. Since the fill load alone is $C_c\gamma_s B_c^2$, the surcharge load will be

Formula 4.17
$$W'_{us} = (C'_c - C_c)\gamma_s B_c^2$$

The addition of the surcharge may change the 'complete' into the 'incomplete' condition. If the conduit is in the incomplete condition without the surcharge, since the curve for C_c is linear, the value of $(C'_c - C_c)$ will be constant for any higher value of $\dfrac{H}{B_c}$, i.e. there is no change in the surcharge load with increase in the depth of cover for a conduit originally in the incomplete condition.

4.6 Wide Trench Fill Loads—Transition Width—Transition Depth (Figs. 4.8, 4.9 (*a*) (*b*) (*c*))

Schlick[19] has shown experimentally that if a trench or subtrench is widened progressively, other conditions being unchanged, the fill load does not continue to increase according to the trench formula $W_c = C_d\gamma_s B_d^2$ but reaches the limiting value of $C_c\gamma_s B_c^2$ as in positive projection conditions under an embankment. The trench width at which this limit is reached (i.e. where $C_d\gamma_s B_d^2 = C_c\gamma_s B_c^2$) is known as the '*transition width*', and any further increase in width produces no further increase in load because the conduit is then in the positive projection (embankment) condition and the load consists of the weight of the prism of fill overlying the conduit plus the frictional forces operating in the fill. Various kinds of wide trench are shown in Fig. 4.8. A trench is said to be

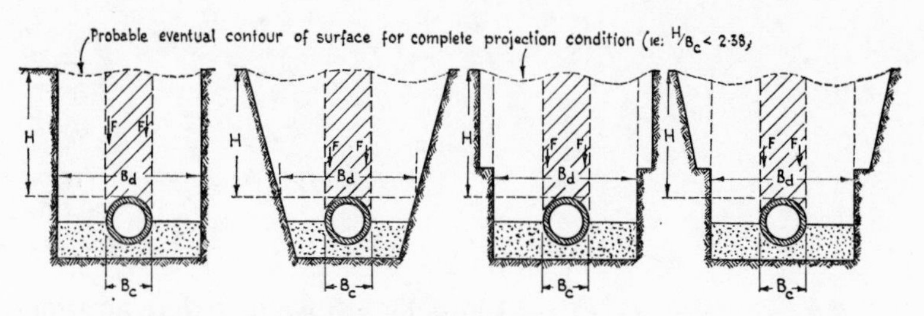

Fig. 4.8 Various 'wide' trench conditions

Crown Copyright—by permission

'narrow' where *Bd* is less than the transition width or 'wide' where *Bd* is greater than the transition width.

The transition width, however, becomes progressively smaller as the depth of the cover over the conduit increases, so that a trench of constant width and varying depth may be 'wide' down to a certain depth and 'narrow' at all greater depths, other conditions being unchanged. The depth at which this change occurs (again where $C_c\gamma_s B_c^2 = C_d\gamma_s B_d^2$) is known as the '*transition depth*' for the particular values of B_d and B_c (Fig. 4.9 (*a*)).

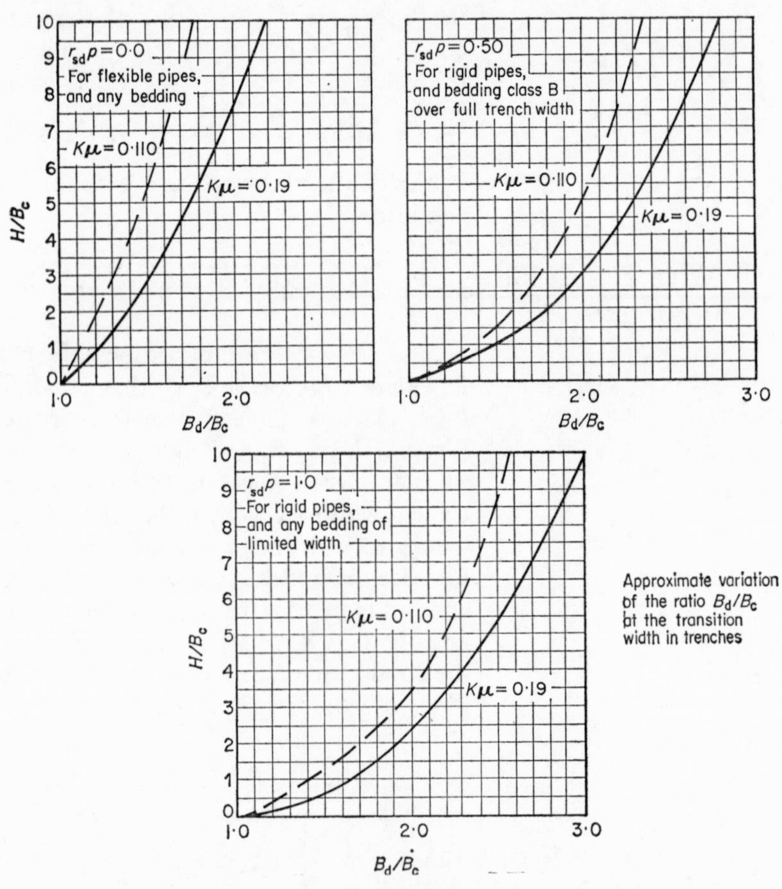

Fig. 4.9 (*a*) Variation of Transition width with H/B_c, $r_{sd}p$ and Ku'

Crown Copyright—by permission

Fig. 4.9 (*a*) shows how the transition width varies with the various values of H/B_c, $r_{sd}p$ and $K\mu'$ likely to be used in trench or 'wide' trench conditions for rigid or flexible pipes.

The exact values of the transition width or depth are most readily obtained graphically by plotting the curves of $C_d\gamma_s B_d^2$

and $C_c\gamma_s B_c^2$ and finding their point of intersection as in Fig. 4.9 (*b*).

Then for a wide trench or for values of *H* less than the transition depth in any trench

$$W'_c = C_c\gamma_s B_c^2 \text{ as in Form. (4.12)}$$

Note that this condition is a special case of the positive projection condition in which the value of r_{sd} is usually assumed to be 1·0 for rigid pipes and 0·0 for flexible pipes, and *p* also is 1·0 for pipes generally but is 0·5 for granular bedding extending over the full

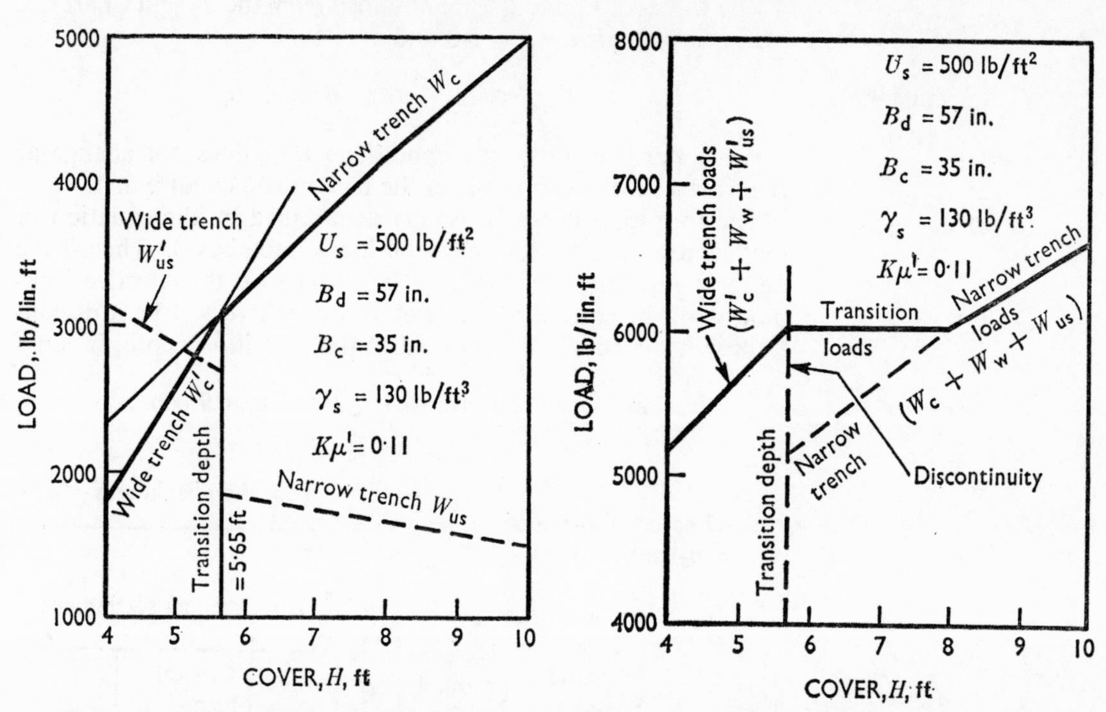

Fig. 4.9 (*b*) Determination of the transition depth

Fig. 4.9 (*c*) Total effective design loads, showing assumed transition loads

width of the trench and up to the middle of the pipe, (see Chart C9). C_c, therefore depends on $r_{sd}p = 1{\cdot}0$ or 0·5 for rigid pipes or 0·0 (when $W'_c = H\gamma_s B_c$) for flexible pipes and its value is obtained from the Design Chart C3. Depending on the ratio $\frac{H}{B_c}$, the condition may be either complete or incomplete. The effective value of B_c for non-circular conduits or for pipes with concrete arch support or concrete surround is the overall width of the completed structure as shown in Fig. 4.3.

4.7 'Wide' Trench Loads Imposed by a Uniform Surcharge of Large Extent

The load imposed on the conduit in a wide trench or subtrench by a uniform surcharge is assumed to be the same as if the conduit were in the positive projection condition under an embankment, i.e. the surcharge is assumed to be frictional and to have the same effect as the equivalent additional height of fill U_s/γ_s ft.

$$\text{i.e. } W'_{us} = (C'_c - C_c)\gamma_s B_c^2 \text{ as in Form. (4.17)}$$

where C'_c depends upon $H'/B_c = (H + U_s/\gamma_s)/B_c$.

The values of C_c and C'_c are obtained from the Design Chart C3.

For flexible pipes $r_{sd} = 0{\cdot}0$ and

Formula 4.17*a*

$$W'_{us} = (H' - H)\gamma_s B_c = U_s B_c.$$

As in positive projection conditions W'_{us} does not change as H increases if the conduit is in the incomplete condition.

The transition depth is usually determined by the equation of the 'narrow' and 'wide' trench fill loads (see above). Then if the uniform surcharge load $W'_{us} > W_{csu}$ and so is the effective component of the total load, W_e (see below), there will be a discontinuity in the curve of total load at the transition depth, because

$$W'_c + W'_{us} > W_c + W_{us} \text{ (see Fig. 4.9 (}c\text{))}$$

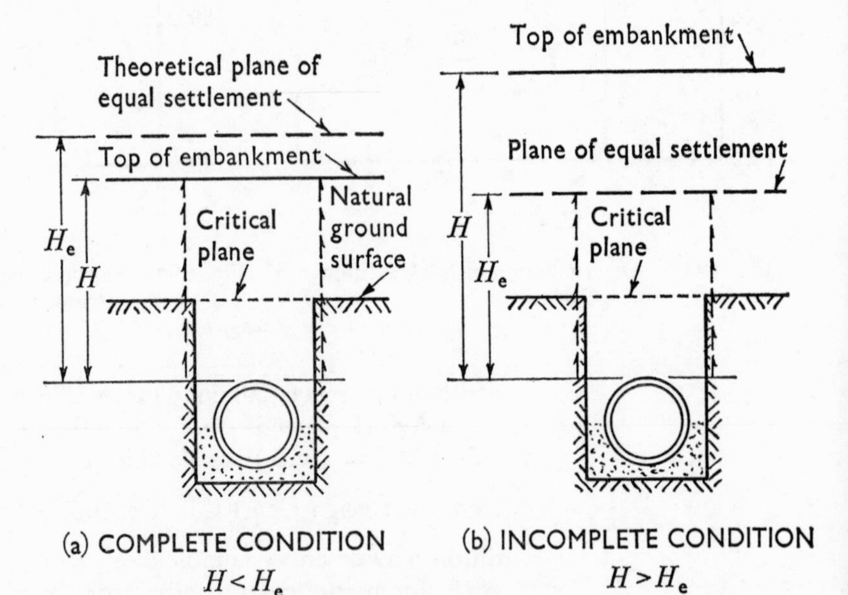

Fig. 4.10 (*a*) Elements of the Negative Projection condition

After Spangler—by permission

To meet this theoretical inconsistency the value of $W'_c + W'_{us}$ at the transition depth may be assumed to remain constant, as H increases, until it is exceeded by $W_c + W_{us}$ at some greater depth of cover, as shown in Fig. 4.9 (*c*).

If however the transition depth is assumed to be determined by the combined loads (i.e. where $(W'_c + W'_{us}) = (W_c + W_{us})$), there will obviously be no discontinuity in the total load curve, and the total load will be lower at values of H in the vicinity of the 'fill load' transition depth. This assumption, however, has not yet been confirmed experimentally. It is not valid, and there is no discontinuity, where the concentrated surcharge load W_{csu} exceeds the uniform surcharge load W'_{us} or W_{us}.

4.8 Negative Projection Fill Loads Under an Embankment or Valley Fill (Fig. 4.10)

4.8 (a) Theory In 1950 Spangler[24] studied the theoretical effect of installing a conduit in a narrow sub-trench excavated in the natural ground underlying an embankment or fill (see Fig. 4.10). He called this the *negative projection condition* to distinguish it from Marston's positive (ditch) projection condition, since it also induces a settlement of the interior prism relative to the exterior prisms and thereby induces beneficial frictional forces in the fill and reduces the fill load imposed on the conduit to the weight of the interior prism minus the sum of the frictional forces in the fill.

The conception of a plane of equal settlement and a height of equal settlement was retained, and with it, the 'complete' and 'incomplete' conditions, but the critical plane was postulated to be the horizontal plane through the top of the sub-trench, i.e. at the natural ground level, and the width of the interior prism was the width of the subtrench.

4.8 (b) The Negative Projection Formulae and Their Derivation

4.8 (b) (i) The Complete Negative Projection Condition ($H_e > H$) (Fig. 4.11 (*a*)) The theoretical derivation of the fill load formula is the same as that for a trench load, except that the coefficient of friction for the fill (μ) is used in place of μ'.

Then

Formula 4.18
$$W_c = \gamma_s B_d^2 \left(\frac{1 - e^{-2K\mu H/B_d}}{2K\mu}\right)$$

and the expression in brackets is known as the load coefficient C_n and the formula is written

Formula 4.19
$$W_c = C_n \gamma_s B_d^2$$

The value of K is Rankine's coefficient of active earth pressure as before, but Spangler assumes the value of $K\mu$ to be not less

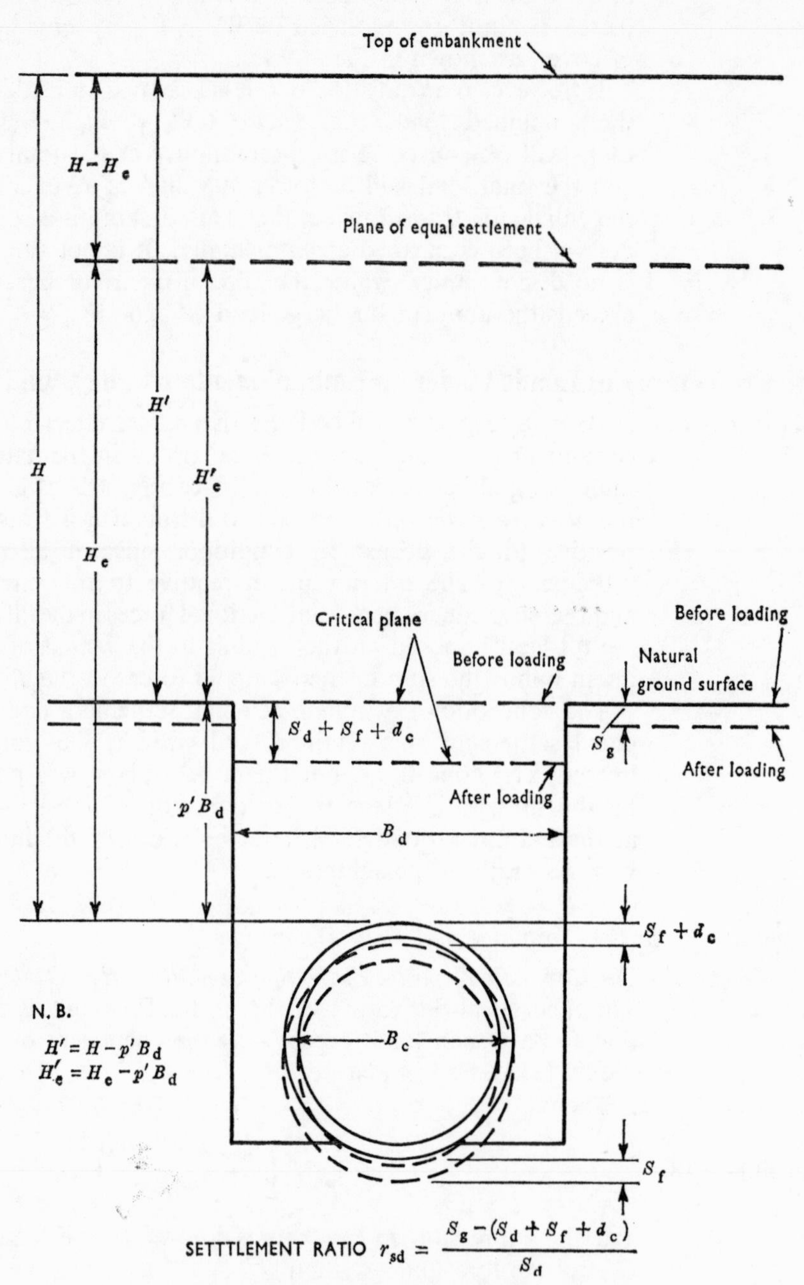

4.10 (*b*) Negative Projection settlement ratio

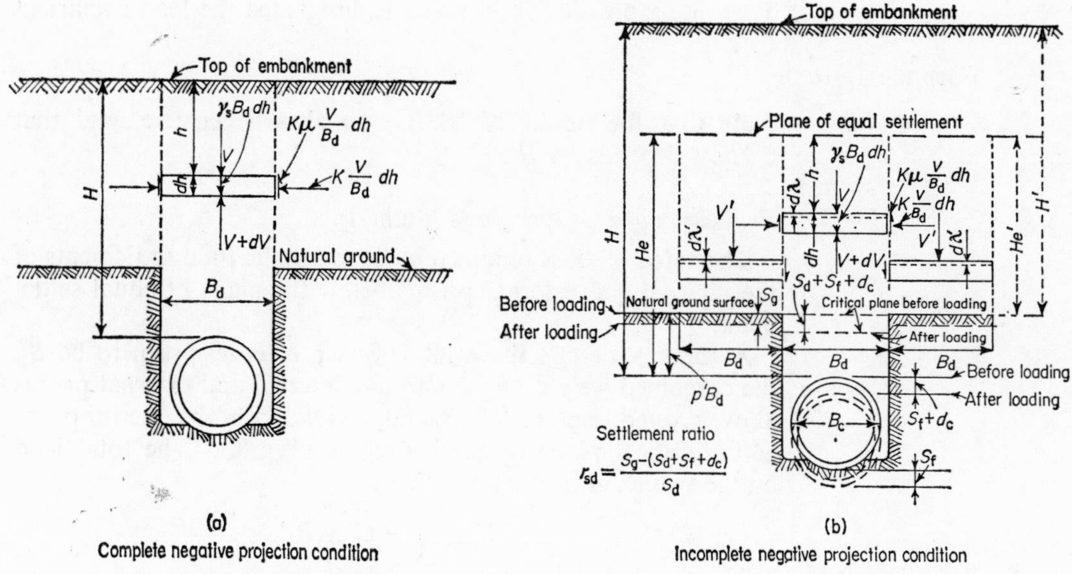

(a) in the complete projection condition (b) in the incomplete projection condition

Fig. 4.11 Forces acting on a conduit in the Negative Projection condition

After Spangler—by permission

than 0·13. For greater values of $K\mu$ his values of C_n are somewhat conservative.

Note that $C_n = C_d$ if $K\mu = K\mu' = 0{\cdot}13$.

Spangler's values of C_n are given in the Design Charts C4 (*a*), (*b*), (*c*), (*d*) for various values of the negative Projection Ratio p'.

4.8 (*b*) (ii) *The Incomplete Negative Projection Condition* ($H_e < H$) (Fig. 4.11 (*b*))

Considering the equilibrium of an element of fill, as in Positive Projection, where h is its depth below the plane of equal settlement

$$V + dV = V + \gamma_s B_d dh - 2K\mu \frac{V}{B_d} dh$$

and when $h = 0$, $V = (H - H_e)\gamma_s B_d$ and the solution of the differential equation is

Equation 4.20
$$V = \gamma_s B_d^2 \left\{ \frac{1 - e^{-2K\mu h/B_d}}{2K\mu} + \left(\frac{H}{B_d} - \frac{H_e}{B_d} \right) e^{-2K\mu h/B_d} \right\}$$

When $h = H_e$, $V = W_c$ and

Formula 4.21
$$W_c = \gamma_s B_d^2 \left\{ \frac{1 - e^{-2K\mu H_e/B_d}}{2K\mu} + \left(\frac{H}{B_d} - \frac{H_e}{B_d} \right) e^{-2K\mu H_e/B_d} \right\}$$

Then the expression in brackets is designated the load coefficient C_n and

Formula 4.21*a*

$$W_c = C_n \gamma_s B_d^2$$

Note that the power of '*e*' is now always negative, and that C_n now depends on H_e.

Evaluation of H_e (Spangler's Method).

The value of H_e is obtained by equating the total settlements of the internal and external prisms below the plane of equal settlement.

Assuming the effective width of each external prism to be B_d, the combined weight of the effective internal and external prisms above ground level is $3H'\gamma_s B_d$ and the load from the interior prism at the critical plane (ground level) is $C'_n \gamma_s B_d^2$. The total load on the two exterior prisms is then:

$$3H'\gamma_s B_d - C'_n \gamma_s B_d^2$$

where C'_n depends on $H' = (H - p'B_d)$ and $H'_e = (H_e - p'B_d)$ (see Fig. 4.11(*b*)). Similarly the sum of the loads on the outer prisms at any depth h below the plane of equal settlement is

$$2V' = 3(H' - H'_e + h)\gamma_s B_d - V$$

where V has the same value as in Equation 4.20.

By the same reasoning as for positive projection, the total settlement of an *exterior* prism between the plane of equal settlement and ground level, i.e. over the height H'_e, is

$$\lambda' = \int_0^{H'_e} \left\{ \frac{3(H' - H'_e + h)\gamma_s B_d - V}{2B_d E} \right\} dh$$

and the settlement of the natural ground is S_g.

Then the total settlement of an *exterior* prism is

$$\lambda' + S_g$$

The settlement of the *interior* prism between the plane of equal settlement and ground level is

$$\lambda = \int_0^{H'_e} \frac{V}{B_d E} dh$$

and denoting the settlement of the fill in the sub-trench between ground level and the top of the conduit as S_d, the settlement of the invert of the conduit as S_f, and the vertical deflection of the conduit as d_c, the total settlement of the *interior* prism is

$$\lambda + S_d + S_f + d_c$$

Equating the *total* settlements

$$\lambda' + S_g = \lambda + S_d + S_f + d_c$$

$$\text{i.e. } \lambda - \lambda' = S_g - (S_d + S_f + d_c)$$

The settlement ratio is now regarded as

$$r_{sd} = \frac{S_g - (S_d + S_f + d_c)}{S_d}$$

and

Equation 4.22

$$\lambda - \lambda' = r_{sd} S_d$$

The *negative projection ratio* is p', where $p'B_d$ is the depth of the top of the conduit *below* the critical plane. The load of the interior prism at the critical plane is $C'_n \gamma_s B_d^2$, where C'_n depends on H' and $H'e$. The value of S_d is then

$$S_d = \left(\frac{C'_n \gamma_s B_d^2}{B_d E}\right) p' B_d$$

Substituting the appropriate values of λ, λ' and S_d in Equation (4.22) and dividing by

$$\frac{3\gamma_s B_d^2}{2E},$$

Equation 4.23

$$\frac{e^{-2K\mu H'_e/B_d} - 1}{2K\mu}\left[\frac{H'}{B_d} - \frac{H'_e}{B_d} - \frac{1}{2K\mu}\right] - \frac{H'_e}{B_d}\left[\left(\frac{H'}{B_d} - \frac{H'_e}{B_d}\right) + \frac{1}{2}\frac{H'_e}{B_d} - \frac{1}{2K\mu}\right]$$

$$= \frac{2}{3} r_{sd} p' \left[\frac{e^{-2K\mu H'_e/B_d} - 1}{-2K\mu} + \left(\frac{H'}{B_d} - \frac{H'_e}{B_d}\right) e^{-2K\mu H'_e/B_d}\right]$$

This equation is solved by trial (or computer) for various values of $\frac{H}{B_d}$, r_{sd} and p' to obtain H_e and thence C_n (Formula (4.21)).

Values of C_n were computed by Spangler(24) for various values of $\frac{H}{B_d}$, r_{sd} and $p' = 0{\cdot}5$, $1{\cdot}0$, $1{\cdot}5$, and $2{\cdot}0$, and a constant value of $K\mu = 0{\cdot}13$ (i.e. $\mu = 0{\cdot}2$) which he considered safe for ordinary fills. They have been recently checked by computer at the Building Research Station and shown to be linear with reference to $\frac{H}{B_c}$ for the incomplete condition. They are plotted in the Design Charts C4, to which the equations of the linear portion of the curves have been added to facilitate extrapolation. These chart values of C_n cannot be used for values of μ less than 0·2. Interpolated values may be used for other values of p' between 0 and

2·0. They are conservative for values of $\mu > 0{\cdot}2$ or of $p' > 2{\cdot}0$.

The validity of the theory was established experimentally by Schlick in 1952.[20]

The *values of Settlement Ratios for Negative Projection* are not yet well established. Spangler's recommended values, which have been adopted by the A.S.C.E.,[1] are given in Table B.

Note from Charts C4 that for equal values of $\frac{H}{B_d}$ the load W_c decreases as the settlement ratio decreases and as the value of p' increases; also that C_n depends on r_{sd} (which is always negative for rigid conduits but may be zero for flexible pipes) and not on $r_{sd}p'$; also that when $r_{sd} = 0$, $C_n < \frac{H}{B_d}$ but approaches it as p' decreases.

For flexible pipes with well compacted side fill the fill load is reduced in a manner similar to that in trench loading

Formula 4.24 i.e. $W_c = C_n\gamma_s B_c B_d$

but for large pipes ($D \geqslant 48$ in.) where the ratio B_d/B_c becomes small, the usual value of $C_n\gamma_s B_d^2$ is considered to be more generally applicable.

4.8 (b) (iii) *Transition Width of the Subtrench*

The conception of a transition width for the sub-trench in line with that for trenches proper, would appear to be reasonable.* Transition would be expected to occur where the negative projection fill load and the positive projection fill load are equal i.e. when

$$C_n\gamma_s B_d^2 = C_c\gamma_s B_c^2$$

This means that the negative projection load can never exceed the positive projection load, however wide the subtrench may be and the inference is that to reduce the fill load as much as possible, the width of the subtrench should be as small as working space permits.

4.8 (b) (iv) *The Neutral Condition*

If $p' = 0$, and therefore $S_d = 0$, the settlement ratio has no real value and the condition becomes 'neutral'. The deformation of the critical plane is assumed to be negligibly small, and the plane of equal settlement coincides with the critical plane. So there are no frictional forces acting in the fill, and the fill load is simply the weight of the interior prism of soil,

Formula 4.25 i.e. $W_c = H\gamma_s B_d$

* In a private communication to the author, Spangler agrees.

This load can be reduced and assimilated to the positive projection neutral condition, with $r_{sd} = 0$ or $p = 0$ provided the side fill in the sub-trench is so compacted as to be equivalent to the undisturbed soil.

Then

Formula 4.26 $$W_c = H\gamma_s B_c.$$

4.9 Negative Projection Loads Imposed by a Uniform Surcharge of Large Extent

In the Complete Negative Projection condition, the surcharge is assumed to be non-frictional as in trench loading and the load imposed on any conduit will therefore have the same value as for trench loading,

i.e. $W_{us} = C_{us}B_d U_s$ as in Form. (4.5)

except that C_{us} depends upon $K\mu = 0{\cdot}13$ instead of $K\mu'$. The value of C_{us} can be obtained from the Design Chart C2.

In the incomplete condition, the addition of the surcharge depresses the plane of equal settlement and so its frictional qualities are of no consequence, and it may be treated as an equivalent additional height of fill of U_s/γ_s ft. The additional load it imposes on any conduit is then

Formula 4.27 $$W_{us} = (C'_n - C_n)\gamma_s B_d^2$$

where C'_n depends upon $(H + U_s/\gamma_s)/B_d$. Its value can be obtained from the same Design Chart C4 and value of p' as that used for C_n.

Note that the addition of the surcharge may convert the complete into the incomplete condition but Formula (4.5) should be used for W_{us} in such borderline cases. Also that since the curves of C_n are linear for the incomplete condition, the value of $(C'_n - C_n)$ is independent of $\frac{H}{B_d}$ and is therefore constant for a given value of U_s/γ_s.

4.10 Induced Trench Condition Fill Load Under an Embankment or Valley Fill

Concerned by the very large loads imposed on culverts laid in the Positive Projection condition, Marston[7,17] suggested in 1922 that the fill load could be greatly reduced by excavating a sub-trench in the fill above the conduit, and then refilling it with some compressible material such as loose fill or, where differential settlement of the surface is of no consequence, straw, before proceeding with the remainder of the filling. He called this the *imperfect ditch condition* because it produced a reversal in the frictional forces by inducing the interior prism to settle relative to the exterior prisms, and thus a condition somewhat similar to that of a ditch (trench). To avoid the confusion arising from the terms 'imperfect' and 'incomplete' it is now known to the author as the '*induced trench condition*. Spangler[24] considered that the induced trench condition was a special case of the negative projection condition with the substitution of the conduit width B_c for the sub-trench width B_d, provided the width of the subtrench above the conduit was B_c and it need not be wider.

Then

Formula 4.28
$$W_c = C_n \gamma_s B_c^2$$

where C_n has the negative projection value for $\frac{H}{B_c}$. The applicable settlement ratio r_{sd} is not well established but values of 0·0 to −0·3 have been tentatively recommended by Spangler. (See Table B). The projection ratio p' is such that $p'B_c$ is the distance from the top of the conduit to the top of the subtrench.

The theory was confirmed experimentally by Schlick in 1952[20].

Note however that this condition, which involves a layer of soil or other yielding material in the subtrench which has different physical characteristics from those of the main fill, will affect the value of S_d and therefore of r_{sd} and hence the value of H/B_c at which $H = H_e$ where the condition changes from

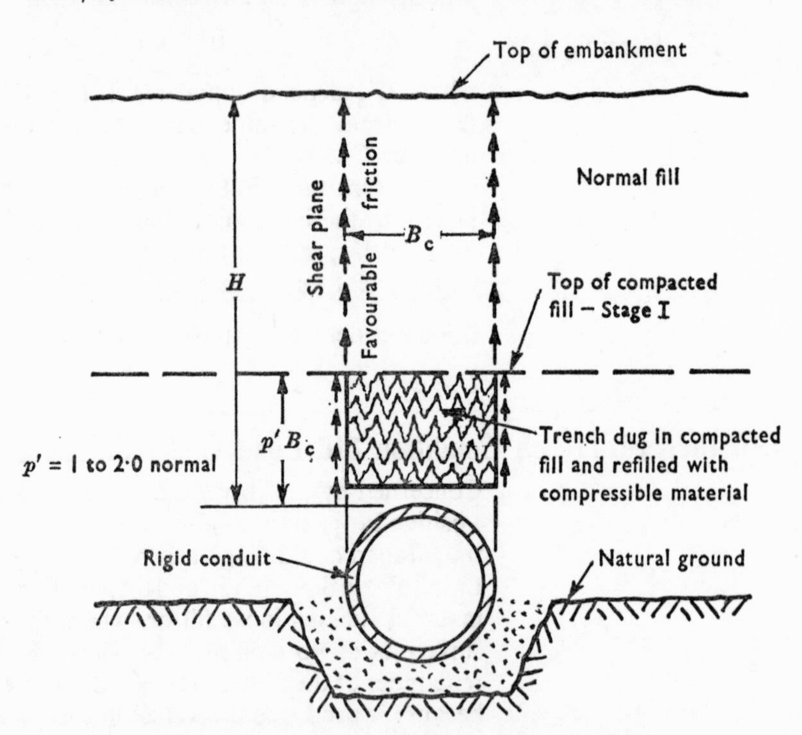

Fig. 4.12 Elements of the Induced Trench condition
After Spangler—by permission

'complete' to 'incomplete'. Thus, excessive settlement in the subtrench may cause excessive settlement of the surface of the fill over the pipeline, e.g. where straw or similar highly compressible material is used in the subtrench. This is of no importance where differential settlement of the surface of the fill is of no consequence, or in a high fill which is unquestionably in the 'incomplete' condition.

Note also that the conduit is installed in the positive projection condition, but loaded in the negative projection condition. The active pressure of the fill is exerted on the conduit and the rigid pipe bedding factor is therefore increased to the positive projection value F_p (see Article 8.2 (*a*)).

4.11 Induced Trench Condition Loads Imposed by a Uniform Surcharge of Large Extent

This load is regarded as being the same as for negative projection conditions except for the use of B_c in place of B_d

Formula 4.29 i.e. $W_{us} = C_{us}B_cU_s$ for the 'complete' or border line condition

or

Formula 4.30 $W_{us} = (C'_n - C_n)\gamma_s B_c^2$ for the 'incomplete' condition

4.12 Summary of the Theoretical Values of the Load Coefficients

The theoretical values of C_d, C_{us}, C_c and C_n are summarised in Table A.

4.13 Probable Fill and Uniform Surcharge Load on Each of Several Pipes in the Same Trench, or Under the Same Embankment

The theory of 'narrow' and 'wide' trench loading and of embankment loads described above and the conception of the deformation of the critical plane as an indicator of the direction of the frictional forces acting in the fill, can be utilised to give approximate estimates of the fill and uniform surcharge loads of large extent imposed on each of several conduits in the same trench or under the same embankment, whatever their disposition, as in the following illustrative examples. The uniform surcharge load is assumed to correspond to the type of fill load imposed on each conduit or half conduit.

4.13 (a) Trench Loading Considering the probable relative settlements of the overlying and intervening prisms of fill and the corresponding directions of the frictional forces acting along the probable shear planes:

Example 1. In the arrangement shown in Fig. 4.13, assuming that B_{dA} and B_{dB} do not exceed the respective transition widths for pipes A and B respectively, the probable relative settlements and frictional planes are shown by the broken lines and the direction of the frictional forces acting on the respective pipes by arrows.

Evidently the left half of pipe B will be subjected to narrow trench loading and the fill load on it will be $\frac{1}{2}C_d\gamma_s B_{dA}^2$. The right half of the pipe however will be subjected to wide trench loading since the intermediate prism of fill to the right of the central prism overlying the pipe will tend to settle more than the central prism. The fill load on this half will therefore be $\frac{1}{2}C_c\gamma_s B_{cA}^2$. The probable total fill load on pipe A will therefore be:

$$W_{cA} = \tfrac{1}{2}(C_d\gamma_s B_{d\,A}^2 + C_c\gamma_s B_{cA}^2)$$

and the corresponding uniform surcharge load will be:

$$W_{usA} = \tfrac{1}{2}\{C_{us}B_{dA}U_s + (C'_c - C_c)\gamma_s B_{cA}^2\}$$

where C_d, C_c, C'_c and C_{us} have the appropriate values for pipe A and the cover depth H_A.

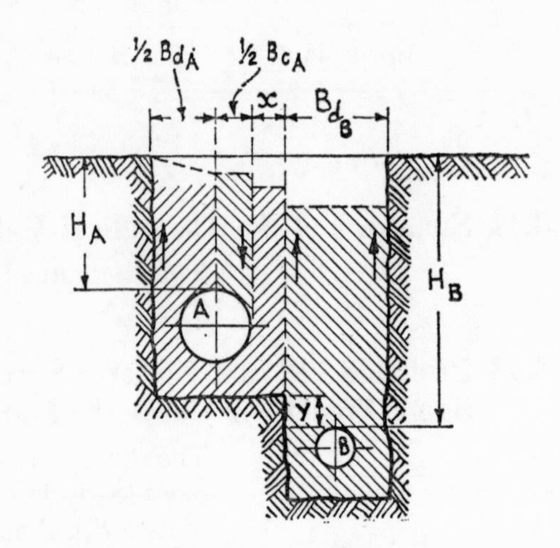

Fig. 4.13 Example of multiple pipes in the same trench

Crown Copyright—by permission

Since the prism of fill of width B_{dB} will tend to settle more than the intermediate prism to its left, pipe B will be subjected to narrow trench loading if the width X of the intermediate prism is less than say B_{dB}, whence

$$W_{cB} = C_d\gamma_s B_{dB}^2$$

and

$$W_{usB} = C_{us}B_{dB}U_s$$

where C_d and C_{us} have the appropriate values for pipe B and the cover depth H_B.

If the width X is large, say equal to or greater than B_{dB}, and the depth Y is small, say approaching zero, the frictional force acting on the left half of pipe B could approach zero, in which event the load on that half of the pipe would approach a maximum value of $\frac{1}{2}H_B\gamma_sB_{dB}$ and the total fill load on pipe B would be

$$W_{cB} = \tfrac{1}{2}(C_d\gamma_sB_{dB}^2 + H_B\gamma_sB_{dB})$$

and the corresponding uniform surcharge load would be

$$W_{usB} = \tfrac{1}{2}(C_{us}B_{dB}U_s + B_{dB}U_s)$$

As the depth Y increases, the fill load on the left half of pipe B will decrease from the maximum value $\frac{1}{2}H_B\gamma_sB_{dB}$ towards the minimum value $\frac{1}{2}C_d\gamma_sB_{dB}^2$. The minimum loading of pipe B for a given value of H_B will occur therefore when the width X is as small as practicable and the depth Y is as large as practicable.

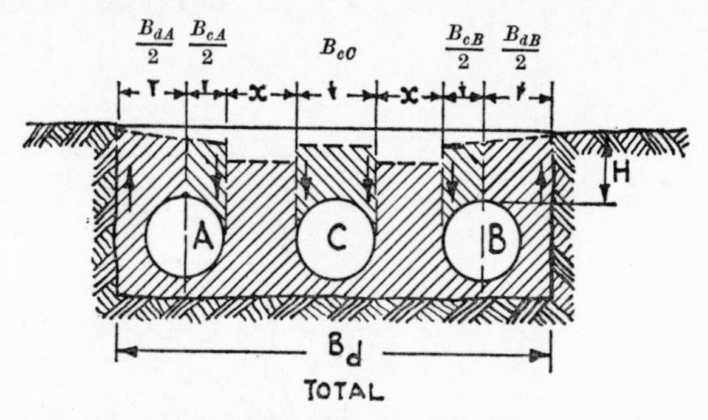

Fig. 4.14 Example of multiple pipes in the same trench

Crown Copyright—by permission

Example 2. Again, in the arrangement shown in Fig. 4.14, assuming that B_{dA} and B_{dB} do not exceed the respective transition widths for pipes A and B and that the width $(2X + B_{cC})$ is equal to or greater than the transition width for pipe C the fill loads on pipes A and B will evidently be similar to that for pipe A in Fig. 4.13, namely

$$W_{cA} = \tfrac{1}{2}(C_d\gamma_sB_{dA}^2 + C_c\gamma_sB_{cA}^2)$$

and

$$W_{cB} = \tfrac{1}{2}(C_d\gamma_sB_{dB}^2 + C_c\gamma_sB_{cB}^2)$$

and pipe C will be subjected to wide trench loading, i.e.

$$W_{cC} = C_c\gamma_sB_{cC}^2$$

If, however, the width $(2X + B_{cC})$ is less than the transition width for pipe C there may be no tendency for the intermediate prisms to settle relative to the central prisms overlying the pipes. The fill load on pipe C will then be $H\gamma_s(B_{cC} + X)$ and on pipes A and B, $\frac{1}{2}\{C_d\gamma_sB_{dA}^2 + H\gamma_s(B_{cA} + X)\}$ and $\frac{1}{2}\{C_d\gamma_sB_{dB}^2 + H\gamma_s(B_{cB} + X)\}$ respectively.

Alternatively when X is small the whole trench load may be shared equally by the three pipes and the fill load on each will be:

$$W_c = (C_d \gamma_s B_d^2 \text{ TOTAL})/3$$

The larger of these alternatives should be adopted for each pipe. The uniform surcharge loads will correspond to the fill load in each case as for the loads of Fig. 4.13, Example 1.

Example 3. In the arrangement shown in Fig. 4.15, pipes A and B are existing pipes crossing a new trench. If the width $X + \frac{1}{2}B_{cB}$ exceeds half the transition width for pipe B, both pipes will be subjected to wide trench loading, i.e. the fill loads will be:

$$W_c = C_c \gamma_s B_{cA}^2 \text{ and } C_c \gamma_s B_{cB}^2 \text{ respectively}$$

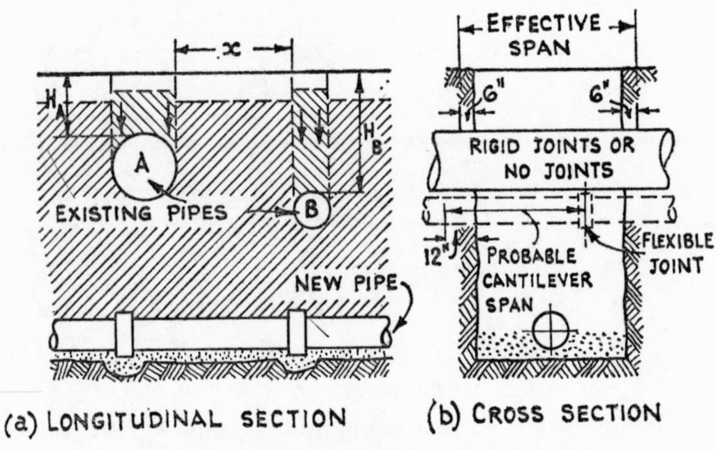

Fig. 4.15 Example of multiple pipes in the same trench
Crown Copyright—by permission

where C_c has a value for each pipe depending on $r_{sd}p$ which may be much greater than unity depending on the care exercised in compacting the fill below the pipes. These loads may be considerably larger than those for which the pipes were designed and the pipes may therefore require strengthening by a higher class of bedding or replacement by stronger pipes. Again, if the fill below the pipes is more yielding than the undisturbed soil in the trench sides the pipes will be constrained to bridge the trench. Their ability to carry their new total effective load as beams or cantilevers, assuming the effective spans shown in Fig. 4.15, is doubtful, especially for unreinforced rigid pipes. Conversely if the new supports are less yielding than the undisturbed soil they will constitute hard spots and may cause fractures of the pipes especially at shallow depths under roads. It may therefore be preferable to relieve such pipes of load rather than to attempt

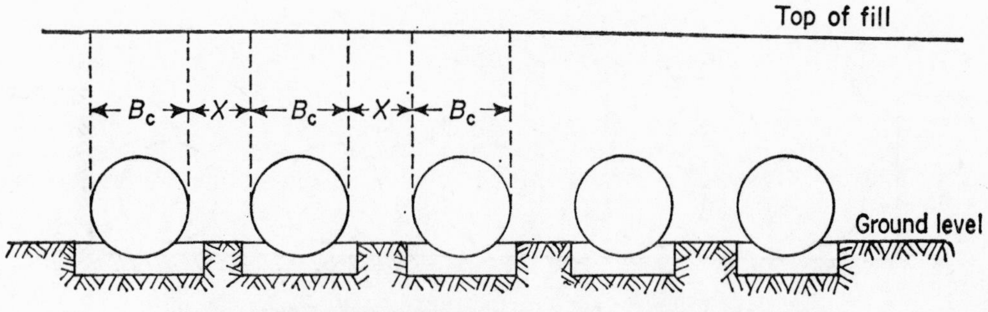

Fig. 4.16 Example of parallel pipes under embankments or fills

to strengthen them, e.g. by the use of ducts, or by ensuring that the fill below the pipes is so treated and compacted as to have the same carrying capacity as the undisturbed soil in the trench sides, or by placing reinforced concrete bridging slabs of adequate strength and length above and clear of the pipes, with ample bearing area in the undisturbed soil of the trench sides.

4.13 (b) Embankment Loading

Example 4. Fig. 4.16.

Any number of parallel pipes all of the same diameter laid under positive projection conditions with inverts at the same level.

For the outside pipes, use the lesser value of

$$W'_c = C_c\gamma_s B_c^2 \text{ or } W_c = \tfrac{1}{2}\{C_c\gamma_s B_c^2 + H\gamma_s(B_c + X)\}$$

For the intermediate pipes, use the lesser value of

$$W'_c = C_c\gamma_s B_c^2 \text{ or } W_c = H\gamma_s(B_c + X)$$

The effective value of W_{us} is obtained by treating it as an equivalent additional height of fill, in a manner similar to that already described for single pipes.

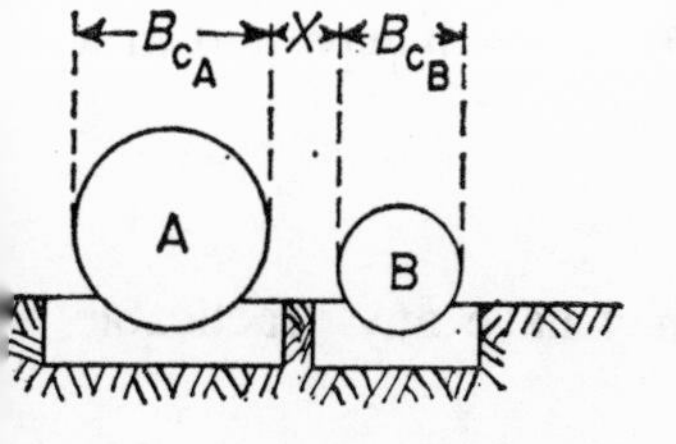

;. 4.17 Example of parallel •es under embankments or fills

Example 5. Fig. 4.17. For two parallel pipes of different diameter, *A* and *B* ($A > B$), laid under positive projection conditions, the inverts being at approximately the same level.

For pipe A,

$$W'_{cA} = C_c\gamma_s B_{cA}^2$$

For pipe B, use the lesser value of

$$W'_{cB} = C_c\gamma_s B_{cB}^2 \text{ or } W'_{cB} = \tfrac{1}{2}\{C_c\gamma_s B_{cB}^2 + C_d\gamma_s(B_{cB} + 2X)^2\}$$

Example 6. Fig. 4.18. Any number of parallel pipes all of the same diameter laid in negative projection conditions with inverts at the same level:

For the outside pipe, use the lesser value of

$$W_c = \tfrac{1}{2}\{C_n\gamma_s B_d^2 + H\gamma_s(B_c + X)\},$$

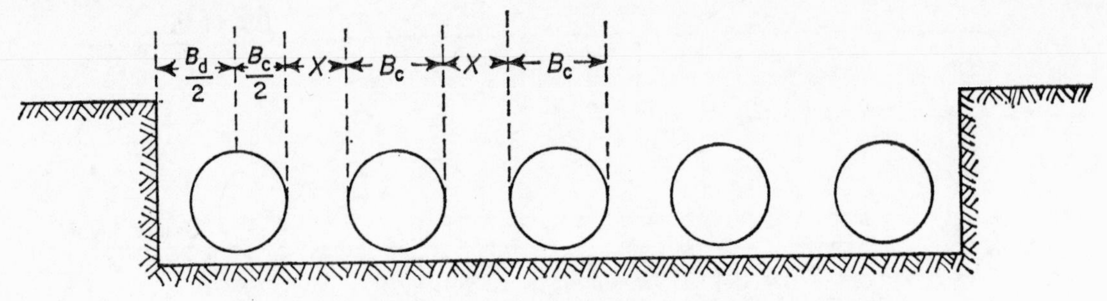

Fig. 4.18 Example of parallel pipes under embankments or fills

or

$$W_c = \tfrac{1}{2}(C_n\gamma_s B_d^2 + C_c\gamma_s B_c^2)$$

For the intermediate pipes, use the lesser value of

$$W'_c = C_c\gamma_s B_c^2 \text{ or } W'_c = H\gamma_s(B_c + X)$$

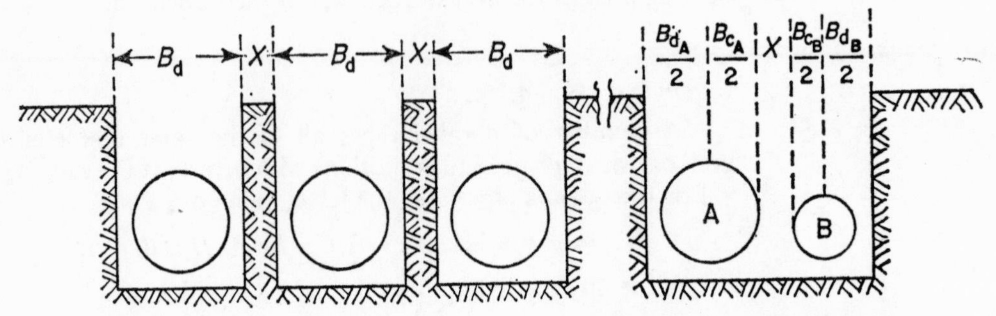

Fig. 4.19 Example of parallel pipes under embankments or fills

Fig. 4.20 Example of parallel pipes under embankments or fills

Example 7. Fig. 4.19. As in Example 6 but each pipe laid in a separate sub-trench of the same width:

$$W_c = C_n\gamma_s B_d^2$$

irrespective of the value of X which may have any practicable value $> B_d$.

Example 8. Fig. 4.20. Two parallel pipes of different diameters A and B ($A > B$) laid in the same subtrench in negative projection conditions with their inverts at approximately the same level:

For the larger pipe A

$$W_{cA} = \tfrac{1}{2}(C_n\gamma_s B_{dA}^2 + C_c\gamma_s B_{cA}^2)$$

For the smaller pipe B use the lesser value of

$$W_{cB} = \tfrac{1}{2}(C_n\gamma_s B_{dB}^2 + C_c\gamma_s B_{cB}^2)$$

or

$$W_{cB} = C_n \gamma_s \left(\frac{B_{dB}}{2} + \frac{B_{cB}}{2} + X \right)^2$$

N.B. p' has different values for pipes A and B.

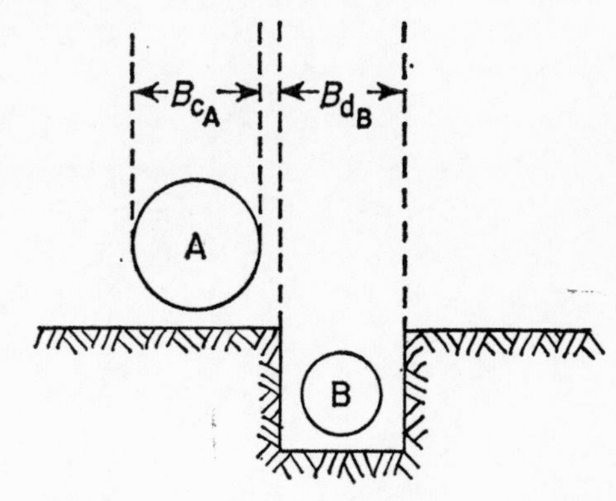

Fig. 4.21 Example of parallel pipes under embankments or fills

Example 9. Fig. 4.21. Two parallel pipes of different diameters A and B where pipe A is laid in the positive projection condition and pipe B in the negative projection condition:

For pipe A,

$$W_{cA} = C_c \gamma_s B_{cA}^2$$

For pipe B,

$$W_{cB} = C_n \gamma_s B_{dB}^2$$

4.14 The Effect of Submergence on Fill Loads

Where the fill is *permanently* submerged in water of density γ_w to a height of H_s feet above the top of the pipe, as in a partially submerged trench (Fig. 4.22) (or embankment), the fill load will be reduced as follows:

4.14 (a) In Narrow Trench Conditions

Formula 4.31
$$W_c = C_d \gamma_s B_d^2 - C_{ds} \gamma_w B_d^2$$

where C_{ds} depends on H_s/B_d (Chart C1).

Formula 4.32 If $H_s > H$, $W_c = C_d(\gamma_s - \gamma_w)B_c^2$ where C_d depends on H/B_d.

4.14 (b) In 'Wide' Trench Conditions or in Positive Projection Conditions

under an embankment:

Formula 4.33
$$W'_c = C_c \gamma_s B_c^2 - C_{cs} \gamma_w B_c^2$$

where C_{cs} depends on H_s/B_c (Chart C3).

Formula 4.34 If $H_s > H$, $W'_c = C_c(\gamma_s - \gamma_w)B_c^2$ where C_c depends on $\frac{H}{B_c}$.

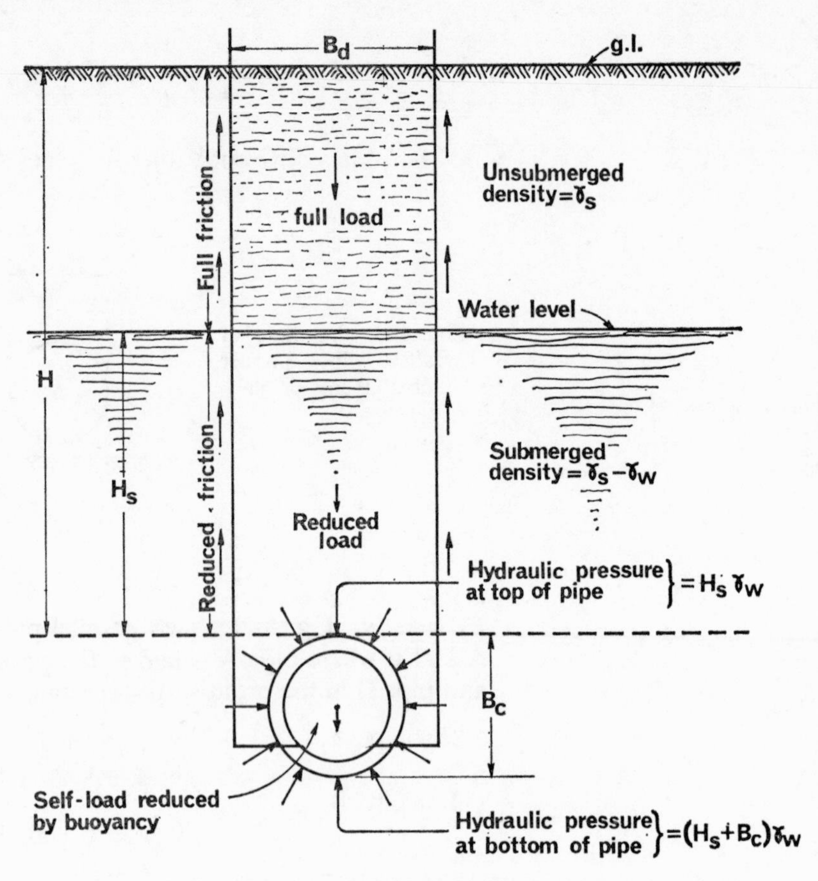

Fig. 4.22 Elements of the submerged condition in trench
Crown Copyright—by permission

4.14 (c) In Negative Projection or Induced Trench Conditions

under an embankment:

Formula 4.35 $$W_c = C_n\gamma_s B_d^2 - C_{ns}\gamma_w B_d^2$$

where C_{ns} depends on H_s/B_d (Chart C4).

Formula 4.36 If $H_s > H$, $W_c = C_n(\gamma_s - \gamma_w)B_d^2$

where C_n depends on $\dfrac{H}{B_d}$.

4.14 (d) For Free Draining Coarse Grained Fills (e.g. gravels, broken stone)

If the drained density of the fill is γ_d and the void ratio is n, the saturated density is

$$\gamma_s = \gamma_d + n\gamma_w$$

and the submerged density is

Formula 4.37 $$\gamma_s - \gamma_w = \gamma_d - (1 - n)\gamma_w.$$

Then in Formulae (4.31) to (4.36) for these soils, substitute γ_d for γ_s, $(1 - n)\gamma_w$ for γ_w and $\{\gamma_d - (1 - n)\gamma_w\}$ for $(\gamma_s - \gamma_w)$.

4.14 (e) *For Flexible Pipes in Wide Trench or Positive Projection Embankment Conditions* (see Chap. 6)

The unsubmerged fill load is $W_c = H\gamma_s B_c$ lb/linear ft (see Table M) and the unit pressure on top of the pipe is $H\gamma_s$ lb/ft² for fine grained (normal) fills.

If the fill is submerged to a depth $H_s < H$ the effective density of the submerged portion of the fill is $(\gamma_s - \gamma_w)$ lb/ft³ and the pressure at the top of the pipe is

Formula 4.38 $$(H - H_s)\gamma_s + H_s(\gamma_s - \gamma_w) \text{ lb/ft}^2$$

Formula 4.39 There is also a mean external fluid pressure of $(H_s + B_c/2)\gamma_w$ lb/ft² acting on the pipe.

Formula 4.40 If $H_s > H$ the fill pressure will be $H(\gamma_s - \gamma_w)$ lb/ft² but the mean external fluid pressure will be $(H_s + B_c/2)\gamma_w$ lb/ft².

For free draining coarse grained fill the corresponding pressures will be:

Formula 4.41 $$\text{if } H_s < H,\ H\gamma_d - H_s(1 - n)\gamma_w;$$

Formula 4.42 $$\text{if } H_s > H,\ H(\gamma_d - (1 - n)\gamma_w);$$

Formula 4.43 and the mean external fluid pressure will be $(H_s + B_c/2)\gamma_w$ as before.

4.14 (f) *Effect of Permanent Submergence on the Contained Water Load*

The external water pressure produces ring bending moments in the pipe which are equal but of opposite sign to the moments produced by the internal water load and therefore neutralises it.

$$\text{i.e. } W_w = 0$$

4.14 (g) *Limitations on the Use of Reduced Loads in Design*

The loads reduced for submergence should never be used in design where there is any possibility of the ground water level ever being lowered, e.g., to facilitate future works in the vicinity, or by natural causes, mining subsidence, land drainage, etc.

4.14 (h) *Surcharges Not to be Reduced*

No reduction in design load or pressure should normally be allowed for accidental submergence of surcharges of any kind.

4.15 Depth of Cover to Prevent Flotation of an Empty Pipeline

Where a pipeline may be submerged when empty, either during construction or after completion, there will be an uplift of $\frac{\pi}{4} B_c^2 \gamma_w$ lb/linear ft acting on the pipe, the fill load will be reduced by submergence, and frictional forces in the fill cannot be relied upon. The minimum submerged cover (H_{min}) required to prevent flotation is then given by the equation

Equation 4.44
$$H_{min} B_c(\gamma_s - \gamma_w) + \text{empty pipe weight} = \frac{\pi}{4} B_c^2 \gamma_w$$

It should be noted that H_{min} is independent of the depth of the water. If the required depth of cover is impracticable, measures must be taken either to keep the pipeline full of water or to anchor it down.

4.16 Inclined Loads

Where a conduit is laid at or near the toe of an embankment, or at the side of a cutting, or where the soil may be surcharged in the vicinity of and above but not over the pipeline, as by material dumps during construction, the maximum resultant fill load on the conduit may be inclined to the vertical and may even approach the horizontal. It is important that such conditions should be recognised and recourse made to an investigation of the magnitude and direction of the maximum resultant load on the pipes. It may be necessary to extend the bedding peripherally as necessary to resist a resultant load which is inclined to the vertical.

For methods of estimating this kind of loading reference may be made to accepted text books on soil mechanics (e.g. Terzaghi[25]) and, where any doubt exists, expert opinion should be obtained.

4.17 Loads Imposed by Concentrated Surcharges for All Installation Conditions

In the period 1920–30 Marston and Spangler[17/21] showed that the vertical load imposed on a conduit (W_{cs}) by a point load of P lb on the surface vertically above the centre of the conduit was practically approximated by an application of the theory of stress in a semi-infinite elastic solid under point loading developed by Boussinesq. Marston's formula was:

$$W_{cs} = C_t P F_i / L \text{ lb/linear ft}$$

The load coefficient C_t was obtained by summing the Boussinesq stresses at a depth H over small areas aggregating the horizontal projection of a conduit of length L, i.e. $B_c \times L$, where L had the arbitrary value of 3 ft or the length of the individual pipe if less than 3 ft, and F_i was an impact factor.

In 1946 Spangler and Hennessey[22] showed that Hall's integration of the Boussinesq stress caused by a point load on the surface and Newkark's[18] integration of the stress at a point on the top of a conduit caused by a rectangular uniformly distributed load on the surface (see Article 4.18), led to influence values which were common to both conditions. They then developed the method described below of evaluating the load imposed on a conduit by a point load in any position on the surface or by any number and arrangement of point loads, using Newmark's table (see Table C) of influence values, or Fadum's graph[25], Chart C5, of these values. It transpired, however, that the distribution of the load across the width of the conduit was not uniformly distributed, which meant that its effect on the ring bending moments of a pipe differed from that caused by the fill load. In order to enable the same bedding factor to be used for all loads, O. C. Young proposed in 1962[11] that an equivalent uniformly distributed load should be substituted for the Boussinesq load, (see Article 4.17 (*b*) (v) 2).

4.17 (*a*) *The Iowa Formula and its Derivation*

In what follows the load acting at the ground surface is termed the concentrated surcharge, and the fraction of this load which is transmitted to the pipe is termed the transmitted load.

For a concentrated surcharge over the centre-line of a pipe, the Boussinesq theory gives a load distribution on the critical horizontal plane through the crown of the pipe, of the form shown in Fig. 4.23, and this distribution is similar in any vertical plane through the point of application of the load. The intensity of this transmitted loading has a maximum value vertically below the point of application of the surcharge and it decreases with the horizontal distance from that point. This variation, and the maximum intensity, decrease as the depth of the critical plane increases.

For purposes of design it is expedient to adopt some effective value of transmitted load which is less than the maximum intensity and which allows for the variation of load intensity along the pipe axis.

Spangler adopts an empirical value obtained by computing the total transmitted load on an individual pipe of width B_c if its length is less than 3 ft, or on a length of 3 ft where the individual pipe length is 3 ft or more, and dividing this load by the length involved. This gives an 'effective' design load per unit length of pipe but the non-uniformity of the load distribution across the diameter of the pipe tends to produce larger maximum circumferential bending moments in the pipe wall than would the same load if uniformly distributed. The difference increases with pipe diameter and decreases with cover depth. To allow for this effect on large pipes at shallow depths, Spangler[8] states that the bedding factor for the transmitted surcharge load under such conditions should not exceed 1·5 whenever the general bedding factor, F_m exceeds 1·5, because the maximum circumferential bending moment occurs at the pipe crown under these conditions. This is approximately true for a single wheel load and 1 ft depth of cover but as already mentioned Young has shown[11] that the transmitted load at cover depths of 2 ft or more caused by a single wheel

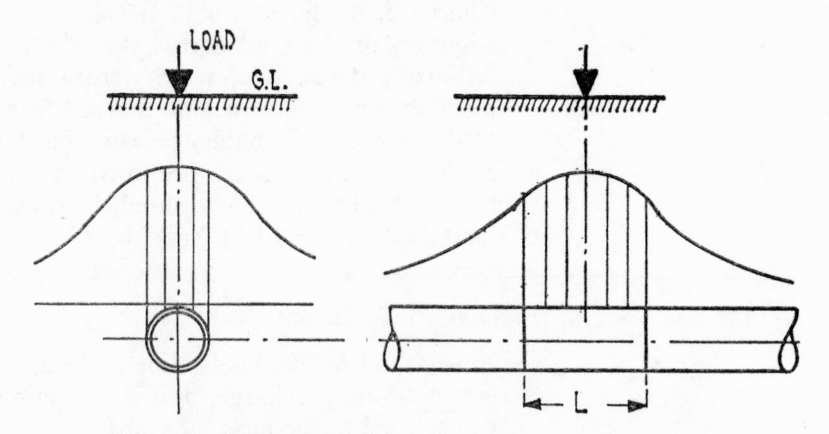

Fig. 4.23 Boussinesq distribution of load on the horizontal plane through the top of a buried conduit caused by a point load on the surface

Crown Copyright—by permission

or by two or more wheels acting simultaneously, is more nearly uniformly distributed. These conditions therefore call for less severe correction. The degree of correction also depends on the type of bedding to be used. It may be made in design by adopting an equivalent uniformly distributed load of the magnitude required to produce the same maximum circumferential bending moment in the pipe as the non-uniform loading (see Charts C6 (*a*) and C6 (*b*)).

The general form of the formula developed by Marston, as modified by Spangler is:

Formula 4.45

$$W_{cs} = \frac{1}{L}(F_{i1}C_{t1}P_1 + F_{i2}C_{t2}P_2 + \ldots F_{in}C_{tn}P_n)$$

where W_{cs} is the uncorrected load per linear ft of pipe; F_{i1}, F_{i2}, etc. are the impact factors appropriate to each individual wheel load, C_{t1}, C_{t2}, etc. are load coefficients computed as described below; P_1, P_2, etc. are the actual individual wheel loads; and L is the length of a pipe or 3 ft, whichever is the less.

W_{cs} must, however, be modified as necessary for non-uniformity of load intensity normal to the pipe axis, and for the type of bedding adopted, by multiplying it by a correction factor F_L (>1), to obtain a corrected value W_{csu}, which may be used as a component of the total effective external design load on the pipe.

Formula 4.46 i.e. $W_{csu} = W_{cs}F_L$

4.17 (b) Notes on the Components of the Formula

4.17 (b) (i) Magnitude of the Point Surcharges, $P_1, P_2, \ldots$

For pipelines under roads, fields and gardens in England and Wales, the recommended minimum concentrated surcharges for which provision should be made in pipeline design are, for the time being, as given in Table D* and the 'worst' positions of the load relative to the pipeline are indicated in Fig. 4.25.

Provision should always be made, however, for the maximum foreseeable surcharges to be imposed if likely to be greater than those given in Table D, e.g. on airfield hard standings, docks, railways, industrial yards, etc.

4.17 (b) (ii) The Impact Factor, $F_{i1}, F_{i2}, \ldots$

Recent experimental work at the Road Research Laboratory has shown that, pending more extensive observations, which are in hand, an impact factor of not less than 2·0† is necessary at any depth of cover for road wheel loads directly above the pipe when the road surface is uneven (e.g. temporary trench reinstatement, resurfacing operations, widenings, etc.). This broadly confirms Spangler's observations(21). The simultaneous impact factors for wheel loads not directly above the pipeline may not be identical but their value will not very materially affect the total load transmitted to the conduit, (e.g. for the B.S. 153 Type HB loading). In this discussion, therefore, a uniform value of $F_i = 2{\cdot}0$ is assumed for each road wheel load. (But see the footnote † below).

4.17 (b) (iii) Computation of the Load Coefficients, $C_{t1}, C_{t2}, \ldots$

With the help of Newmark's Table C, or the graph, Chart C5, derived therefrom by Fadum, influence values, I_σ, may be obtained

* Table D has been revised in accordance with Ref. 10*a*, para. 235.

† As a result of later tests (27*a*, 27*b*) the value of the Impact Factor has been reduced to 1·3 on the 20,000 lb wheel loads of BS 153 HB loading used in Charts C8 (*a*) and C8 (*b*). Lighter wheel loads with a higher speed limit require a higher Factor (see Table D). It was also found that the Factor does not vary with the depth of cover.

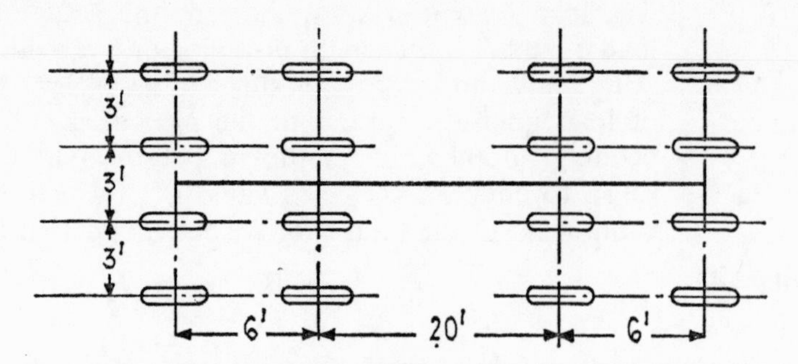

Fig. 4.24 B.S. 153, Type HB road wheel spacing

Crown Copyright—by permission

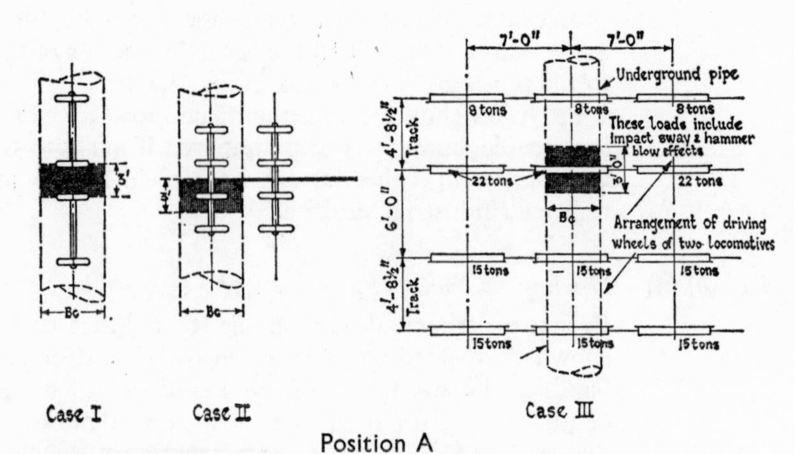

Position A

Case I. Fields and gardens, private carriageway, access roads and other roads not in Case II
Case II. Main traffic routes
Case III. Railways

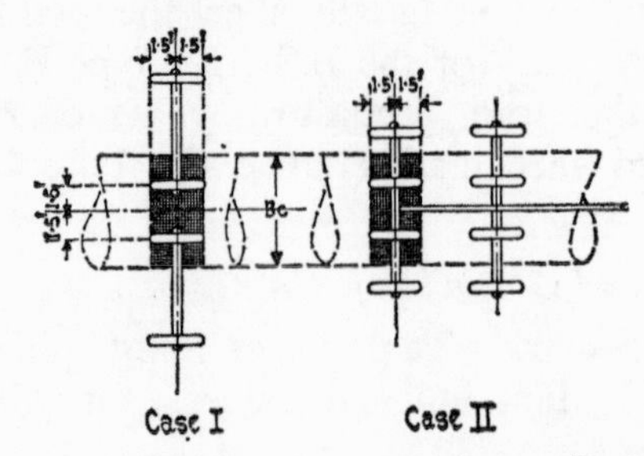

Position B

Case III. Not applicable since British Railways require that pipe crossings be approximately at right angles to the tracks

Fig. 4.25 'Worst' wheel load position

Crown Copyright—by permission

which, when algebraically summed, give the value of C_t for a wheel load as follows:

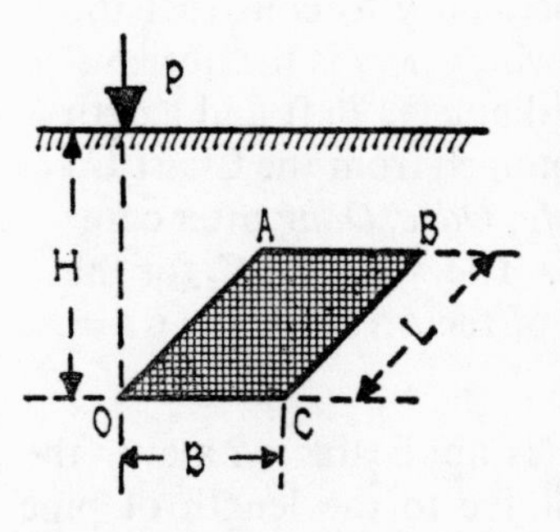

Fig. 4.26 Position of point load P relative to a rectangle at depth H on which the influence value I is computed
After Spangler—by permission

Referring to Fig. 4.26, let P be a point surcharge on the surface vertically above the point O of a rectangle $OABC$, of width B and length L, lying in a horizontal plane at depth H below the surface. The values of m and n in the Newmark table or graph are than $m = B/H$ and $n = L/H$, for the simultaneous values of which the value of I_σ is obtained from Table C, or the Chart C5.

Referring now to Fig. 4.27 for a point surcharge P on the surface vertically over the mid-point, O, of a pipe of outside diameter B_c and length L ft, the horizontal projected area of the pipe is divided into four equal rectangles, as $OABC$, each of width $B_c/2$ and length $L/2$. The values of m and n are then $m = \dfrac{B_c}{2H}$ and $n = \dfrac{L}{2H}$ Then if I_{ob} is the influence value obtained from the Chart C5 for these values of m and n, the value of C_t will be $4 \times I_{ob}$.

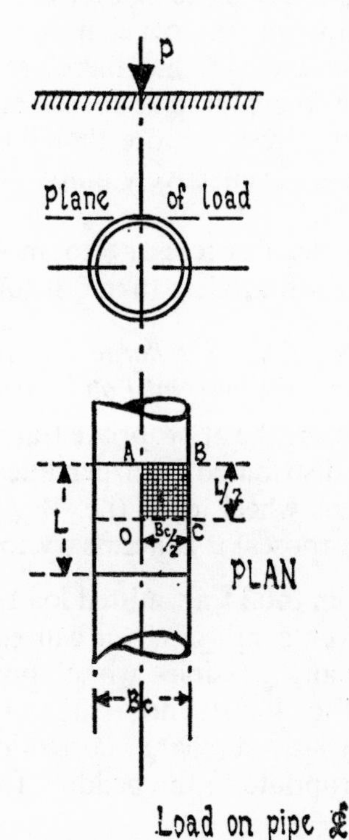

Fig. 4.27 (a) Point load P vertically above conduit centre line
After Spangler—by permission

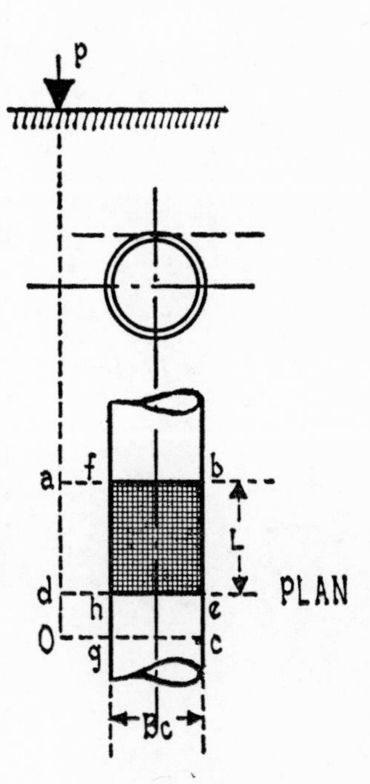

Fig. 4.27 (b) Point load P not vertically above conduit centre line
After Spangler—by permission

Referring to Fig. 4.27 (*b*), in which the surcharge P is vertically above a point O, which is not directly over the pipe, and again with a cover of H ft, it is necessary to construct the virtual rectangles *Oabc*, *Oafg*, etc., where *hebf* is the horizontal projection of the pipe of outside diameter B_c ft and length L ft. The value of I_σ must now be obtained from the Chart C5 for each of the rectangles *Oabc*, *Oafg*, *Odec*, *Odhg* after computing m and n for each rectangle. The value of C_t for the area *hebf* is then the algebraic sum of these values, i.e. $Ct = Iob - Iof - Ioe + Ioh$.

A similar method of computing C_t is applicable whatever the position of the point surcharge P relative to the length of pipe under consideration. These coefficients govern the total load transmitted to a pipe of length L ft and outside diameter B_c ft, at the horizontal plane through the top of the pipe.

4.17 (*b*) (iv) *The Pipe Length L*

The only pipes with British Standard lengths less than 3 ft are the small bore clayware pipes to B.S. 65 (1966) and BS 540 (1966). In the preparation of the concentrated load graphs Charts C7 and C8 a value of $L = 3$ ft has therefore been adopted throughout. The use of these graphs for small bore pipes 2 ft long underestimates W_{cs} by not more than about 7% and the value of F_L for such pipes is effectively unity. The small correction can usually be ignored.

It is not considered necessary to modify W_{csu} for make-up lengths of pipe which are less than 3 ft long.

4.17 (*b*) (v) *The Effects of Multiple Surcharge Position and the Correction Factor F_L for Non-uniformity of Load Distribution*

In order to determine the appropriate transmitted equivalent maximum uniformly distributed load per linear ft of pipe for one or more simultaneous wheel loads (i.e. W_{csu}, the corrected value of W_{cs} for design purposes) it is necessary to compute:

The maximum total transmitted load on a 3 ft length of the pipe at any given depth of cover caused by the simultaneous surcharges in any possible 'worst' position relative to the pipeline, (see Fig. 4.25): and

The concentrated surcharge correction factor F_L, referred to above, appropriate to the bedding to be employed.

4.17 (*b*) (v) 1. *The Transmitted Load*

The rather tedious problem of determining the maximum total transmitted load may be solved by constructing influence grids of the type shown in Fig. 4.28 for each wheel and for each of several values of the cover depth, by means of Table C or Chart C5. Then, by superimposing the grids for a given depth, one grid over the other, but displaced by the distance between wheels, the pattern of load intensity on a pipe of any diameter in any particular position relative to the wheels is obtained for the given depth, by summing the load intensities in each of the overlapping grid units which lie over a 3 ft length of the pipe. The sum of all the grid unit values overlying the pipe multiplied by the wheel load P gives the total transmitted load on the pipe. The variation in the intensity of load across the pipe diameter may be obtained in a similar manner. (N.B.—If the wheel loads vary in magnitude the grid values may be obtained in terms of PI_σ instead of I_σ).

This process is repeated for each depth of cover in turn, and the whole computation repeated for each critical position of the pipe relative to the wheels.

6"

6"

4·5	5	5·5	5·5	6	6				
5	5·5	6	6·5	7	7				
5·5	6	7	7·5	8	8				
5·5	6·5	7·5	8	9	9				
6	7	8	9	10	10	9	8	7	6
6	7	8	9	10	10	9	8	7	6
				9	9				
				8	8				
				7	7				
				6	6				

TYPICAL INFLUENCE GRID
(values are illustrative only.)

Fig. 4.28 Typical influence grid

Crown Copyright—by permission

4.17 (*b*) (v) 2. *The Correction Factor* F_L

By taking a range of pipe diameters and depths of cover and calculating for each case

$$F_L = \frac{\text{Total equivalent uniformly distributed transmitted load}}{\text{Total actual transmitted load}}$$

the curves of Charts C6 (*a*) and (*b*) were obtained by Young[11] for Class A 120° R.C. bedding and Class B bedding respectively. Values for special higher strength bedding have also been computed by Young[39].

Within the range of conditions considered, the values of F_L, as found for the two-wheel loading of Fig. 4.25, position '*a*', *Case 1*, do not differ materially either from those for a single wheel load placed centrally over the pipe or from those for the eight-wheel loading of Fig. 4.25, position '*a*', *Case 2*, and hence Charts C6 (*a*) and (*b*) may for practical purposes be used for F_L for any one of those wheel arrangements.

For the lower part of the curves in Chart C6 (*b*) the maximum moment occurs at the invert, and for the upper part at the crown. The kink in the curves marks the transition from one condition to the other.

4.17 (*b*) (v) 3. *Summary of Conditions Affecting* W_{csu}

The results of the theoretical investigation of the value of W_{csu} for the surcharges referred to in Table D may be summarised as follows:

Case 1: *Two equal wheel loads of 7000 lb, 3 ft apart* and track widths of 6 ft or more, (i.e. for pipes under access roads, fields, and gardens.)*

1. Two 'worst' wheel positions are possible, viz: (*a*) vehicle axles parallel to the pipe axis and all wheels vertically over the pipe centre-line, *see position* '*a*', *Case 1*, in Fig. 4.25, and (*b*) axles normal to the pipe axis and wheels 1·5 ft from the pipe centreline, see *position* '*b*', *Case 1*, in Fig. 4.25.
2. The outer wheels of the vehicles do not materially affect the value of W_{csu}, for these loads and wheel track widths of 6 ft or more.
3. The variation in loading intensity across the pipe diameter is always greater for position '*a*' than for position '*b*', and it increases as the cover decreases and as the pipe diameter increases.
4. The required correction to W_{cs} increases as the bedding factor F_m increases.
5. The variation of load intensity across the pipe diameter is negligible for position '*b*', but the actual transmitted load on the pipe may be greater in this position than in position '*a*', depending on pipe diameter. Thus, for values of B_c less than 3 ft, position '*a*' always produces a more severe actual transmitted load than position '*b*', whilst for values of B_c greater than 3 ft, the reverse is true. The most severe circumferential bending moment may therefore occur in either position, depending upon the

* Also applicable to 2/16,000 lb loads 3 ft apart, see Table D.

particular conditions of diameter, cover depth, and bedding class.

6. For Class A 120° R.C. beddings and other concrete beddings subtending angles of 120° or less, and 2 ft or more of cover, position '*a*' governs and W_{cs} must be corrected.
7. For Class B or weaker beddings and pipes up to 3 ft diameter, position '*a*' governs, and theoretically W_{cs} must be corrected, but the correction is so small as to be negligible for cover depths of 2 ft or more.
8. For Class B bedding and pipes of over 3 ft diameter, position '*b*' governs and correction to W_{cs} is not required.

Case 2: *B.S. 153, type HB Road loading* (i.e. for pipes under public roads), Fig. 4.25.

The same corrections as for *Case 1* are required for this loading, but for the total load, all eight wheels of a group must be considered. The second group of eight wheels does not materially affect the value of W_{csu}.

Case 3: *British Railways loading*, Fig. 4.25.

For pipes laid approximately normal to the tracks the same value of W_{csu} as for road loading in position '*a*', *Case 2* with an impact factor not less than 2·0 may be used for cover depths of 3 ft or more, provided the locomotive loads do not exceed those given in Fig. 4.25.

4.17 (*b*) (vi) *Load Charts C7* (*a*) *and* (*b*), *and C8* (*a*) *and* (*b*)

Computations have been carried out for the surcharges discussed above and the resulting corrected values are given in the Charts C7 (*a*) and (*b*) and C8 (*a*) and (*b*). Charts C7 (*b*) and C8 (*b*) should be used for concrete arch support with Class B bedding, and for flexible pipes for which concrete beddings are inappropriate.

1. Charts C7 (*a*) *and* (*b*) *for access road, field and garden loading: two equal 7000 lb wheel loads 3 ft apart in the 'worst' position.**

These graphs give the corrected effective maximum resultant values of the equivalent transmitted load W_L for the two 7000 lb wheel loads for a pipe of external diameter B_c in. and length 3 ft, or more, at various depths of cover H ft in the 'worst' position, excluding impact effects. They may be used for wheel loads of any equal magnitude P lb greater or less than 7000 lb by multiplying W_L by $P/7000$. They are not applicable to wheel track widths

* These charts may also be used for 2/16,000 lb loads 3 ft apart, see Ref. 10*a*, para. 235(*b*), note 2.

less than 6 ft for which condition the value of W_L must be computed as in Article 4.17 (*b*) (iii) above.

Assuming a uniform impact factor F_i, and the same wheel spacing and pipe bedding the effective equivalent uniformly distributed load transmitted to the pipe is then:

Formula 4.47

$$W_{csu} = \frac{F_i W_L P}{7000} \text{ lb/linear ft}$$

2. Charts C8 (a) and (b) for road loading: B.S. 153, type HB loading in the 'worst' position

These graphs cover the special case of road loading in which $P_1 = P_2 = P_8 = 20{,}000$ lb* and give the effective maximum resultant equivalent transmitted load W_o for the wheels in the 'worst' position on a pipe of external diameter B_c in. and length 3 ft, or more, at various depths of cover H ft, excluding impact effects.

Assuming a uniform impact factor F_i, the effective equivalent uniformly distributed load transmitted to the pipe is then

Formula 4.48

$$W_{csu} = F_i W_o \text{ lb/linear ft}$$

For the time being F_i should be assumed equal to 2·0 at all depths†.

For the same wheel spacing and pipe bedding but values of P other than 20,000 lb:

Formula 4.49

$$W_{csu} = \frac{F_i W_o P}{20{,}000} \text{ lb/linear ft}$$

3. British Railway loading, (Fig. 4.25)

Charts C8 (*a*) and (*b*) may be used with an impact factor of 2·0

Formula 4.50

for pipes laid approximately normal to the tracks, i.e. $W_{csu} = 2W_o$. N.B. Any change in the impact factor of B.S. 153 H.B. Road Loading is not applicable to this loading.

4.17 (*b*) (vii) *Notes on the Theory*

1. The assumption of a uniform elastic material as the medium between the surcharge and the pipe is rarely or never achieved in practice. It appears, however, that clay more normally approaches this ideal condition, for instantaneous loading, than granular soils.

2. As the theory stands, the load transmitted to a pipe from the concentrated surcharge is assumed to be independent of the stiffness of the pipe, and of the nature of the fill or of the road structure, except as modified by impact effects. In other words the transmitted load on a pipe of given diameter laid at a given depth will be the same irrespective of trench width (and is therefore the

* Derived from the B.S. 153 load of 25,000 lb per wheel by allowing for the overstress of 25 per cent permitted by the Standard.

† But see Table D.

same for embankments as for trenches) and of whether or not other pipes are laid alongside it. It is known that a rigid pavement does in fact modify the transmission of the load, but since such rigidity may at any time be diminished or destroyed by roadworks, it would seem advisable, for pipe design purposes, to ignore any relief of load or impact factor from this cause.

4.18 Loads Imposed by Uniformly Distributed Surcharges of Small Extent for ALL Installation Conditions

If the contact area of a surcharge such as a column footing or a vehicle track has a *largest dimension greater than half the individual cover depth H/2* it is regarded as, and termed, 'a uniformly distributed load of small extent'. For static surcharges, its position relative to the conduit must be predetermined. For mobile surcharges, the position producing the maximum imposed load on the conduit must be considered.

The Boussinesq theory of stress in a semi-infinite elastic solid caused by a surface load has again been invoked to give an approximate estimate of the *vertical* load imposed on a conduit by a rectangular uniformly distributed surface load. The integration of the stresses at a point vertically below one corner of the rectangle was achieved by Newmark[18] who also computed the influence values I_σ given in Table C for the imposed load. These values were adopted by Terzaghi[25] and reproduced in graphical form by Fadum (Chart C5).

Thus in Fig. 4.29 for a rectangle of dimensions L ft $\times$ B ft uniformly loaded with U_{sus} lb/ft² the stress at a point 0 vertically below a corner of the rectangle at a depth H is

$$\sigma = U_{sus} I_\sigma \text{ lb/ft}^2$$

where I_σ is the influence value for L/H and B/H.

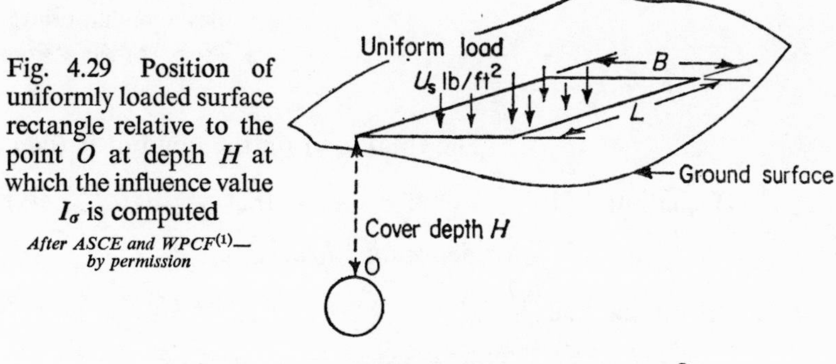

Fig. 4.29 Position of uniformly loaded surface rectangle relative to the point O at depth H at which the influence value I_σ is computed

After ASCE and WPCF[1]—by permission

In applying this theory to a conduit, a system of virtual rectangles (Figs. 4.30 (*a*) and (*b*)) is set up in a similar manner to that used for a concentrated load but having one common corner vertically above the centre line of the conduit and at the point nearest to the c.g. of the surcharge. The influence value for each such rectangle is then obtained from Table C or Chart C5. The algebraic sum of these influence values is then used to determine the resultant stress at the top mid-point of the conduit, e.g. in Fig. 4.30 (*a*)

$$\sigma = U_{sus}\Sigma I_{\sigma} = U_{sus}4I_{\sigma} \text{ lb/ft}^2$$

in Fig. 4.30 (*b*)

$$\sigma = U_{sus}\Sigma I_{\sigma} = U_{sus}(I_{ob} - I_{od} - I_{og} + I_{oh}) \text{ lb/ft}^2$$

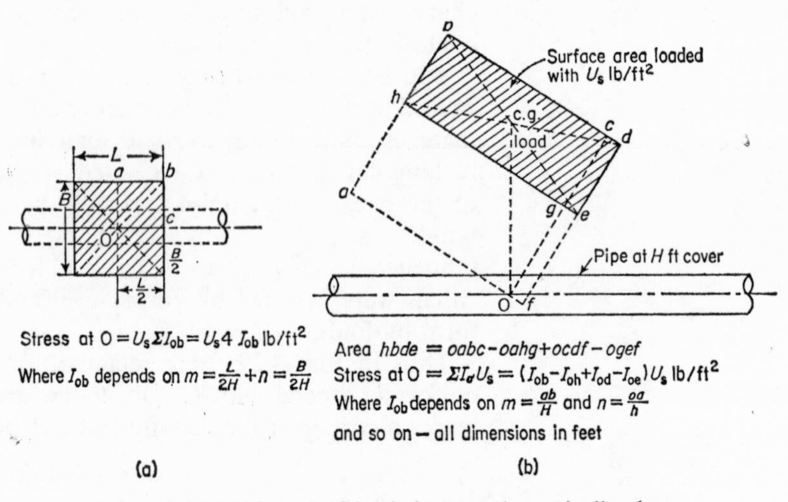

Fig. 4.30 (*a*) Centre of loaded rectangle vertically above conduit centre line

Fig. 4.30 (*b*) Centre of loaded rectangle not vertically above conduit centre line

The total load on the conduit is then assumed to be

Equation 4.51
$$W_{sus} = \Sigma I_{\sigma} B_c U_{sus} \text{ lb/linear ft}$$

or denoting ΣI_{σ} as C_{sus}

Formula 4.52
$$W_{sus} = C_{sus} B_c U_{sus} \text{ lb/linear ft}$$

This is not strictly true since the stress will vary across the width of the conduit, and the resultant stress may be inclined, and greater than its vertical component. For permanent *static* loads the assumption of elasticity in the fill is dubious and it would be preferable to support them independently of the conduit, e.g. by piles, and especially so where their exact magnitude and location are not precisely foreknown.

For mobile and transient loads such as tracked construction vehicles, the maximum load on the conduit will occur when the centre of gravity of the vehicle lies vertically above the centre line of the conduit. The load may be 'worst' when the tracks are normal to the conduit axis. Impact effects, depending on the roughness of the surface must be considered. Their value is not well established and the estimation of F_i is a matter of experience. The value of the load transmitted to the conduit is then

Formula 4.53

$$W_{sust} = C_{sus} B_c U_{sust} F_i \text{ lb/linear ft}$$

If in doubt, and especially with shallow cover, a value of F_i not less than 2·0 should be adopted.

Since this type of load is assumed to be independent of frictional forces in the soil, the formulae are applicable to conduits of any kind, installed by any method, in any kind of soil, and are unaffected by submergence or by the proximity of other conduits.

4.19 Fluid Pressures

All fluid pressures are independent of the nature, function, shape or method of installation of the conduit.

4.19 (a) External Fluid Pressures,

Caused by submergence or vacuum, induce ring compression stresses in thick walled *rigid* pipes and are usually beneficial. They may be ignored in design, however, especially where they are temporary. Under premanent submergence, they may be the major loads, e.g. on pipes laid with little or no cover under lakes, rivers or land subjected to flooding. In all such cases and in tidal waters, the maximum height of the water level above the top of the conduit ($H_{s_{max}}$ ft) should be determined from local records if possible. Then the external pressure

$$p_{g_{max}} = (H_{s_{max}} + B_c/2)\gamma_w/144 \text{ lb/in.}^2$$ (see also Article 4.14).

These pressures affect the deflection of a circular flexible pipe and they may contribute largely to the compressive stress in the pipe wall and therefore to the possibility of yield or buckling of the pipe wall especially in thin walled pipes.

Partial vacuum pressures may occur in pressure pipes owing to faulty operation or failure of air valves during filling, or in the event of a burst, whether or not the pipeline is buried, and in siphons. For such pipelines both the maximum and the minimum pressures must be determined from observations of water levels if possible or, if not, they must be estimated. The maximum external pressure is then

Formula 4.54 $$p_{e_{max}} = p_{g_{max}} + p_{vac} \text{ lb/in.}^2$$

As already mentioned in Article 4.14 (*f*), permanent submergence of the top of the pipe neutralises the internal water *load* in either pressure or non-pressure pipes.

4.19 (*b*) *Internal Fluid Pressures*

In aqueducts and sewers, internal pressure may occur in inverted siphons or, relatively slightly, where a pipeline is, or may be, hydraulically surcharged. Its value may be assumed as the mean static head at zero velocity p_h,

Formula 4.55 $$\text{i.e. } p_{i_{max}} = p_h \text{ lb/in.}^2$$

In gravity pressure pipelines it is also the maximum static head at zero velocity plus any allowance for surge (see Chapter 7 of Ref. 41) which may be considered necessary to provide for faulty operation of valves, or the maximum test pressure to which the pipeline is to be subjected, whichever is the greater,

Formula 4.56 $$\text{i.e. } p_{i_{max}} = p_h + p_s \text{ or } p_{t_{max}} \text{ lb/in.}^2$$

In pumping (rising) mains it will be the sum of the static head (lift) (p_h) plus the friction head (p_f) plus the velocity head (p_v), or the surge pressure (p_s) (unless surge is effectively suppressed), or the maximum test pressure $p_{t_{max}}$ whichever is the greatest,

Formula 4.57 $$\text{i.e. } p_{i_{max}} = p_h + p_f + p_v \text{ or } p_s, \text{ or } p_{t_{max}} \text{ lb/in.}^2$$

Internal pressure causes tensile ring stress in the pipe wall and therefore tends to weaken rigid pipes and reduce their external load carrying capacity. In flexible pipes, however, it tends to reduce the deflections caused by external loads, to reverse the ring stress and reduce the bending moments in the pipe wall, and to suppress buckling, *but only whilst it is operative*. It also produces unbalanced end thrust, or axial tension, in the pipeline depending upon the type of joints used.

In rising mains the internal pressure may be partially or wholly neutralised by permanent submergence of the pipeline, but in pressure mains proper, external pressures afford little significant reduction of the effective internal pressure.

Effective design pressures are summarised in Table E.

4.20 The Total Effective External Design Load—W_e or W'_e

The maximum resultant external load on which the crushing strength of rigid pipes, or the wall thickness of flexible pipes is based, is the sum of the effective component loads which can be imposed simultaneously on the conduit. Because of the major assumptions:

1. *that in any kind of conduit concentrated surcharges will not be imposed simultaneously with distributed surcharges of large extent after completion of the backfill.*

The effective value of the total external design load *for narrow trench conditions or for negative projection or induced trench conditions* in a completed embankment will be *the greatest value of*:

Formula 4.57

$$W_e = W_c + W_w + W_{us} + W_{sus}$$
$$\text{or } W_c + W_w + W_{csu} + W_{sus}$$
$$\text{or } W_c + W_w + W_{sust} + W_{sus}$$

and *for wide trench, or positive projection conditions* in an embankment, the greatest value of

Formula 4.58

$$W_e = W'_c + W_w + W'_{us} + W_{sus}$$
$$\text{or } W'_c + W_w + W_{csu} + W_{sus}$$
$$\text{or } W'_c + W_w + W_{sust} + W_{sus}$$

During the construction of an embankment or fill, where heavy mobile construction equipment is used, it will be necessary to estimate the depth of cover, below which it would not be safe to operate the equipment over the conduit

i.e. at any depth, $W_c + W_{csu}$ (or W_{sust}) must be less than the design value of W_e. (See worked examples 15, 16.)

2. *That in pressure pipes the maximum external surface surcharge loads will not be imposed on the pipeline simultaneously with the maximum internal test pressure.*

For narrow trench conditions or negative projection or induced trench conditions under an embankment, the effective value of the total effective external design load will then be the greater value of:

Formula 4.59

$$W'_e = W_e \text{ (Form. 4.57)}$$

when p_i = normal working pressure p_w:

Formula 4.60

$$\text{or } W'_e = W_c + W_w + W_{sus} \; (+W_{us} \text{ if present during a test})$$

when p_i = the maximum field test pressure p_t.

For wide trench conditions or positive projection conditions under an embankment, it will be

Formula 4.61 $W'_e = W_e$ (Form. 4.58)

when p_i = normal working pressure p_w

Formula 4.62 or $W'_e = W'_c + W_w + W_{sus}$ ($+W'_{us}$ if present during a test)

when p_i = the maximum field test pressure p_t.

Section C

Pipe Strength

Chapter 5
The Load Carrying Capacity of Rigid Conduits

5.1 Classification of Rigid Conduits

Conduits may be broadly classed as circular or non-circular, depending on their cross-sectional shape. The circular pipe is the most commonly used and most adaptable type and is the only one standardised in the U.K. The classification of conduits of various materials is given in Table F.

Some special non-standard circular conduits which are initially rigid, may crack or yield under external loading, and then deform sufficiently to become effectively flexible (e.g. Deckon pipes).

Non-circular rigid conduits are not normally used as pressure conduits. They require a different method of design from that used for pipes. Precast concrete oval and egg-shaped tubes are included in this category.

Each type and variety of conduit needs individual consideration in design as noted below.

5.2 Rigid Circular Non-pressure Pipes (e.g. concrete, reinforced concrete, asbestos cement, clayware, cast or spun iron).

This type of pipe is characterised by a high resistance to crushing loads but it is brittle and cracks under very small deformation of the vertical diameter. Its load carrying capacity depends on (i) its elastic resistance to circumferential bending moments and (ii) the manner in which the bending moments which vary around the pipe periphery are affected by the distribution of the load and the reaction from the bed.

If the load and reaction are diametrically opposed line loads, the circumferential bending moment is a maximum and the load required to fracture the pipe (i.e. the ultimate load) under these conditions is known as the *laboratory strength*. This is the basic strength on which the structural design of an unreinforced pipe depends and, if the tensile modulus of rupture of the material and the dimensions of the pipe are precisely known, it can be computed by established theory. The theoretical bending moments, direct compressive and shear forces in a weightless ring are given by Roark[33]. In view, however, of the unavoidable variations in the modulus and in the wall thickness of commercial pipes, Marston considered and the author would confirm, that the experimental determination of this two-edge crushing strength is a more reliable basis of design than that given by the theoretical approach.

If the extension of the horizontal diameter could be greatly restrained, (e.g. in a rock trench), the bending moments would be reduced and the load carrying capacity increased. However, in most soils the deformation before cracking would be too small to produce any reliable restraint by the passive pressure of the soil. This effect therefore is usually ignored in the design of rigid pipes. Active soil pressure is exerted, however, on conduits in positive projection and induced trench conditions under embankments or fills, and in very wide trench conditions, and it reduces the circumferential bending moment in the pipe and therefore increases its load carrying capacity.

5.2 (*a*) *Safe Crushing Test Strength, W_T, and Laboratory Strength, $C_{BS}W_T$*

This basic strength is determined by standard crushing tests as specified in the British Standard for the particular pipe in question. It is designated W_T and is defined as the statistical minimum 2 or 3-edge *ultimate* test load for all unreinforced rigid pipes; or the one-hundredth inch ($\frac{1}{100}$ in.) *cracking load*, or 80 per cent of the ultimate, whichever is the less, for reinforced concrete pipes; or 90 per cent of the one-thousandth inch ($\frac{1}{1000}$ in.) *cracking load* for prestressed cylinder type concrete pipes. It is not yet defined for non-cylinder type prestressed concrete pipes.

The nature and arrangement of the British Standard bearing strips for supporting and loading the pipe in the crushing test machine differ, however, some being 2-edge bearings and others 3-edge bearings (see Fig. 5.1). The older B.S. 556 type of 2-edge bearing employing rubber bearing and loading strips 6 in.

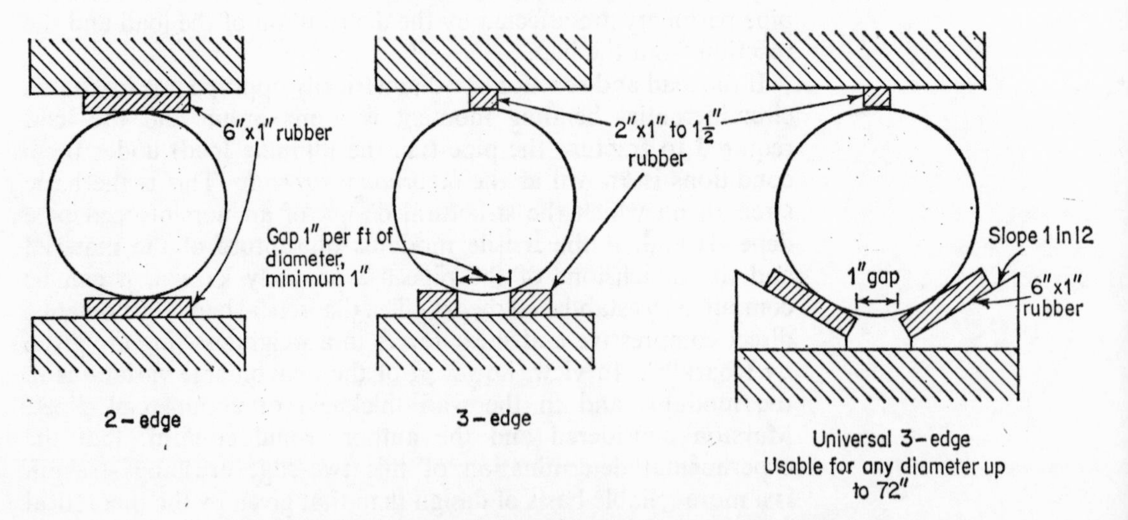

Fig. 5.1 Various standard crushing test bearings for rigid pipes

wide, introduces an optimistic error into the test which increases with the load and decreases with the pipe diameter. A correction factor ($C_{BS} < 1{\cdot}0$) is therefore needed to convert test results obtained with this type of bearing strip to the equivalent laboratory strength. Thus, if the minimum crushing test load, either ultimate for a plain pipe, or $\frac{1}{100}$ in. crack load for a reinforced concrete pipe, is W_T lb/linear ft, its *laboratory strength* will be

$$C_{BS}W_T.$$

The approximate values of the factor for this kind of bearing have hitherto been taken as:

For pipes of 33 in. internal diameter or less, $C_{BS} = 0{\cdot}90$
For pipes of 36 in. diameter or more $C_{BS} = 0{\cdot}95$

Marston[17] however defined the laboratory strength as the 3-edge bearing crushing test strength. So for the more recently standardised 3-edge bearings in the British Standards for various pipes, the value of C_{BS} may be taken as 1·0.

5.2 (*b*) *Field Strength*

When a pipe is buried the loads imposed upon it are assumed, or adjusted, to be uniformly distributed over the upper half of the pipe, both axially and transversely. As the reaction from the bedding is progressively spread around the lower half of the pipe, the load carrying capacity, or '*field strength*', of the pipe progressively increases, because of the reduction of the bending moments, up to a limit imposed by the degree of resistance of the bedding to lateral deformation.

5.2 (*c*) *Bedding Factor*

The ratio of the field strength for trench conditions to the laboratory strength is known as the '*trench*' *bedding factor* (F_m). Values of this factor for specific methods of constructing the bedding of the pipe have been established experimentally by Marston and Schlick[17,35,37] and the usually accepted values are given in Chart C9, together with the appropriate construction details. Higher values of F_m for specially designed concrete beddings have been investigated by Young.[39]

These factors are also used for 'wide' trench conditions, for heading and thrust bore conditions, and for negative projection conditions under embankments, i.e. F_m is always associated with pipes laid in a trench or subtrench (see Figs. 3.1, 4.2, and 5.2). Thus the field strength of a pipe in all the above conditions may be expressed as:

$$W_f = C_{BS}F_mW_T$$

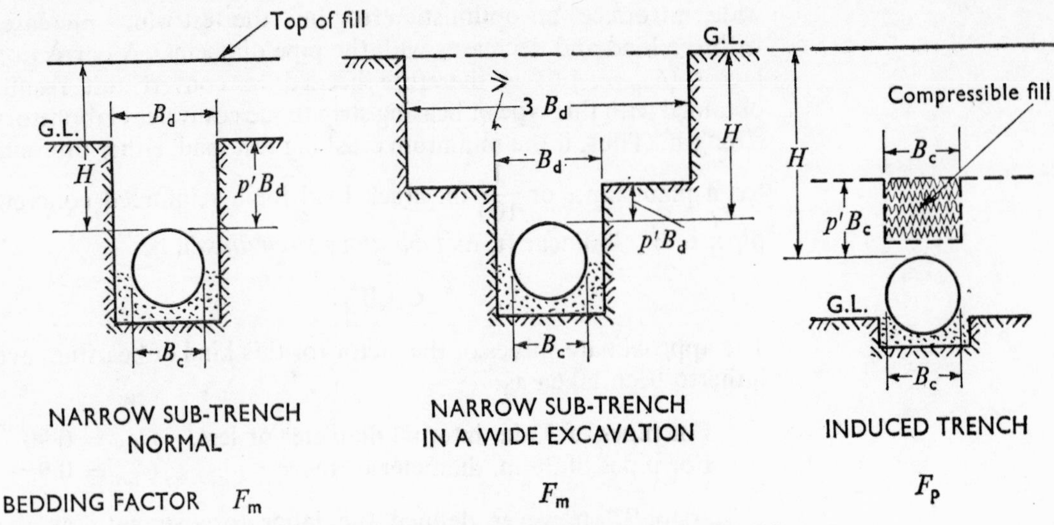

Fig. 5.2 Various negative projection laying conditions
Crown Copyright—by permission

In positive projection and induced trench conditions under embankments or fills, and in '*very wide*' *trench* conditions, the field strength of a pipe is increased by the active soil pressure exerted upon it laterally by the fill, and its bedding factor is therefore increased to a value F_p which is greater than F_m. The effective active soil pressure, however, is caused only by the fill; transient or localised surcharges and the water load do not contribute to it. Values of this '*embankment*' *bedding factor* for the specific methods

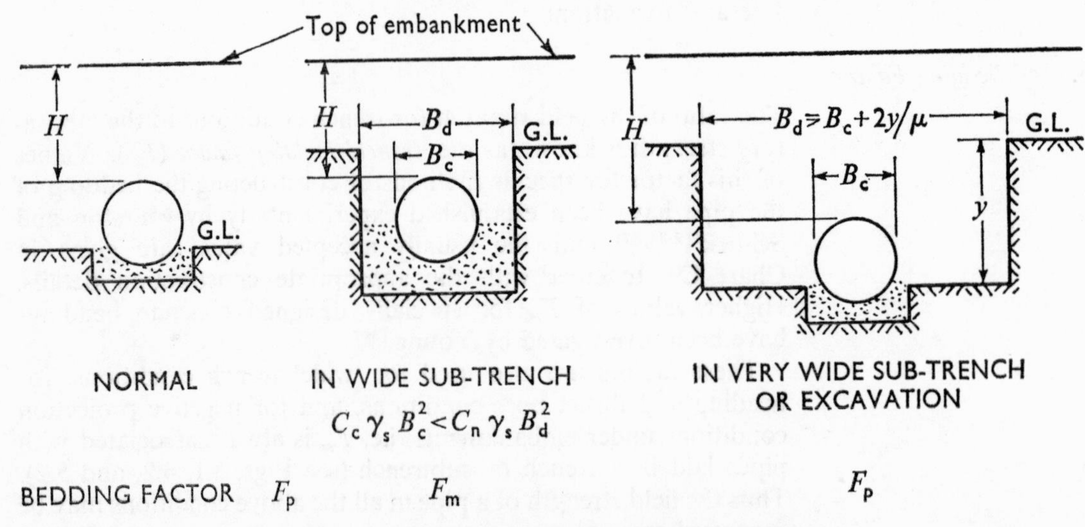

Fig. 5.3 Various positive projection laying conditions
Crown Copyright—by permission

of constructing the bedding given in Chart C10, are computed as in the note on F_p in Article 8.2. So for these conditions:

$$W_f = C_{BS}F_pW_T$$

and F_p is always associated with a pipe which is *not* laid in a trench or sub-trench (see Figs. 3.1 and 5.3).

5.2 (*d*) *Factor of Safety* (F_s)

For pipes having a standard safe crushing test strength (W_T) a factor of safety on pipe strength is not needed. But, as evidenced by the bedding factor, about ½ to ⅔ of the field strength of a pipe depends upon the proper construction of the bedding. Since the quality of the site work may vary along the length of the pipeline, and especially so in adverse weather conditions, a factor to cover these risks is usually applied. The value of this factor should logically increase as the depth of fill increases and as the *permanent* fill load becomes the preponderant component of the total load.

The factor should be increased further to cover the uncertainties of the strength of non-standard pipes, or where there is any doubt as to the quality of the site work, e.g. for an inexperienced contractor.

Suggested values of the factor are given in Table G, but final selection must be a matter of good judgement of local conditions and of experience. A value of not less than 1·5 for all unreinforced rigid pipes is recommended by the A.S.C.E.[1] and a similar value should be adopted for all rigid pipes laid under embankments or high fills where the fill load is preponderant.

The effective field strength of the pipe is then:

Formula 5.1
$$W_f = \frac{C_{BS}F_mW_T}{F_S}$$

for the 'narrow' and 'wide' trench, heading and thrust bore, and negative projection conditions.

For positive projection, induced trench and 'very wide' trench conditions it is:

Formula 5.2
$$W_f = \frac{C_{BS}F_pW_T}{F_S}$$

5.2 (*e*) *The Required Value of the Safe Crushing Test Strength*, W_T

The value of W_T required to carry the loads which may be imposed upon the pipeline is obtained by equating the effective field strength of the pipes to the total external design load W_e, determined as in Chapters 4 and 8.

Thus for all trench and subtrench conditions:

$$\frac{C_{BS}F_m W_T}{F_S} = W_e$$

whence:

Formula 5.3 $$W_T = W_e F_s / C_{BS} F_m$$

and for all positive projection conditions:

Formula 5.4 $$W_T = W_e F_s / C_{BS} F_p$$

5.2 (f) Applicability of the Formula

These formulae for W_T are only applicable to circular rigid pipes which at any given depth of cover are uniformly loaded over their width and length and similarly uniformly supported by one of the specific beddings illustrated in Charts C9 and C10.

5.2 (g) British Standard Safe Crushing Test Loads, W_T

In the recent revision of the British Standards for rigid pipes of plain or reinforced concrete, B.S. 556 (1966), asbestos cement, B.S. 3656 (1963), and clayware, B.S. 65 and B.S. 540 (1966), minimum values of the safe crushing test load have been specified for pipes of various standard sizes and for each of two or more strength classes. These B.S. values are summarised for convenience in Table H.

The A.S.T.M. and I.S.O. standards for similar pipes are listed in Table J. The strength classes and W_T values vary between the respective Standards but their use corresponds with that of the B.S. pipes as described herein.

For cast and spun iron pipes and asbestos-cement pressure pipes the value of W_T can be derived from the results of the tests on the materials specified in the respective British Standards 78, 1211, and 486, or from the values guaranteed by reputable pipe makers.

5.3 Rigid Circular Pressure Pipes

5.3 (a) Effect of Internal Pressure on External Load carrying Capacity

In a pipe ring subjected to an external load and a simultaneous internal pressure, the maximum tangential tensile stress will be the sum of the maximum ring bending tensile stress induced by the external load and the direct ring tensile stress induced by the internal pressure. For a given pipe the external load carrying capacity is therefore reduced by the internal pressure. Conversely for the same external load a pipe must have a higher safe crushing test strength when subjected to internal pressure than when carrying the external pressure alone.

5.3 (*b*) *Required Value of the Safe Crushing Test Load,* W_T

For reinforced concrete pipes subjected to the combined loading Heger[32] has shown theoretically that an equation of the following form is valid:

$$\frac{\text{Total External Load}}{\text{Field Strength}} + \frac{\text{Internal Pressure}}{\text{Bursting Pressure}} = \text{Unity}$$

$$\text{i.e.}\ \frac{W'_e}{W_T C_{BS} F_m} + \frac{p_i}{p_{ult}} = 1$$

Introducing factors of safety against crushing and bursting

$$\frac{W'_e F_{se}}{W_T C_{BS} F_m} + \frac{p_i F_{si}}{p_{ult}} = 1$$

then for trench, subtrench, heading, thrust bore, and negative projection conditions:

Formula 5.5 $$W_T = W'_e F_{se} / C_{BS} F_m \left(1 - \frac{p_i F_{si}}{p_{ult}}\right)$$

Shlick[37] has shown that for cast iron pipes the corresponding formula is:

Formula 5.6 $$W_T = W'_e F_{se} / C_{BS} F_m \left(1 - \frac{p_i F_{si}}{p_{ult}}\right)^{\frac{1}{2}}$$

This formula is also considered to be valid for spun iron and asbestos cement pipes.

For prestressed *cylinder type* concrete pipes Kennison[62] gives:

Formula 5.7 $$W_T = W'_e F_{se} / C_{BS} F_m \left(1 - \frac{p_i F_{si}}{p_{oc}}\right)^{\frac{1}{2}}$$

The corresponding formulae for non-cylinder type prestressed concrete pipes have not yet been established authoritatively.

The bedding factor F_p should be used in place of F_m for positive projection, induced trench and very wide trench conditions in formulae 5.5, 5.6 and 5.7.

These formulae are all based on the supposition that failure of the pipe will occur when $W'_e F_{se}$ and $p_i F_{si}$ are imposed upon it simultaneously.

In all types of rigid pressure pipe the value of W_T is as defined in Article 5.2 (*g*) above.

p_{ult} is the bursting pressure, with zero external load

p_{oc} is the pressure required to produce zero compression in the pipe wall, with zero external load.

W'_e is as defined in Article 4.2 (ii) and depends on the operative value of p_i.

5.3 (c) Applicability of the Formulae

These formula are applicable to rigid pressure pipes of the specified materials when used as water or gas mains or as rising (pumping) mains, or as inverted siphons or other hydraulically surcharged sewers, or otherwise when subjected to internal pressure.

They assume that the pipes are uniformly loaded externally over their length and width and that they are similarly uniformly supported by one of the specific beddings illustrated in Charts C9 and C10.

Pressure pipes under railway embankments are usually required to be placed inside a protective sleeve pipe which relieves them of external load, other than vacuum, and provides a drainage channel and protection for the embankment in the event of a burst. This sleeve pipe, however, must be designed to carry the external load. (See Article 6.4 for steel sleeves or Article 5.2 for rigid pipe sleeves).

5.4 Deckon Pipes

There is no B.S. for these pipes which are proprietary and patented. The Deckon pipe consists of a plain concrete pipe with an external sheath of glass fibre reinforced plastic in the form of a double helical wrapping of glass fibre rovings impregnated with polyester resin which constitutes an impervious, strengthening and chemically resistant membrane. The number of fibres per unit length of pipe must be such as to ensure that, after a long period (40–50 years) underground they do not fail in tension under the imposed bending moments and it is simply adjusted by varying the number of layers or 'winds' of roving. The tensile strength of glass fibre is high but it decreases with time under sustained stress, and more so in moist conditions. The greater the stress the shorter the time to reach failure. The statistical extrapolation of long term tests has shown that to ensure the required moment of resistance in the pipe wall after 50 years, the sustained tensile stress in the glass fibre laminate should not exceed 33 per cent of the immediate ultimate stress.

The *safe crushing test load* depends on the moment of resistance of the reinforced section of the pipe wall at the springings, instead of on the tensile modulus of the concrete at crown and invert as in a plain pipe, and the test performance specification has been modified from that laid down in B.S. 556, 1966, by replacing the criterion of a maximum crack width of 1/100th in. at proof load to one of a maximum tensile strain in the laminate at the springings.

Thus the Ministry of Housing and Local Government requires these pipes to comply with the following specification:

1. *Crushing Test-proof Load*

   When tested in accordance with Appendix 'F' of B.S. 556, 1966, a 'Deckon' pipe shall sustain the minimum proof test load specified in Table 2 of the above British Standard. The tensile strain in the external laminate, measured at the horizontal diameter of the pipe when under this load, shall not exceed 33·3 per cent of the ultimate laminate strain.

   At this load, tensile cracking of the concrete core pipe at crown and invert is normal and is permissible, there being no limit on crack width.

   The tensile strain in the laminate at the springings, shall be measured with the use of demountable strain gauges at a minimum of four to six points, dependent on pipe length, or other suitable means. The strain measurement shall be directly related to the established ultimate strain value previously measured by testing and statistical analysis, using methods acceptable to the Purchaser's Engineer.

2. *Crushing Test—Ultimate Load*

   When tested in accordance with Appendix 'F' of B.S. 556, 1966, a 'Deckon' Pipe shall sustain without collapse or signs of compression failure, a load which is not less than the minimum ultimate load specified in Table 2 of the above British Standard.

3. Deckon Pipes shall comply in all other respects, with B.S. 556, 1966, including the selection of pipes for test.

Whilst this type of pipe normally has several times the crushing strength of a plain pipe of the same concrete and wall thickness, it will crack at crown and invert under approximately the same load as the plain pipe, which is only a small fraction of the safe working load. The skin, however, is watertight so that these cracks are of no consequence provided the joints are effectively sealed. This is achieved by extending and thickening the reinforced plastic skin to form the socket of the pipe and using a rubber sealing ring in the joint.

5.4 (a) Non-pressure Pipes

When the core pipe cracks at crown and invert the critical circumferential bending moment increases from $W_T R/\pi$ at the crown or invert, to $W_T R/2$ at the springings. There is also a tangential compressive load of $W_T/2$ at the springings. The resulting combined compressive stress determines the minimum wall thickness of the core to carry the given B.S. 2 or 3-edge test load W_T. But practical considerations will usually call for a thicker wall. The sectional area of the glass must be sufficient to ensure that throughout the useful life of the pipeline, failure will not occur in the glass.

If the ultimate load is accidentally exceeded and the concrete fails in compression at the springings, two more hinges are formed and the horizontal deflection of the pipe increases until the passive soil pressure is sufficient to balance the vertical load. The pipe thus becomes 'flexible' and it now consists essentially of four two-pinned arches in equilibrium. It will not collapse so long as the horizontal deflection is restrained by the soil, and the glass fibre does not fail under local stress at the hinges or because of strength regression.

It should be noted that at present Deckon pipes are designed as rigid pipes similarly to steel reinforced concrete pipe, i.e. having insufficient lateral deflection to mobilise lateral support from the soil under normal working conditions, and this design technique ensures slightly greater safety margins for Deckon.

Investigations are in hand to develop a design technique for 'flexible' pipes which will take advantage of lateral pressures mobilised by buried Deckon pipes.

5.4 (*b*) *Field Strength and Bedding Factor*

Since these pipes are cracked at crown and invert under working external loads, the beddings and bedding factors for ordinary rigid pipes are not applicable.

If the pipe is laid on a flat trench bottom, as with an ordinary Class D bed, the bending moment at the *springings* of the pipe will not differ materially from $W_T R/2$ as for the test load and the effective bedding factor will therefore be 1·0.

If, however, the load and reaction are uniformly distributed across the pipe diameter, as with say a normal Class B bedding, the bending moment at the springings will be halved relative to the test load moment and the effective value of the bedding factor will be 2·0.

Then for all trench and sub-trench conditions:

$$W_T = W_e F_s / C_{BS} F_m \text{ as in Formula (5.3)}$$

where $C_{BS} = 1{\cdot}0$ and $F_m = 1{\cdot}0$ for a Class D bed or 2·0 for a Class B bed.

Concrete beddings are not appropriate for these pipes.

In positive projection conditions under an embankment or valley fill, active earth thrust will be exerted on the pipe and will modify the bending moment at the springings. Thus if the lateral thrust is simplified to $K\left(H + \frac{Bc}{2}\right)\gamma_s B_c$ distributed uniformly over the vertical projection B_c of the pipe, the reverse moment at the springings will be $-K\left(H + \frac{B_c}{2}\right)\gamma_s B_c R/4$ and the combined moment will be $\frac{C_c \gamma_s B_c^2 R}{4} - K\left(H + \frac{B_c}{2}\right)\gamma_s B_c R/4$ and since the springing moment under the 3-edge test load is $\frac{C_c \gamma_s B_c^2 R}{2}$ the positive projection bedding factor F_p will be:

Formula 5.8

$$F_p = \frac{C_c \gamma_s B_c^2 R}{2} \Big/ \frac{R}{4}\left\{C_c \gamma_s B_c^2 - K\left(H + \frac{B_c}{2}\right)\gamma_s B_c\right\}$$

$$= 2 \Big/ \left\{1 - \frac{K}{C_c}\left(\frac{H}{B_c} + \frac{1}{2}\right)\right\}$$

This theory appears to be reasonable but has not yet been confirmed experimentally.

5.4 (c) Factor of Safety

Since this factor will tend to extend the life of the glass fibre, if over-loads do not occur accidentally, it is suggested that a factor of 1·25 on the ultimate for all depths should be used for non-pressure pipes.

5.5 Non-circular Rigid Conduits

Equations 5.1 and 5.2 above are not applicable to conduits having oval, egg-shaped, rectangular, or other non-circular cross-sections, the field strength of which must be individually ascertained either by theory or suitable loading tests, preferably the latter.

Chapter 6
The Load carrying Capacity and the Wall Thickness of Flexible Pipes
(Steel, ductile iron, pitch fibre, and some plastics)

6.1 General Notes

The essential difference between flexible and rigid pipes is that in the absence of side support the former will deform progressively without cracking under a slowly increasing crushing load. In practice when supported by surrounding soil and subjected to external pressure they may fail ultimately either by plastic yield of the wall material if relatively stiff, or by local buckling if lacking in stiffness (see Fig. 6.1). Failures of similar kinds may occur in flexible pipelines which are subjected to pure external fluid pressure, e.g. when submerged in water but not buried, or when laid above ground and subjected to partial vacuum conditions. Under any of these conditions internal fluid pressure reduces the compressive stress in the pipe wall and so reduces the probability of failure up to the point where plastic yield in tension occurs. If, as is usual in water mains, the internal pressure is greater than the external pressure, it tends to restore the circularity of pipes which have been deformed by external loading. But, since the internal pressure may on occasion be reduced to zero, or may even be negative (partial vacuum), pressure pipes must be separately designed to withstand the worst conditions of either internal or external loading. Non-pressure pipes in which the internal

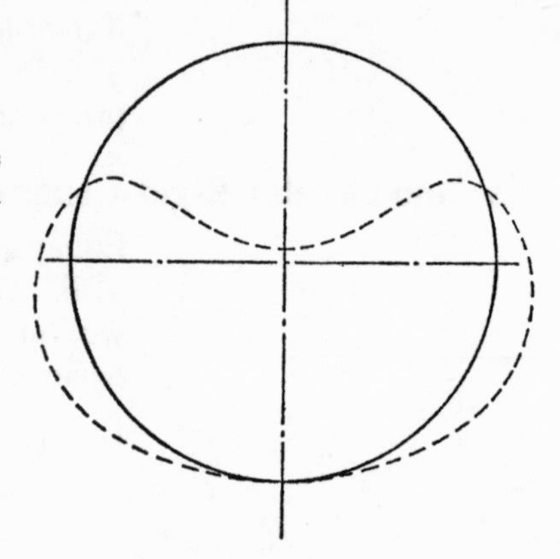

Fig. 6.1 Buckling mode of a flexible pipe with excessive vertical deflection
After Spangler—by permission

pressure rarely or never exceeds atmospheric pressure are designed to withstand the worst conditions of external loading only.

Evidently the design of flexible pipes differs fundamentally from that of rigid pipes. Crushing tests are inappropriate and are not specified for flexible pipes (except in B.S. 2760, 1956, where they occur purely as quality control tests for pitch fibre pipes). Wall thickness now depends on the elasto-plastic properties of the pipe material, and on those of the supporting soil if the pipes are buried, and on the permissible safe limit of vertical deflection, as well as on the magnitude of the loading and the diameter of the pipe.

Thus, with due allowance for practical imperfections, the wall thickness must be largest of the values required:

1. to maintain the radial deflection of the pipe within safe specified limits under extreme conditions of external loading when the internal pressure is atmospheric or below, and depending on whether or not the pipe is permanently supported by earth pressure;
2. to prevent critical compression failure either by plastic flow or elastic buckling under the same conditions as in (i) above; and
3. to prevent tensile (bursting) failure in pressure pipes under the maximum internal fluid pressure and minimum *permanent* external pressure.

Precise methods of design to meet the radial deflection limitations for buried pipes are not yet thoroughly well established and, to some extent, are still empirical. American experience in the use of steel pipes extends over the past 50–100 years however and a recent report by the American Waterworks Association[41], Chapter 8, states that steel pipes are virtually uncrushable under any existing fill or transient wheel loads when the cover is at least equal to the pipe diameter and the average deflection of the pipe under maximum external load is limited to about 2 to 3 per cent of its nominal diameter.

Spangler's classical work[46] on flexible pipe deflection has not been superseded, but Meyerhof and Fisher[44] have recently contributed very useful clarification to the estimation of Spangler's 'Modulus of Soil Reaction', the value of which has hitherto proved difficult to predetermine.

Recent experimental work by Whitman and Luscher[49] and independently by Meyerhof, Blaikie, and Fisher[43] has indicated that owing to the composite action of the pipe-soil system the critical (or collapse) pressures of buried flexible pipes may be much larger than has hitherto been anticipated.

For smooth walled pipes which are laid above ground or submerged but not buried the collapse pressures proposed by Stewart,[41] Chap. 6, based on Timoshenko's value

$$p_c = \frac{2E}{(1 - m^2)} \left(\frac{t}{2R}\right)^3 \text{lb/in}^2.$$

adjusted for practical imperfections and ovality in the pipes are presumably still valid (see Eqns. (6.2) and (6.3) below).

The design of *ductile iron pipes* will no doubt follow the method adopted for smooth walled steel pipes allowing for the differences in the properties of the metal[45,45a] but it remains to be firmly established, meanwhile the advice of the makers should be obtained.

For pitch fibre and plastic pipes no similar methods of estimating wall thicknesses have yet been proposed. For the small sizes of pipe with relatively thick walls in present use the current recommendations, based on limited loading tests, regarding adequate bedding and proper compaction of the surrounding soil, are regarded as essential however.

6.2 Basic Principles of Design of the Wall Thickness of Steel Pipes

6.2 (a) Installation or Working Conditions

For design purposes the following conditions under which a pipeline is to operate are to be distinguished as each requires a different treatment (see Fig. 6.2).

1. *Pipelines above ground* (see Articles 6.4 (*a*) and 6.5 (*a*)). The external load is normally zero but external fluid pressure may occur owing to partial vacuum in pressure pipes, (see Plate 5), or the accidential submergence of any pipe by floods.
2. *Pipelines under water but not buried* (see Articles 6.4 (*a*) and 6.5 (*a*)). Again there is no external fill load but the external fluid pressure is permanent and may vary.

In conditions 1 and 2 deflection is of no consequence and design is based on buckling considerations only. The effective critical pressure is that for the pipe alone. There is no help from composite action with soil.

3. *Pipelines buried in uncompacted backfill* (see Articles 6.4 (*b*) and 6.5 (*b*)). Lateral soil pressure is dubious and therefore of no consequence. Deflection must be controlled entirely by the elastic resistance to bending of the pipe wall without assistance from lateral soil pressure. Buckling can be ignored if the deflection is limited to a maximum of 5 per cent of diameter.[46]

4. *Pipelines buried in properly controlled compacted backfill* (see Articles 6.4 (*c*) and 6.5 (*c*)). The resistance to deflection and compression failure are now both modified by the composite action of the pipe and the soil ring which surrounds it. Relative to condition (3) the deflection is controlled by the lateral soil pressure and, relative to (1) and (2) the effective critical pressure is increased, i.e. the permissible compressive stress in the pipe wall is increased.

Careful attention to site work, particularly as regards adequate compaction of the backfill within a radial width or depth of at least one pipe diameter, or more in embankment positive projection conditions, is essential.

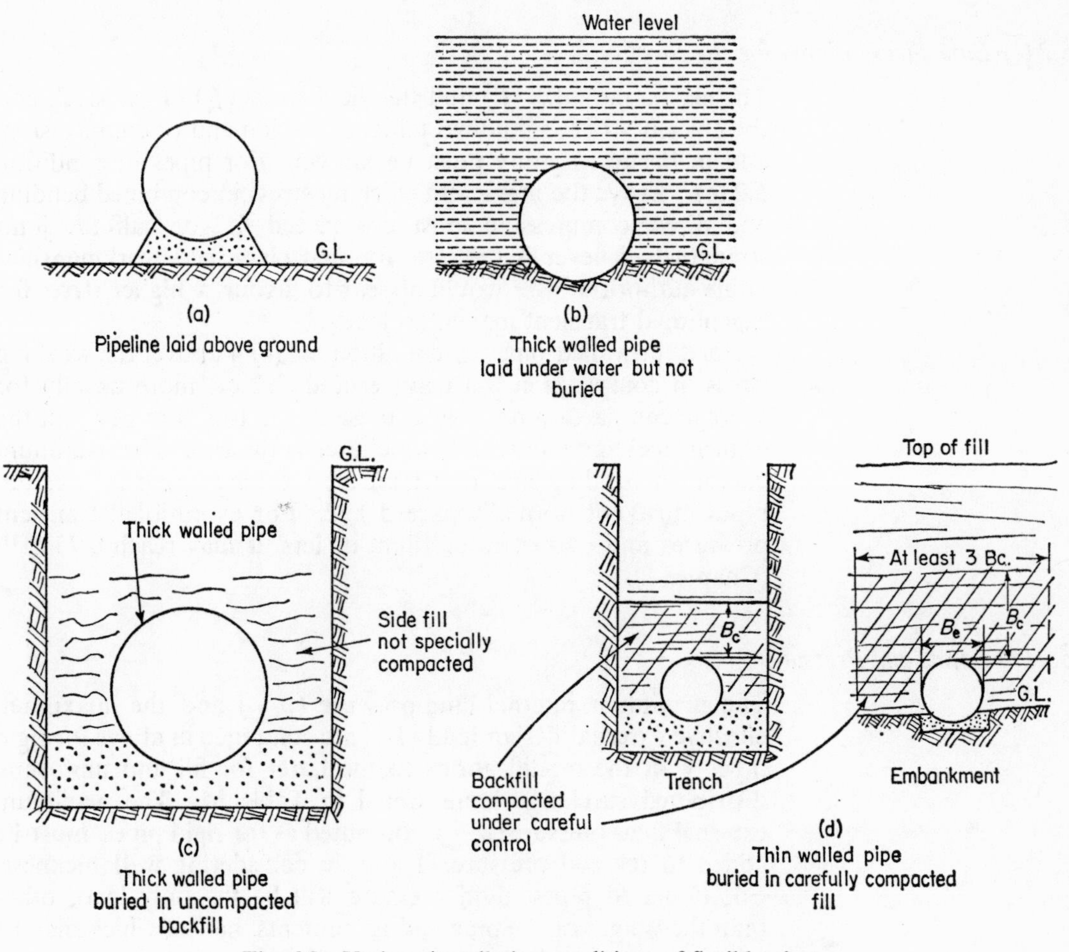

Fig. 6.2 Various installation conditions of flexible pipes

6.2 (b) Safe Limits of Deflection

Experience has shown that a steel pipe when unsupported laterally tends to become unstable and buckle inwards when the deflection of the vertical diameter reaches about 20 per cent of the diameter. It is therefore usually considered[41,44,46,47] that the vertical (or horizontal) deflection, under the maximum external design load and atmospheric internal pressure, should not exceed 5 per cent of the pipe diameter for unlined or flexibly lined pipes, whilst for pipes with cement mortar or other brittle linings it should not exceed 2 per cent to avoid cracking the linings. The horizontal deflection is approximately equal to the vertical deflection. Some authorities apply a safety factor of 1·25 on deflection[44] or a so called 'deflection lag' factor of 1·5 depending on the degree of compaction of the surrounding soil[46] (see Article 6.6).

6.2 (c) Safe Stress Limits for the Steel

The elastic modulus (E) and the yield stress (f_y) of the steel, and the efficiency of longitudinal joints in tension and/or compression, and in flexural rigidity must be known. For pipes in condition 6.2 (*a*) 3 above the maximum working stress in combined bending and direct compression must not exceed $f_y/2$ or half the joint strength whichever is the less, for first class field workmanship. Some authorities[41,46] would appear to favour a higher stress for exceptional transient maximum loads.

For thin walled pipes in condition 6.2 (*a*) 4 above, the working stress in compression must not exceed $f_y/2$ or, more usually for average compaction practice at present, $f_y/3$ to $f_y/4$ or one half the critical buckling stress, $f_c/2$, whichever is the less. The maximum combined bending and direct tensile working stress in pressure pipes must not normally exceed $f_y/2$. For exceptional transient pressures and diameters of 18 in. or less, it may reach $0{\cdot}75 f_y$[41] (Chapter 6).

6.2 (d) Loads and Pressures

The maximum internal fluid pressure ($p_{i_{max}}$) and the maximum effective external design load (W_e) are computed as above for rigid pipes with the modifications to the latter for fill and uniformly distributed surcharge loads noted in Table M. The maximum external fluid pressure ($p_{e_{max}}$), computed as for rigid pipes, must be added to the soil pressure, if any, in considering wall thickness. For unburied pipes, fluid pressure will be the only load, other than the weight of the pipe and its contents, half of which may be taken as an equivalent external load for large pipes.

For pressure pipes the wall thickness (t_i) for the internal fluid pressure, whether or not modified by the external load, is computed separately from that for the external load and external fluid pressure (t_e), and the larger value of the two is required. For non-pressure pipes the computation of t_e is identical with that for pressure pipes except that partial vacuum pressures do not occur.

6.2 (e) *Practical Limitations to Thinness of Wall*

To meet the practical requirements in fabricating, transporting, and installing *smooth walled* steel pipes an empirical rule[41] (Chap. 8) requires that the wall thickness should not be less than 1 per cent of the pipe radius, i.e. $t/R \geqslant 0{\cdot}01$.

6.2 (f) *Allowance for Corrosion*

Any necessary sacrificial extra wall thickness for corrosion must be added to the thickness derived by any of the following design procedures. In general prevention of corrosion by suitable effective protective coatings and linings is necessary whether or not extra thickness of metal is provided.

6.2 (g) *Pipe Beddings and Bedding Constants* (K_e)

Spangler[46] has shown that the length of the arc of the pipe circumference over which the bedding extends affects the radial deflection of the pipe under external loading, and has established the bedding constants (K_e) for various values of the bedding angle (θ) given in Table L. The bedding factors relating to bending moments in rigid pipes (F_m or F_p) are not appropriate for flexible pipes and cannot be used.

The construction of the beddings for flexible pipes has not been specified with the same precision as for rigid pipe beddings. Of the latter, Classes B and C (see Charts C9 and C10) are considered to be suitable, but the laying of smooth thick walled steel pipes on a flat trench bottom in soil, (not rock), with due regard to uniformity of support and the avoidance of hard and soft spots in the bed is not unusual, using a value of $K_e = 0{\cdot}108$. Concrete beddings are not usually appropriate for flexible pipes.

6.3 Applicability of the Design Methods

The methods are applicable to mild steel drawn (seamless) or welded smooth walled pressure or non-pressure pipes and to non-pressure steel pipes with various types of corrugated wall, when laid with various degrees of compaction of the surrounding soil, whether in trench or heading or under an embankment or fill or, for smooth walled pipes, by thrust bore.

6.4 Wall Thickness Required for External Loads and Pressure (t_e)

6.4 (a) For Working Conditions 1 and 2 in Article 6.2 (a), i.e. for Flexible Pipes Installed above Ground or Submerged but not Buried and having no Lateral Soil Support

Under these conditions the external fluid pressure $p_{e_{max}}$ (including partial vacuum pressure where applicable) is the only external load and it must not exceed half the critical external pressure for the unsupported pipe. The wall thickness (t_e) will therefore be determined by substituting the value of $2p_{e_{max}}$ for p_c in equations (6.1) or (6.2) below, depending on its magnitude (see Appendix A, Ex. 10).

Critical (or collapse) pressure

The external fluid pressure at which a thin walled circular flexible pipe unsupported by surrounding soil becomes unstable elastically and starts to buckle or yield plastically is known as the 'critical' or collapse pressure (p_c). For a long unreinforced circular smooth wall steel pipe having $E = 30 \times 10^6$ lb/in.^{2} and Poisson's ratio $m = 0{\cdot}30$ its value according to Timoshenko assuming perfect circularity and uniform wall thickness is

$$p_c = 66 \times 10^6 \left(\frac{t}{2R}\right)^3 \text{lb/in}^2. \text{ (see Article 6.1)}$$

For commercial pipes, however, to allow for variations in wall thickness, out of roundness and other manufacturing tolerances, Stewart[41] (Chapter 6) has proposed for steel having $E = 30 \times 10^6$ lb/in.^{2} and $f_y \geqslant 27{,}000$ lb/in.^{2} and an external radius R*.

For buckling failure:

Formula 6.1
$$p_c = 50{\cdot}2 \times 10^6 \left(\frac{t}{2R}\right)^3 \text{lb/in.}^2$$

where

$$\frac{t}{2R} \leqslant 0{\cdot}023 \quad \text{and} \quad p_c \leqslant 581 \text{ lb/in.}^2$$

and for plastic failure:

Formula 6.2
$$p_c = 86670 \left(\frac{t}{2R}\right) - 1386 \text{ lb/in.}^2$$

where
$$\frac{t}{2R} > 0{\cdot}023 \text{ and } p_c > 581 \text{ lb/in.}^2$$

The corresponding *critical stress* is $f_c = p_c R/t$.

* For thin smooth walled pipes there is little difference between the internal, mean and external values of R.

6.4 (b) For Working Condition 3 in Article 6.2 (a): Self-supporting Smooth or Corrugated Wall Pipe Installed in a Trench and Acting as an Elastic Ring

When the pipe deflection is limited to not more than $R/10$ (5 per cent of diameter)

Equation 6.3

$$\Delta_x = \frac{K_\theta W_e R^3}{12EI} \text{ in.} = \Delta_y \text{ approx.}$$

where W_e is the total effective external load in lb/linear ft and K_θ is the bedding constant (see Table L).

But in the presence of an external fluid pressure p_e the deflection caused by the fill load will be increased to Δ'_x by the excess of the vertical pressure load over the horizontal pressure load, caused by the difference in the vertical and horizontal projected areas of the pipe[(46a)]

i.e. $p_e(2R + \Delta'_x)$ vertical load $- P_e(2R - \Delta'_x)$ horizontal load $= 2p_e\Delta'_x$ lb/lin. in.

Then the total effective vertical load is:

$$\frac{W_e}{12} + 2p_e\Delta'_x = \frac{\Delta'_x}{\Delta_x}\frac{W_e}{12} \text{ lb/linear in.}$$

since, from Equation (6.3) the ratio of load to deflection is constant.

Whence, substituting the value of Δ_x from Equation (6.3),

Equation 6.4

$$\Delta'_x = \frac{K_\theta W_e R^3}{12(EI - 2p_e K_\theta R^3)} \text{ in.}$$

For smooth walled pipes $I = t_e^3/12$ and

Equation 6.5

$$\Delta'_x = \frac{K_\theta W_e R^3}{Et_e^3 - 24p_e K_\theta R^3} \text{ in.}$$

and if $p_e = 0$

$$\Delta'_x = \frac{K_\theta W_e R^3}{Et_e^3} = \Delta_x \text{ as in Equation (6.3)}$$

The value of t_e or I may then be derived from these equations (see App. A, Ex .11). Values of I for corrugated pipes are given in Table K.

Check on compressive stress in the steel

It can be shown that the maximum bending moment and fibre stress in the steel ring occurs at the invert and the compressive stress is increased by the external fluid pressure.

Thus substituting the value of $\Delta'x$ from Equation (6.4) in Equation (iii) of Table L, viz.

$$M_b = \frac{K_b}{K_\theta}\frac{\Delta'_x EI}{R^2} = M_{max} = \frac{K_b W_e REI}{12(EI - 2K_\theta p_e R^3)}$$

where K_b is the moment coefficient at the invert (see Table L). The maximum *bending stress* is then

Equation 6.6
$$f_{cb} = \frac{M_{max}d}{2I} = \frac{K_b W_e REd}{24(EI - 2K_\theta p_e R^3)} \text{ lb/in.}^2$$

where d is the depth of the corrugations in a corrugated pipe and $d = t$ for a smooth walled pipe.

For a smooth walled pipe laid on the trench bottom, the bedding angle will be about 30° and from Table L, $K_\theta = 0{\cdot}108$ and $K_b = 0{\cdot}235$. Then from Equation (6.6)

$$f_{cb} = \frac{0{\cdot}117\ W_e REt_e}{Et_e^3 - 2{\cdot}592 p_e R^3} \text{ lb/in.}^2$$

The compressive ring stress is

Equation 6.7
$$f_{cr} = \frac{p_e R}{A} \text{ lb/in.}^2$$

where A = the sectional area of wall, in.^{2} per inch of length and $A = t_e$ for a smooth walled pipe.

The maximum total compressive stress is then

Equation 6.8
$$f_{c\,max} = f_{cb} + f_{cr} \text{ lb/in.}^2 \text{ (approx.)}$$

which must not exceed $f_y/2$.

Then substituting t_e for d, $t_e^3/12$ for I and t_e for A in Equations (6.6) and (6.7), from Equation (6.8)

Equation 6.9
$$\frac{K_b W_e REt_e}{2(Et_e^3 - 24K_e p_e R^3)} + \frac{p_e R}{t_e} = \frac{f_y}{2}$$

Whence t_e may be obtained; but its practical value cannot be less than t_i obtained as in Article 6.5 (*b*) for the condition where $p_e = 0$ in pressure pipes (see App. A, Ex. 11).

N.B. W_e is measured in lb/linear ft. The remaining quantities are measured in lb/in. units. The coefficients in Equations (6.3) to (6.6) and (6.9) have been adjusted accordingly.

6.4 (c) Working Condition 4 in Article 6.2 (a): Thin-walled Smooth or Corrugated Pipe Installed with Controlled Compaction of the Soil Surrounding the Pipe (see App. A, Ex. 12, 13, 14)

The method of design described below is a combination of Spangler's method of estimating pipe deflection under external loading, with Meyerhof's more recent suggestions[44,46] regarding the 'Modulus of Soil Reaction' (i.e. Spangler's value 'e' which Meyerhof calls the 'coefficient of soil reaction' or 'subgrade modulus' 'k'. These values ('e' and 'k') are identical). Both methods postulate controlled compaction of the bedding and fill around the pipe to 95–100 per cent Procter density (see B.S. 1377: 1961–3).

6.4 (*c*) (i) *Pipe Deflection*

Spangler's basic equation for long-time deflection under these conditions is[46]

Equation 6.10
$$\Delta_x = \Delta_y = \frac{F_{lag} K_\theta W_e R^3}{12(EI + 0{\cdot}061eR^4)}$$

For the composite action of pipe and soil Meyerhof[44] postulates a ring of soil concentric with the pipe and of an annular width of at least the diameter of the pipe (see Fig. 6.3), and having approximately uniform stiffness. The bedding angle is therefore 180° and $K_\theta = 0{\cdot}083$ (Table L).

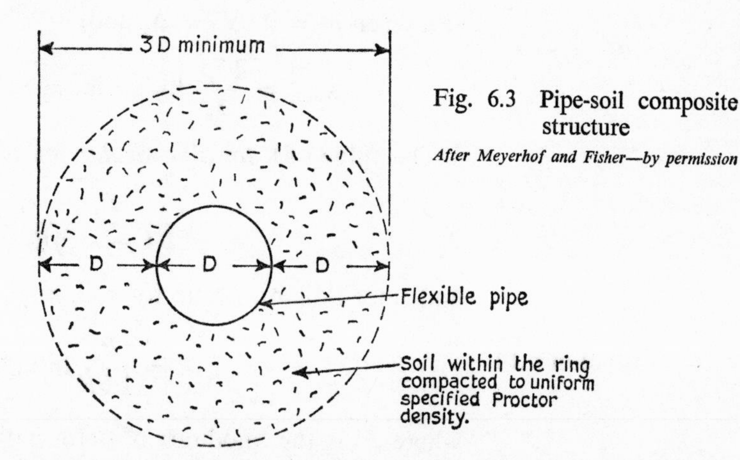

Fig. 6.3 Pipe-soil composite structure

After Meyerhof and Fisher—by permission

The average unit vertical pressure at the top of the pipe is

$$p_o = \frac{W_e}{(12)(2)R} \text{ lb/in.}^2$$

Substituting $p_o(24)R$ for W_e and 'k' for 'e' in Equation (6.10) and ignoring F_{lag} at this stage (see Article 6.6), for $K_\theta = 0{\cdot}083$,

$$\Delta_x = \frac{0{\cdot}167 p_o R^4}{EI + 0{\cdot}061kR^4} \text{ in.}$$

and since EI is small for thin walled pipes compared with $0{\cdot}061kR^4$ it may be safely ignored.

Then

Equation 6.11
$$\Delta_x = \frac{2{\cdot}7p_o}{k}\text{ in.}$$

But, as in Article 6.4 (*b*) above, the effect of external fluid pressure is to increase the deflection of the pipe to Δ'_x by the addition of a load $2p_e\Delta'_x$ lb/linear in.*

Then

$$p_o2R + 2p_e\Delta'_x = \frac{\Delta'_x p_o 2R}{\Delta_x}$$

whence, substituting $\frac{\Delta_x k}{2{\cdot}7}$ for p_o (Equation (6.11))

Equation 6.12
$$\Delta'_x = \frac{2{\cdot}7p_oR}{kR - 2{\cdot}7p_e}\text{ in.}$$

then

Equation 6.13
$$k_{min} = \frac{2{\cdot}7(p_oR + p_e\Delta'_x)}{R\Delta'_x}\text{ lb/in.}^2\text{/in.}$$

and when $p_e = 0$, $\Delta'_x = \Delta_x$ and

$$k_{min} = \frac{2{\cdot}7p_o}{\Delta_x}\text{ lb/in.}^2\text{/in. as in Equation (6.11)}$$

The value of *k for a particular soil fill* is given as[44]

$$k = \frac{E_s}{2(1 - m_s^2)R}\text{ lb/in.}^2\text{/in.}$$

and since Poisson's ratio for soil, $m_s = \frac{1}{2}$ approximately

Equation 6.14
$$k = \frac{E_s}{1{\cdot}5R}\text{ lb/in.}^2\text{/in.}$$

where E_s is the 'modulus of deformation' of the compacted soil (i.e. stress/strain) as determined by triaxial tests on samples of the fill which have been compacted to a known, or required, density and then saturated under a transverse fluid pressure equivalent to the minimum overburden pressure at the pipe axis over the length of pipeline under consideration. This test will give the immediate value of E_s but the value may change with time under load for some soils (see Article 6.6).

Note that for a particular soil, overburden pressure, and degree of compaction, Equation (6.14) indicates that k is inversely proportional to R. The design operations will be facilitated by the early determination of the required degree of compaction and

* See Article 6.4 (*c*) (ii) on p. 101 for the appropriate values of p_e.

of E_s for the proposed fill, preferably by competent soil technologists.

For pipelines up to about 60 in. diameter Meyerhof[44] states that the value of k to be used in design should not be less than 20 lb/in.²/in., and that E_s should not be less than about 1000 lb/in.² and that it should increase by about 200 lb/in.² per foot of pipe diameter for diameters greater than five feet.

In weak natural soils it may be necessary to import a stronger soil complying with the above requirements for the fill surrounding pipes in a trench, and to widen the trench (so increasing the fill load) sufficiently to ensure that the natural soil is not overstressed by the lateral pressure from the pipes.

6.4 (*c*) (ii) *Ring Stress*

For flexible pipelines in this condition (Article 6.2 (*a*) (iv)) the 'compression ring' theory of White and Layer[48] is adopted. This theory postulates that the compressive ring stress in the pipe wall is uniform around the periphery and that the external soil pressure at any point on the periphery is inversely proportional to the radius of curvature of the wall at that point (i.e. that pR is constant) provided that the minimum cover is sufficient (i.e. about $B_c/4$) to balance the active pressure imposed on a thin walled pipe. A corrugated pipe has considerable bending resistance and it may tolerate a minimum cover of about $B_c/8$[48].

Since the radius at the top of the pipe tends to increase with the application of external load, the soil pressure tends to decrease and the compressive ring stress is therefore not likely to exceed

Equation 6.15
$$f_c = \frac{(p_o + p_e)R}{A} \text{ lb/in.}^2$$

where $p_o = W_e/(12)(2)(R)$ lb/in.² when the pipeline is not submerged;

or $p_o = W_{es}/(12)(2)(R)$ lb/in.² when reduced for partial or complete submergence;

A = the sectional area of the pipewall in.² per in. of length;

$A = t$ for smooth walled pipes, in.²/in.

and p_e
- $= 0$ for non-pressure pipes not submerged;
- $= p_g$ lb/in.² for non-pressure pipes partially or wholly submerged;
- $= p_{vac}$ lb/in.² for pressure pipes not submerged;
- $= p_g + p_{vac}$ lb/in.² pressure pipes partially or wholly submerged.

The allowable ring stress f_a must not exceed the lesser value of:

1. $f_y/2$ to $f_y/4$ lb/in.^{2} depending on the expected efficiency of the compaction of the fill, or the same fraction of the axial seam strength of the pipe if less than f_y.
2. One-half the critical stress computed from equation (6.17) below, or the allowable stress obtained from the appropriate Chart 12 (*a*), (*b*) or (*c*).

6.4 (*c*) (iii) *Critical (or Collapse) Stress*

As mentioned earlier it has been shown that the critical stress for thin walled pipes when buried and supported by properly compacted soil differs considerably from that of naked or unsupported pipes.

Meyerhof[44] postulates that providing

Equation 6.16

$$R \geqslant 2\sqrt[4]{\frac{EI}{(1-m^2)k}} \text{ in.}$$

the critical ring stress may, allowing for practical imperfections in the pipe, be taken as

Equation 6.17

$$f_c = \frac{f_y}{1 + (f_y A/2)(\sqrt{(1-m^2)/kEI})}$$

and the permissible stress is $f_a = f_c/2$.

Permissible stresses of $f_c/2$, and the lower limit of R, have been computed for steel having $E = 30 \times 10^6$ lb/in.^{2}, $m = 0{\cdot}3$ and $f_y = 40{,}000$ lb/in.^{2} and are given in the Charts C12 (*a*) and (*b*) for two types of corrugated wall, and in Chart C12 (*c*) for smooth wall pipes having $f_y = 40{,}000$ lb/in.^{2} and 36,000 lb/in.^{2} respectively. The upper limit of f_c is taken as f_y. The lower limit of R is shown on the same charts.

The values of A, I and t for commercial *corrugated* pipes are given in Table K for sections in common use. For smooth walled pipes $A = t$ in.^{2}/linear in. of pipe and $I = t^3/12$ in.^{4}/linear in. of pipe.

6.4 (*c*) (iv) *Selection of Wall Thickness*

For thin walled pipes which depend for their stability on composite action between the soil and the pipe the selection of the pipe wall thickness (t_e) is a process of successive approximation as follows:

Given the pipe radius R in., the external design load W_e lb/linear ft. of pipe, the maximum external fluid pressure p_e lb/in.^{2}, and the limiting vertical deflection of the pipe Δ_x in:

Step 1. Obtain the vertical pressure at the top of the pipe caused by the external load

$$\text{i.e. } p_o = \frac{W_e}{(12)(2)R} \text{ lb/in.}^2$$

Step 2. Obtain the required minimum value of 'k' from Equation (6.13), and if less than 20 lb/in.^{2}/in. adopt a value of 20 lb/in.^{2}/in. or assess the value from known test values of E_s, by Equation (6.14).

Step 3. Assume an allowable ring stress f_a of say $0{\cdot}25 f_y$ and from Equation (6.15) obtain a value for A, the area per in. length of wall section, i.e.

$$A = (p_o + p_e)R/0{\cdot}25 f_y \text{ in.}^2$$

Step 4. Adopt the nearest standard plate thickness to give the area A for smooth walled pipe or, for corrugated pipes, select a gauge thickness 't' from Table K giving the nearest value to A, and obtain its I value. Then check that the value of R exceeds the minimum permissible value given by Chart C12 (*a*), (*b*) and (*c*) or by Equation (6.16) i.e. that

$$R \geqslant 2\sqrt[4]{\frac{EI}{(1 - m^2)k}} \text{ in.}$$

Thus taking $E = 30 \times 10^6$ lb/in.^{2} and $m = 0{\cdot}3$ and k as obtained or assumed in Step 2;

$$R \geqslant 2\sqrt[4]{\frac{33 \times 10^6 I}{k}} \text{ in.}$$

Step 5. If Equation (6.16) is not satisfied reduce I or increase k as necessary and convenient.

Step 6. Obtain the ratio t/k for a corrugated wall, or kt for a smooth wall, and from Chart C12 (*a*), (*b*) or (*c*) as appropriate obtain the allowable stress f_a.

Step 7. With this value of f_a, if greater than $0{\cdot}25 f_y$, obtain a revised value of $A(=(p_o + p_e)R/f_a)$ and select a corresponding new value of t giving the nearest lower value of A from Table K. N.B. $t = A$ for smooth walled pipes.

Step 8. Obtain the new ratio of t/k or kt and repeat steps 6 and 7 to check that the value of A is satisfactory.

Step 9. Repeat step 8 if necessary to revise A.

Step 10. Check that the value of Δ_x by Equation (6.12), (i.e. $\Delta_x = 2{\cdot}7 p_o/R$ in.) does not exceed the specified deflection.

N.B. 1. If the compaction of the fill can be increased to give a higher value of 'k' the wall thickness of the pipe can be reduced by repeating steps 6 to 10. This will also reduce Δ_x.

2. For smooth walled pressure pipes t_e may be greater but cannot be less than t_i (see Article 6.5 (*c*)). (See also App. A, Ex. 14.)

6.5 Wall Thickness Required by Internal Pressure in Smooth Walled Pipes (t_i)

6.5 (*a*) *For Working Conditions 1 and 2 in Article 6.2* (*a*), *i.e. For Pipes Installed above Ground or Submerged but not Buried and having no Soil Support* (see also App. A, Ex. 10)

The effective internal pressure is $p_{i_{max}}$ and the permanent external pressure is $p_{e_{min}}$ (excluding p_{vac} which is not operative simultaneously with p_i). The effective resultant pressure is $p_{i_{max}} - p_{e_{min}}$ and the tensile ring stress is

$$f_{sr} = (p_{i_{max}} - p_{e_{min}})\frac{R}{t_i}$$

and since this stress must not exceed $f_y/2$

Equation 6.18
$$t_{i_{min}} = \frac{2(p_{i_{max}} - p_{e_{min}})R}{f_y}$$

where R is the external radius of the pipe (see Note to Article 6.4 (*a*)).

6.5 (*b*) *For Working Condition 3 of Article 6.2* (*a*), *i.e. For Self-supporting Pipe Installed in a Trench and acting as an Elastic Ring having no Effective Lateral Soil Support* (see also App. A, Ex. 11)

For an unstrutted pipe when the trench is backfilled before the application of internal pressure.

The effect of the internal pressure is to reduce the deflection of the pipe, because of the excess of vertical over horizontal internal pressure load (the reverse of the effect in Article 6.4 (*b*)),

$$\text{i.e. } \frac{W_e}{12} - 2p_{ie}\Delta'_x = \frac{\Delta'_x}{\Delta_x}\frac{W_e}{12}$$

whence

$$\Delta'_x = \frac{K_\theta W_e R^3}{12(EI + 2K_\theta p_{ie} R^3)} \quad \text{Cf. Equation (6.4)}$$

where $p_{ie} = (p_{i_{max}} - p_{e_{min}})$.
Substituting $t_i^3/12$ for I for a smooth walled pipe

Equation 6.19
$$\Delta'_x = \frac{K_\theta W_e R^3}{Et_i^3 + 24K_\theta p_{ie} R^3}$$

For a pipe laid on a flat trench bottom the bedding angle θ is assumed as 30° and $K_\theta = 0{\cdot}108$ (Table L).
Then

Equation 6.20
$$\Delta'_x = \frac{0{\cdot}108 W_e R^3}{Et_i^3 + 2{\cdot}592 p_{ie} R^3}$$

Check on maximum tensile stress

As in Article 6.4 (*b*) *the maximum bending stress* occurs at the invert where $M_b = K_b WR/12$ and since $\Delta'_x = K_\theta W_e R^3/12EI$ (see Table L),

$$M_b = \frac{K_b}{K_\theta} \cdot \frac{\Delta'_x EI}{R^2} \text{ as in Table L, and } f_{sb} = \frac{M_b t}{2I} = \frac{6M_b}{t_i^2}.$$

Substituting the values of M_b, and Δ'_x from Equation (6.19),

Equation 6.21
$$f_{sb} = \frac{K_b}{2} \cdot \frac{W_e R E t_i}{(Et_i^3 + 24K_\theta p_{ie} R^3)}$$

For $\theta = 30°$, $K_\theta = 0{\cdot}108$ and $K_b = 0{\cdot}235$ (Table L),

$$f_{sb} = \frac{0{\cdot}117 W_e R E t_i}{Et_i^3 + 2{\cdot}592 p_{ie} R^3}$$

The tensile ring stress is

Equation 6.22
$$f_{sr} = \frac{p_{ie} R}{t_i}$$

and the combined total maximum stress is

$$f_s = f_{sb} + f_{sr} \text{ (approx.)}$$

which must not exceed $f_y/2$,

Equation 6.23
$$\text{i.e. } \frac{K_b}{2} \frac{W_e R E t_i}{(Et_i^3 + 24K_\theta p_{ie} R^3)} + \frac{p_{ie} R}{t_i} \leqslant f_y/2 \text{ Cf. Equation (6.9)}$$

t_i is then obtained to satisfy Equation (6.23) but it cannot be less than t_e obtained as in Article 6.4 (*b*) for the condition where $p_i = 0$.

N.B. W_e is measured in lb/linear ft. The remaining quantities are in lb-in. units. The coefficients in Equations (6.19) to (6.23) have been adjusted accordingly.

6.5 (*c*) *Working Condition 4 of Article 6.2 (a), i.e. For Thin Smooth-walled Pipes Installed with Controlled Compaction of the Soil Surrounding the Pipes* (see also App. A, Ex. 12, 13)

Whether or not the pipe is predeformed before or during backfilling the internal pressure tends to equalise the vertical and horizontal deflections. During normal working the tensile ring stress is reduced by the external load and pressure. During a pressure test before backfilling, or after repairs, however, when the pipe is uncovered, it must carry the full internal pressure. The ring stress is then

Equation 6.24
$$f_{sr} = \frac{p_{i_{max}} R}{t_i} \text{ lb/in.}^2$$

which must not exceed $f_y/2$ lb/in.^{2}.
Then

Equation 6.25
$$t_{i_{min}} = \frac{2p_{i_{max}} R}{f_y} \text{ in.}$$

NOTE that t_i may be greater but cannot be less than t_e (see Article 6.4 (*c*)).

6.6 Practical Considerations

It is evident that precision in the estimation of wall thickness or deflection is not to be expected. In view of the inevitable variation in soil properties, conservative values of k should be used. Moreover, wall thickness may be dictated by consideration of corrosion and the probability of accidental mechanical damage to the pipe wall rather than by the loading, and the stiffness of the pipes may be adversely affected by inadequacy in any longitudinal joints or seams.

The use of large buried pipes with uncompacted fill is inappropriate for use under roads and verges because of the unavoidable subsequent settlement of the fill and the variation in deflection of the pipe with consequential damage to the road bed.

If there is any possibility of subsequent excavation alongside a soil supported pipe the composite action may be destroyed. It would therefore appear advisable to design the pipe to withstand the fill load, at least, without soil support and to prohibit the passage of heavy vehicles over the pipeline during the danger period (see also App. A, Ex. 13).

Spangler's 'lag factor', (F_{lag}) = 1·25 to 1·50, is intended to allow for the slow increase of deformation of the soil under sustained lateral pressure in non-pressure pipes. It is not normally applicable therefore to high pressure pipes or to the temporary pressures induced in the soil by transient wheel loads and it need only be applied, if at all, to the deflection caused by fill and permanent or long-term uniform surcharge loads, i.e.

$$\Delta_{y\ total} = F_{lag}\Delta_y \text{ for fill and other permanent loads} + \Delta_y \text{ for transient loads.}$$

6.7 The Use of Strutted or Tied Predeformed Flexible Pipes

Extending the vertical diameter of a steel pipe during manufacture or prior to, or during, its installation and before placing the backfill, by means of vertical struts, or horizontal ties and walings, is a device frequently used to reduce inconveniently large or impermissible eventual subsidence of the trench backfill particularly in road trenches, for large pressure or non-pressure pipes. This device may prove very useful in water mains, laid under roads, where the internal pressure considerably exceeds the external pressure caused by the fill load, i.e. (by W_c not W_e). Under such conditions the pressure tends to restore the pipe to its circular shape or to maintain that shape. If the pipe is laid in its undeformed state its vertical diameter will be reduced by the fill load, and heave of the fill will occur when the line is under pressure. Conversely, if the pipe is appropriately predeformed, by strutting or tying, it will be restored to the circular form, approximately, by the fill load and there will be little or no further deformation when the internal pressure is applied. A similar result can be achieved for small bore pipes by applying the internal pressure before placing compacted backfill.

In a non-pressure pipe the same device reduces the lateral pressure applied to the soil and so increases the load carrying capacity of the pipe or, theoretically, permits a thinner pipe to be used.

If either type of pipe is lined with mortar or other brittle protection material, predeforming the pipe prestresses the lining so that in the working position it is not subjected to tensile stresses caused by pipe deformation and so is less liable to crack so long as the bond to the steel remains sound during the prestressed period.

By lengthening the vertical diameter of a pipe by a distance Δ_s the final vertical deformation of the pipe diameter can be restricted to zero or to a permissible degree of shortening, by making

$$\Delta_s = \Delta_y - \Delta'_y$$

where Δ_y = the permissible limit of deformation

and Δ'_y = the final deformation desired.

Thus, if Δ_y is 5 per cent, and if $\Delta'_y = 0$, $\Delta_s = 5$ per cent; or if $\Delta'_y = 2$ per cent, $\Delta_s = 5 - 2 = 3$ per cent.

The struts or ties may be removed before completing the backfill provided there is a sufficient depth of fill to ensure adequate composite action with the soil and no premature release of the pre-deflection, and provided that construction traffic is so controlled as to avoid excess deflection of the unstrutted pipe. Equation (6.11) is not applicable whilst the struts or ties are in place.

6.8 Pitch Fibre and Plastics Pipes (B.S.2760; B.S.3505, 3506)

These pipes, at present of small diameter (about 8 in. bore maximum), are relatively thick walled. They are therefore unlikely to fail under uniform ring compression but may fail by plastic yield under sustained vertical load unless given adequate lateral support by the soil and bedding. As yet, there is no generally accepted rational method of determining their wall thickness but the same basic principles as for steel pipes are applicable, viz.

(i) That for non-pressure or pressure pipes the deformation of the vertical diameter should not exceed 5 per cent under the maximum external load W_e with atmospheric internal pressure.

(ii) That for plastic pipes the combined compressive direct and bending stresses under atmospheric internal pressure, should not exceed the long term ultimate compressive stress reduced by an adequate factor of safety.

(iii) That for plastics pressure pipes the tensile ring stress under maximum internal pressure and zero external load or the combined bending tensile and direct stress under working conditions should not exceed the long-term ultimate tensile strength of the material reduced by an adequate factor of safety.

Condition (i) can be achieved by providing fully and uniformly compacted bedding, side fill and initial overfill, preferably of stable granular material. Short-term tests have shown that a B.S. 3505 Class AA uPVC pipe embedded in fully compacted sand did not exceed 5 per cent deformation under the probable maximum loads likely to be imposed on the pipe up to a cover of 24 ft.

For condition (iii) B.S. 3505 and 3506 specify the permissible working pressure for the several classes of uPVC pipe, (see Table H) (presumably under zero external load), but they do not specify either the long term Young's Modulus or the yield stresses of the material.

Concrete beddings are not recommended for either kind of pipe.

Section D

Construction Methods
Materials
Computation

Chapter 7
Choice of Construction Method and Materials

7.1 Construction Below Ground

7.1 (*a*) *Choice of Method* (see Fig. 3.1)

The choice lies between narrow trench, narrow sub-trench in V trench, wide trench, wide sub-trench in V trench, very wide trench, heading (or tunnel), or thrust bore. It will depend upon the existing site conditions such as the presence of existing underground works or other obstructions, the contour of the route, the laying depth and soil conditions, traffic conditions on roads, length of haul for spoil or fill, and on the favourable balance of the costs of labour and materials. Thus comparative estimates are certainly advisable and for large schemes, are necessary in the interests of economy (see App. A, Summary of worked examples Nos. 1–6).

7.1 (*a*) (i) *The Narrow Trench*

This requires the minimum of excavation for exposed work but, at depths over about 5 ft, the trench will usually require timbering. In urban roads, traffic congestion and consequent delay are practically unavoidable. The loads imposed by the fill are relatively small at depths less than, say, 7 ft cover, but become preponderant at greater depths in roads. Strict control of trench width is essential, working space is thereby limited, and there is a weather risk, all of which slow the work and increase labour costs and overheads.

7.1 (*a*) (ii) *V Trench with Narrow Sub-trench*

This method is only suitable for open country in the absence of existing underground works. It facilitates mechanical excavation and requires no timbering except in a deep sub-trench. There are risks of weather and slips in clay soils, especially at overall depths of cut exceeding about 12 ft, otherwise the work is straightforward.

7.1 (*a*) (iii) *The Wide Trench*

This involves somewhat more excavation than a narrow trench. Its main advantages are the relief from strict control of width and greater working space, both of which tend to reduce working time. The load on the pipe is increased and more bedding material may be required, but, on balance, the easier working conditions may be preferred. Traffic and weather hazards are much the same as for the narrow trench. If the depth varies considerably, a trench which is 'wide' at the shallow end may become 'narrow' as the depth increases, so far as the load on the pipes is concerned.

7.1 (*a*) (iv) *V Trench with Wide Sub-trench*

This has much the same advantages and disadvantages as the wide trench relative to the narrow trench, except that the sub-trench will always be 'wide' and the load therefore always greater than with a narrow sub-trench.

7.1 (*a*) (v) *The Very Wide Trench*

This only occurs where the main excavation is made for some purpose other than pipe installation. If the fill load is operative, it will be high but it can be reduced by laying the pipe in a narrow sub-trench if possible (negative projection condition), or in the induced trench condition if a sub-trench is inappropriate (see Embankment loading, Article 7.2 (*a*)).

7.1 (*a*) (vi) *In Heading or Tunnel*

This involves much less excavation, is largely free of weather hazards, and largely avoids traffic congestion, and if deep enough avoids obstructions from existing underground works. However, timbering is usually essential, the working space is cramped and handwork in the removal of spoil, and in careful repacking is usually necessary and slow, all of which increase labour costs. In a tunnel 24-hour working is possible and is completely independent of the weather. The effective design load is not greatly affected relative to narrow trench conditions. Groundwater and unstable soil may be even more troublesome than in a trench. To provide working space the minimum cross section of the heading is about 3 ft 9 in. by 2 ft 6 in. even for the smallest pipe.

7.1 (*a*) (vii) *By Thrust Boring*[87,87a]

This requires the minimum of excavation, no timbering other than in the back and front pits and is independent of the weather but it cannot be completely relied upon as regards the line and level of the pipe. Handwork is usually needed for the excavation, working from inside the pipe if it is large enough, but 24-hour working is possible. It is not suitable for any situation where underground works may be encountered, but it works well under busy streets and in bad ground, the pipes acting as a shield. Steering is difficult in coarse or cemented gravels. No repacking is necessary but pressure grouting may be needed in some soils. Special concrete pipes with flush joints are needed and are available. Welded joints are used for steel pipes.

7.1 (*b*) *Choice of Bedding*

In any of the above methods of installation other than thrust boring the choice of bedding will affect the strength and cost of the pipes (see App. A, Worked examples—1–6).

7.1 (*c*) *Minimum Cover Depth*

The minimum cover over drain or sewer conduits in roads should not normally be less than 4 ft in order to avoid interference with other underground utility pipes and cables; in fields or gardens it should not normally be less than 3 ft to avoid interference with agricultural and horticultural operations. These depths are also believed to be sufficient to prevent ice formation in pipes in the British Isles. Where submergence is possible the cover should always be sufficient to prevent the flotation of an empty pipeline (see Article 4.15, p. 60). Water mains are not usually laid with less cover than 3 ft. Gas mains are not subject to the same restriction but pipes with very shallow cover may need special protective measures as suggested below.

7.1 (*d*) *Protection of Shallow Pipelines*

7.1 (*d*) (i) *Under Roads and Verges*

1. *Rigid Pipes*

Where the cover depth under roads or verges is less than the 4 ft minimum recommended in Article 7.1 (*c*) above, the load transmitted to the pipes from the wheels of heavy vehicles rises rapidly as the cover decreases. Special measures are needed to resist these loads and the vibration which frequently accompanies them. Under macadam or other flexible pavements, 4 in. and 6 in. rigid gully connections are highly vulnerable in this respect, and especially so if rigidly jointed, since they may be subjected to high shear stress. Rather than attempt to make rigid pipes strong enough to withstand these loads it is preferable either to relieve them of load completely or to substitute flexible pipes of steel or ductile iron.

Thus, where possible, shallow trenches should be backfilled with compressible material and bridged by a reinforced concrete slab which extends over a sufficient width to transmit all wheel loads directly to the undisturbed soil. Alternatively, rigid pipes may be encased in reinforced concrete having adequate longitudinal steel, or they may be protected by placing them inside a larger, and preferably flexible, pipe (see Fig. 11.2). Under concrete pavings it may only be necessary to increase the reinforcing steel in the paving sufficiently to enable it to bridge the pipe trench effectively.

If the pipeline lies within the road bed and is so shallow that any bridging would be in contact with the pipes, steel or ductile iron pipes, suitably protected against corrosion, should preferably be used. They should be encased in salt-free sand, preferably not less than 4 in. thick. In new roads under these conditions the road bed should be fully compacted before installing the pipes, otherwise the rolling operations may damage or displace the pipes.

Road gullies and manholes should either be firmly supported, so that they do not sink under wheel loads and so tend to shear the pipe connection, or the gulley gratings or manhole frames should be supported by a protective reinforced concrete slab which should not be in contact with the top of the gulley or manhole shaft respectively (see Fig. 11.2).

2. *Flexible Pipes*

Where large flexible pipes cannot be relieved of transient surcharge loads and where the maximum achievable value of the coefficient of soil reaction 'k' (i.e. k_{max}) would be inadequate to satisfy Equation (6.13), the wall thickness of the pipe must be increased sufficiently to satisfy Equation (6.10), taking $F_{lag} = 1{\cdot}0$ and $e = k$ (i.e. the value EI becomes significant). Alternatively k may be taken as zero as in Article 6.5 (*b*) and 6.4 (*b*) above.

Where subsequent excavation in the vicinity of the pipeline, at any time or by any body, may remove or disturb the side supporting soil, its reinstatement is likely to be ignored or inadequate and it would therefore be advisable to use Equation 6.4 for the design of the pipes (see App. A, Ex. 13).

7.1 (d) (ii) *Under Fields and Gardens: Rigid or Flexible Pipes*

Where the cover is less than 3 ft the pipes may require protection against mechanical damage by agricultural and horticultural operations, e.g. ploughing and digging. In gardens, covering the pipes with loose concrete or stone slabs, or tiles, placed about 3 in. above the pipes may serve the purpose. In fields, similar but heavier slabs may be used and manholes might be extended above the surface by 12 to 18 in., especially at changes in direction, to give permanent warning of the presence of the pipeline, and avoid their being covered with soil and lost.

7.2 Construction Under Embankments or Valley Fills

7.2 (a) Choice of Method

The choice now lies between positive or negative projection conditions or the induced trench condition. For the same size and shape of rigid or flexible conduit installed under identical conditions of height and density of fill, these three methods of construction have the following characteristics:

1. *The positive projection condition* involves the largest fill and uniform surcharge loads but requires the least site work and permits the use of the higher bedding factors (F_p) for pipes.
2. *The negative projection condition* involves a considerable reduction in loading but requires additional site work in the excavation of the sub-trench, and the lower bedding factors (F_m) must be used.
3. *The induced trench condition* involves the lowest loading of all and permits the use of the higher bedding factors (F_p) but it requires the most site work. At depths imposing the 'complete' condition this method is open to objection under roads, railways and industrial yards, or in other situations where differential settlement of the surface of the fill over the pipeline would be unacceptable. At greater depths, imposing the incomplete projection condition, this objection does not hold provided care is taken to avoid excessive settlement in the sub-trench (see Article 4.10).

The choice of method will therefore depend on (*a*) practicability under local conditions, (*b*) the cost of reducing the load by extra site work relative to that of providing stronger pipes or other conduits (see App. A, Ex. 7).

7.2 (b) Choice of Bedding

This again will affect the required strength and cost of the pipes. Certain difficult soil conditions may necessitate the use of concrete beddings for rigid pipes but, where possible, they should be avoided in the interests of the speed of working and cost. Concrete bedding should not be used for flexible pipes when composite action with the soil is utilised as in Article 6.4 (*c*).

7.3 Construction Above Ground

7.3 (a) Choice of Method

It is occasionally necessary to lay pipes above ground either to facilitate inspection and adjustment after large ground movements, as in some mining conditions, and very unstable ground, or to maintain a gradient and avoid an inverted siphon, e.g. at river or valley crossings. There are no external loads, other than partial vacuum in pressure pipes, and the weight of the pipe and its contents. The choice of method therefore lies between various means of providing for the relatively large thermal and moisture changes which may be expected to occur. These movements may be additional to any subsidence movements of the pipelines and will depend largely on the thermal characteristics of the pipes (see Table S).

Metal pipes may be supported, preferably clear of the ground, on separately supported saddles or rollers to avoid axial restraint of the pipeline. The supports should have provision for vertical adjustment to compensate for ground movement. Non-metallic pipes, having little beam strength, are probably better laid on the ground on a uniform granular bed which can be adjusted as necessary for ground movements. See also Appendix B, for conditions in Mining Areas.

7.3 (b) Protection of Pipelines Above Ground

All rigid pipes should have flexible-telescopic joints. For flexible pipes, telescopic joints, or expansion-contraction joints at intervals, are necessary to avoid thermal buckling or axial tensile stresses. Pitch fibre and plastics pipes need thermal insulation by soil or other material as well as protection against mechanical damage. Concrete and clayware pipes require a generous protective covering of soil or other shielding material to prevent cracking by differential thermal or moisture effects. Steel and cast, spun or ductile iron, may be exposed to the air but require an anti-corrosive coating which will withstand the extremes of temperature which are likely to occur. Positive type joints (Table Tb) (with expansion joints at intervals may be needed in pressure lines). Reinforced concrete and asbestos cement pipes are less subject to cracking than plain concrete pipes but should preferably be shielded or covered by a layer of soil. Pipes laid across bridges, unless enclosed in ducts, will require similar protection against temperature variations. Any protective coatings applied directly to the pipes should be either ductile or have thermal coefficients as nearly as possible the same as those of the pipe material, otherwise they may crack or spall. Natural rubber joint rings should not be exposed to sun and air and, for work above ground, synthetic rubber joint rings are preferable (see Table Q).

7.4 Choice of Materials for Underground Construction

7.4 (a) Chemical Stability

The first essential requirement of all materials used in the construction of buried conduits is that they shall be, or shall be so treated as to be, chemically stable in both the external and internal environments to which they may be subjected. The broad suitability of the available B.S. pipe and pipe jointing and bedding materials for various probable environments is given in Table Q.

7.4 (b) Choice of Pipe Material

This is affected to some extent by the limitation in the available sizes of pipes in some materials, and by relative cost.

The hydraulic properties of a sewer are probably affected much more by local irregularities at the pipe joints than by the material of which the pipes are made since the slimy coating which eventually forms in most sewers tends to obliterate the relative surface roughness of the pipe walls.[116,119,120]

Further points requiring attention and affecting the choice of pipe material are as follows:

1. Clayware pipes are brittle and should not be laid at very shallow depths under roads or above ground, or in the close vicinity of strongly rooted trees above the lowest water table unless specially protected.

2. Unreinforced concrete pipes are made in longer lengths than clayware pipes but they also are brittle and pipes in short lengths are to be preferred for the smaller sizes (4 in. and 6 in.) when laid with shallow cover under roads, or in other situations where high dynamic loading is possible. Like clayware they may be broken by tree roots above the lowest water table.

3. Reinforced concrete pipes are now available and in the same lengths as unreinforced pipes. They are more resistant to tensile stresses and therefore more suitable for use under roads than unreinforced pipes.

4. Asbestos-cement pipes are made in various lengths and are more resistant to longitudinal bending than unreinforced concrete pipes but short lengths are to be preferred for the smaller sizes (up to 12 in.) when laid with shallow cover under roads.

5. Unplasticised PVC (uPVC) pipes are also made in long lengths and are not brittle at 50°F but, when laid with very shallow cover, or near trees, they may be deformed excessively by dynamic loads or roots unless specially protected. They are very thin, light and flexible and therefore suitable for use in boggy and made ground or in peaty soils.

6. Pitch fibre pipes have properties somewhat similar to those of PVC as regards flexibility and lightness. They should not be used for effluents or ground conditions which are predominantly hot, and not with very shallow cover without special protection against mechanical damage.

7. Cast and spun iron pipes, are more robust than other rigid pipes. They are made in various lengths but, as with asbestos-cement pipes, the short lengths should be preferred for the smaller sizes when laid with shallow cover under roads.

8. *Steel and ductile cast iron pipes* are both flexible and are the best suited for high dynamic loading. These pipes are resistant to damage by tree roots. They are made in long lengths but need special protection against corrosion.

9. *Pipes for pressure and rising mains* may be of steel, cast or spun or ductile iron, asbestos cement, prestressed or reinforced concrete, or uPVC (see Tables F and H).

10. *Protective coatings and linings.* Pipes with protective coatings require special care in handling to avoid damage to the coatings and if such damage does occur it should be made good thoroughly before the pipes are finally laid or buried. Under adverse conditions such as clay soils infected with sulphate reducing bacteria, saline or acid ground water, soil containing aggressive salts, or the presence of stray electric currents, iron and steel pipes may require *cathodic protection* in addition to protective coatings.

Coal tar pitch has been found preferable to bituminous compositions for the normal protective linings of iron and steel pipes. For external coatings, glass fibre reinforcement which does not rot is preferable to organic fabrics such as hessian or cotton. Plastics linings and coatings for metal and concrete pipes are now available and, despite their relatively high cost, they, or plastics pipes of small diameter, may, under chemical advice, be necessary to resist certain trade effluents or very aggressive soil conditions, e.g., when pipes are laid in made ground containing industrial waste products, in soil badly polluted by liquid wastes, or in highly sulphated or acid soils.

7.4 (c) Choice of Joint Materials

The types of flexible joint at present in use for various kinds of pipes and their size limits are listed in Table T (*a*) and (*b*).

Provided the essential requirement of chemical stability is met (see Table Q) the main choice lies between *rigid joints* made with materials such as cement mortar, or special chemically resistant compounds which set and harden, and *flexible-telescopic joints* which depend for their watertightness on non-hardening materials such as natural and synthetic rubbers, rubber-bitumen compositions, and some plastics. The following observations should be borne in mind when making this choice viz:

1. That it is always better, if possible, to eliminate a cause of stress than to provide resistance against it.

2. Whenever soil movements, relative settlements, or considerable variations in temperature are probable, (e.g. in all shrinkable clay soils above the lowest water table, in mining or other areas liable to subsidence, at junctions with manholes or other structures, or for hot or abnormally cold industrial effluents, or in the vicinity of furnaces or other soil heating appliances) the flexible-telescopic type of joint is to be preferred for rigid pipes since it enables the pipes to yield or move under the forces imposed on them by such conditions without fracture.
3. Since flexible-telescopic joints are usually made more easily and quickly than rigid joints they are preferable to the latter in difficult laying conditions, (e.g. under water, or where speed is essential, or where the working space is limited).
4. Cement mortar and some plastic compositions which set and harden have a high coefficient of thermal expansion relative to clayware and, as regards plastics, relative to concrete (see Table S). The moisture movement of cement mortar is also much greater than that of clayware: and it increases with increase in cement content.

   Under conditions where considerable variations of temperature or moisture may occur either before or after backfilling these rigid jointing materials therefore tend to cause burst sockets or collars particularly in clayware pipes and should be avoided (see Table R).
5. Hot or cold mastics which harden or shrink with age or which creep or flow at the extremes of temperature to which they may be subjected should not be used for jointing. When making hot run bituminous joints dry conditions are essential and overheating the composition should be carefully avoided as it may cause subsequent embrittlement of the material.
6. The life of natural rubber in exposed joints above ground, or in the presence of petroleum products, oils, and fats in the effluent is doubtful. Synthetic rubbers such as neoprene and butyl rubber are considered to be more suitable for such conditions (see Table Q).
7. Run lead and caulked leadwool joints for cast iron or steel pipes provide slight flexibility but are not suitable when lengthening and shortening or vibration of the pipeline may occur, since these conditions adversely affect their watertightness.

8. The ogee or rebated type of joint for concrete pipes cannot readily be made water-tight, especially for thin walled pipes, by means of mortar joints.
9. Pipes which are themselves flexible do not require flexible joints but in subsidence areas or in variable temperature conditions where draw can occur telescopic joints are preferable to rigid joints.
10. All plastics pipes have a very high coefficient of thermal expansion and they therefore require telescopic joints, or contraction joints, at all possible anchorages such as valves, branches, or bends, especially if laid when their temperature considerably exceeds 50°F or where they may be subjected to considerable variation in temperature.
11. Vibration of the ground by traffic or machinery adversely affects all joints and particularly rigid joints.
12. Tunnel segments of cast iron or concrete should be flexibly jointed thereby making the rings flexible and reducing or eliminating circumferential bending stresses.

7.4 (d) Choice of Bedding Materials

1. *Site Concrete.* Corrosive conditions in the external environment which affect the choice of pipe material also affect the choice of bedding material. Thus if a special cement is required for concrete pipes, the same cement will be necessary for concrete beddings or surrounds. Similarly if external conditions are so severe as to exclude unprotected concrete pipes, they would also exclude concrete beddings or surrounds. In any sulphate condition where good quality dense concrete is acceptable, special care should be exercised to ensure the same quality in the site concrete.

2. *Granular Material.* In acid groundwater conditions limestones are unsuitable for granular beddings but they, or chalk, are sometimes used as a buffer protective surround for cast iron pipes in aggressive clay soils. Subject to this condition granular bedding material should comply with the requirements for Class B beddings (see Charts C9 and C10), but some latitude may be allowed in the maximum grain size for large pipes. It should not exceed say D/40 and the properties of free drainage and ready compactibility must be preserved by suitable grading. For pitch fibre, plastics and Deckon pipes, sharp edged material which could cut into the pipe wall should be avoided.

7.5 Pipes in Made or Other Unstable Ground

7.5 (a) In made ground or similar conditions where further consolidation and uneven settlement are expected, the pipes should be either flexible, or flexibly jointed. Sewer and surface drainage pipes should be laid at the greatest slope permitted by the local conditions, so ensuring as far as possible that a back fall is avoided as settlement occurs. Where the practicable slope is too small to accommodate the probable settlement it may be necessary to carry the pipe line on beams which are supported by piles driven through the made ground into the firm ground below. This latter method increases the load on the pipes and requires special attention to pipe strength and bedding (see Article 8.1 (*a*) (iii) and the note on the projection ratio '*p*').

7.5 (b) In peat, boggy ground, or unstable waterlogged fine sands, the pipes may tend to rise or sink, depending on their weight relative to that of the material displaced. Under these conditions a light flexible pipe (e.g. pitch fibre or plastics) should preferably be used. Exceptionally for small pipes, and usually for large pipes, it may be preferable to use a heavier kind of pipe supported on beams and piles, as in (*a*) above.

Where the ground surface is reasonably uniform and firm enough to support a tractor, it may be possible to lay small bore galvanised steel or plastics pipes with a mole plough and so avoid the cost of trenching. The pipes should be positively jointed above ground prior to ploughing in.

7.5 (c) In areas subject to mining subsidence, pipelines are stretched and shortened as well as bent during the passage of a subsidence wave. The pipes should therefore be either flexible or flexibly jointed and the joints for either type should also be telescopic. Short length pipes should be used in order to provide an axial movement of not less than ± 1 per cent of the pipe length in each joint.[83] (See also App. B.)

7.5 (d) In extreme conditions of any kind it may be preferable, where possible, to lay suitable pipes above ground and, for sewers and surface drainage pipes, to support them in such a manner that the fall can be re-adjusted as necessary from time to time.

Chapter 8

Summary of the Formulae, with Notes on their Components and Detailed Methods of Computation of the Required Strength of Rigid Pipes or the Wall Thickness of Flexible Pipes

(See also the worked examples in Appendix A)

8.1 Load Formulae

Summary of formulae

The various load formulae and their specific applicability are summarised below and, more briefly, in Table M. In using these formulae the following notes on their components should be borne in mind.

8.1 (*a*) *Loads Imposed by Fill and Uniformly Distributed Surcharge Loads of Large Extent*

8.1 (*a*) (i) *In 'Narrow' Trench Conditions*

Pipe Type	*Fill load* (W_c)	*Uniform surcharge of large extent* (W_{us})
Rigid	$C_d \gamma_s B_d^2$	$C_{us} B_d U_s$
Flexible	$C_d \gamma_s B_c B_d$	$C_{us} B_c U_s$

8.1 (*a*) (ii) *In Heading or Thrust Bore Conditions*

Rigid or Flexible	$C_d \gamma_s B_c^2$ (Solid packing)	$C_{us} B_c U_s$
Rigid	$C_d \gamma_s B_t^2$ (Compressible packing)	$C_{us} B_t U_s$
Flexible	$C_d \gamma_s B_c B_t$ (Compressible packing)	$C_{us} B_c U_s$

These formulae (i) and (ii) are applicable wherever W_c is less than $C_c \gamma_s B_c^2$, and to rigid or flexible conduits, subject to the distinction made in col. 4 of Table M, of any material or shape, and for pressure or non-pressure service, when laid in a narrow trench, or heading, or a narrow sub-trench in a wide parallel or V trench (see Fig. 3.1 and 4.1). Corrections are necessary for permanent submergence (see Article 4.14) or for the loads on each of several pipes in the same trench (see Article 4.13 (*a*)).

Notes on the components of the formulae

C_d — Depends on H/B_d or H/B_c or H/B_t and on $K\mu'$ (see chart C1). Use B_d, B_t or B_c as shown above or in Table M.

B_d — Is the width in feet of the trench measured over all timbering at the level of the top of the pipe. It should be kept as small as is consistent with practicable working space alongside the pipes, allowing for timbering where necessary (see Fig. 4.2).

B_c — Is the maximum overall outside diameter or finished width in feet of the conduit above the horizontal diameter including any *in-situ* concrete (see Fig. 4.3 and Chart C9). For the O.D. of pipes, in the absence of precise values see Table N.

B_t — Is the width of heading in feet measured similarly to that of B_d.

H — Is the depth in feet of soil cover. Estimates of W_c for both H_{max} and H_{min} should be made. H_{max} should include any anticipated raising of the ground surface.

$K\mu'$ — The lowest probable value of the coefficient μ' of sliding friction of fill on natural soil should be used (see Table O).

γ_s or γ_d — The highest probable value in lb/ft^3 of the saturated density γ_s or the drained density γ_d of the fill should be used above the lowest permanent ground water level, and the submerged density $(\gamma_s - \gamma_w)$ or $\{\gamma_d - \gamma_w(1 - n)\}$ below that level. For free-draining coarse granular fill the drained density $\gamma_d = \gamma_s - n\gamma_w$, where n is the void ratio (see Table O).

C_{us} — Depends on H/B_d or H/B_c or H/B_t and on Ku' (see Chart C2). It assumes that frictional or cohesive forces are inactive in the surcharge material. Use B_d, B_c or B_t as shown above or in Table M.

U_s — The highest anticipated uniformly distributed surcharge load of large extent in lb/ft^2, or, in the absence of specific requirements, a minimum value of 500 lb/ft^2, should be used in design. It is assumed that the surcharge may be applied temporarily throughout any particular length of the pipeline from time to time.*

8.1 (a) (iii) *In Embankment and 'Wide' Trench Positive Projection Conditions*

Pipe type	*Fill load,* $(W'c)$	*Uniform surcharge load of large extent,* (W'_{us})
Rigid	$C_c\gamma_s B_c^2$	$(C'_c - C_c)\gamma_s B_c^2$
Flexible	$H\gamma_s B_c$	$B_c U_s$

* But see Ref. 10*a*, para. 235 (*d*) or Table D.

These formulae are applicable wherever W'_c is less than $C_d \gamma_s B_d^2$ and for rigid or flexible conduits, subject to the distinctions made in col. 6 of Table M; of any material or shape, for pressure or non-pressure service; when laid in a 'wide' parallel sided or V trench with or without a 'wide' sub-trench, (see Fig. 4.8), or in a 'very wide' trench without sub-trench; and in all positive projection conditions under embankments or valley fills (see Figs. 4.5 and 5.3) or in negative projection conditions with a 'wide' sub-trench (see Fig. 5.2). Frictional forces are assumed to be operative and cohesive forces inoperative in both the fill and the uniform surcharge.

Corrections are necessary for permanent submergence or for closely spaced parallel pipes (see Article 4.14 and 4.13 (*b*)).

Notes on the components of the formulae

C_c & C'_c Depend upon H/B_c and H'/B_c respectively and on the settlement ratio r_{sd} and the positive projection ratio p. The values given in Chart C3 have been computed for $K\mu = 0{\cdot}19$ when r_{sd} is positive, or $K\mu = 0{\cdot}13$ when r_{sd} is negative, but they are valid for all lower or higher values of $K\mu$ respectively without serious error. The curves of Chart C3, other than those for the 'complete' projection condition and the 'complete' trench condition, are linear and may be extrapolated by means of the appropriate equations given on the chart for higher values of H/B_c than those given by the graph.

H Is the depth of soil cover in feet, above the top of the conduit. Estimates of W'_c and W'_{us} should be made for both H_{max} and H_{min}, and H_{max} should include any anticipated future raising of the surface.

H' $= H + U_s/\gamma_s$, i.e. the uniform surcharge is assumed to be an equivalent increase in cover depth with the same frictional characteristics as the fill. Estimates of the load should be made for both H'_{max} and H'_{min}.

U_s, γ_s or γ_d, B_c Have the same value, as above, for trench loading.

r_{sd} The recommended values of the settlement ratio to be assumed under various conditions for rigid and flexible pipes are given in Table B. Note that the value of C_c and therefore of W_c will differ,

p depending on whether rigid or flexible pipes are to be used. Also that W_c increases as $r_{sd}p$ increases. pB_c is the actual or average projection of the top of the conduit above the natural ground level (see Figs. 4.4 and 4.7). For oval or egg-shaped conduits or for conduits in soft weak soils, or for those laid wholly above natural ground and supported by piles, p will be greater than 1·0 since pB_c will be greater than B_c. For 'wide' sub-trenches in pseudo negative projection, or 'wide' trenches proper, in which the bedding extends over the full width of the trench, pB_c is the projection of the top of the conduit above the top of the bedding.

Where the value of $r_{sd}p$ is large or indefinite, W_c should be based on the formula for the complete projection condition. In extrapolating this part of the C_c curve beyond the range of H/B_c given in Chart C3 the value of C_c may be computed from the formula for complete projection given in Table A. This formula may yield a very high load and it should not be used without careful judgement of the conditions and consideration of the practicable alternatives which would reduce the load, e.g. induced trench conditions.

N.B. It is not necessary to limit the trench or sub-trench width under 'wide' trench conditions.

8.1 (a) (iv) *In Negative Projection and Induced Trench Conditions Under Embankments or Fills*

Pipe Type	*Fill load (W_c)*	*Uniform surcharge load of large extent (W_{us})*
Rigid or large flexible	$C_n\gamma_s B_d^2$	$(C'_n - C_n)\gamma_s B_d^2$ Incomplete Projection $C_{us}B_dU_s$ Complete Projection
All thin steel and flexible $\not>$ 48 in I.D.	$C_n\gamma_s B_c B_d$	$(C'_n - C_n)\gamma_s B_c B_d$ Incomplete Projection $C_{us}B_cU_s$ Complete Projection
For Induced trench conditions		
Rigid or flexible	$C_n\gamma_s B_c^2$	$(C'_n - C_n)\gamma_s B_c^2$ Incomplete Projection $C_{us}B_cU_s$ Complete projection

These formulae are applicable to rigid or flexible conduits subject to the distinction made in col. 6 of Table M; of any material or shape; and for pressure or non-pressure service; when laid in a narrow sub-trench below the original ground level, or in positive projection conditions with an induced trench above the conduit. They are always associated with a narrow sub-trench (see Figs. 4.11 and 5.2). If the induced trench condition is used under roads or railways, unless the value of H/B_c is large enough to produce the incomplete negative projection condition, differential settlement over the pipeline may be troublesome. Correction for permanent submergence may be needed (see Article 4.14 (*c*)).

Notes on the components of the formulae

C_n & C'_n Depend on H/B_d or H/B_c and H'/B_d or H'/B_c respectively, and on r_{sd}, p', and $K\mu$. The values given in Chart C4 have been computed for $K\mu = 0{\cdot}13$ but they are valid (with varying safe error) for higher values of $K\mu$ but not for lower values, i.e. they should not be used for soft clays in which ϕ is less than 11°. The curves given in the Charts C4, other than those for the complete condition, are linear and may be extrapolated by means of the appropriate equations, given on the charts for values of H/B_d and H'/B_d greater than those given by the graphs.

H Is the depth of soil cover in feet above the top of the conduit. Estimates of W_c and W_{us} should be made for both H_{max} and H_{min}. H_{max} should include any anticipated future raising of the surface.

$H' =$ $H + U_s/\gamma_s$ as for positive projection, above.

C_{us}, U_s, γ_s or γ_d and B_c Have the same values as above for trench loading.

B_d Is the width of the sub-trench over all timbering measured at the original around level. It should be as small as the practicable working space alongside the pipes will permit allowing for timbering only where necessary. In induced trench conditions its minimum effective value is B_c providing the compressible fill is limited to that width, which is advisable.

r_{sd} The settlement ratio has no well-established values at present. Spangler's recommended values are given in Table B.

p' $p'B_d$ is the actual or average minimum distance between the top of the pipe in the sub-trench and the natural ground see Fig. 4.7 and 4.10. As p' increases C_n decreases for equal values of H/B_d or H/B_c. It is not to be confused with p for positive projection. The values of C_n have been checked by computer for values of $p' = 0{\cdot}5$, $1{\cdot}0$, $1{\cdot}5$, and $2{\cdot}0$.

Intermediate values may be interpolated, or the graph for the next lower value of p' may be used. For the induced trench condition p' should be reasonably large, say 1·0 to 2·0.

N.B. When $p' = 0$, $C_n = H/B_d$ and $W_c = H\gamma_s B_d$ and $W_{us} = B_d U_s$, but if the fill in the sub-trench is made as resistant to settlement as the natural ground, $W_c = H\gamma_s B_c$ and $W_{us} = B_c U_s$., i.e. the condition is neutral.

8.1 (b) Loads Imposed by Concentrated Surcharges for all Conditions

Pipe type	*Load*		*Concentrated Surcharge load* (W_{csu})
Any	Single point load		$F_i C_t P F_L / 3$
	2/7000 lb loads 3 ft apart		$F_i W_L$*
	B.S. 153 Type HB Road loading	In England and Wales	$F_i W_o$
	Railway loading		$2\ W_o$

* Applies also to 2/16,000 lb loads when W_L is multiplied by 16,000/7,000 [10a].

These formulae are applicable to all temporary or transient 'point' load surcharges, e.g. vehicle wheels, tower crane or pile driver pads, etc., applied over or in the vicinity of the conduit, provided that the maximum dimension of the area of contact is less than $H/2$ (for larger contact areas see Article 8.1 (*c*) below). Also to rigid or flexible conduits of any shape, installed in any of the load conditions, in any sort of soil, and with any kind of paving (it is assumed that temporary reinstatement of trenches or repair work may occur in reinforced concrete pavings, thereby reducing them temporarily to the same condition as flexible pavings. No reduction in these loads is permissible for submergence of the conduit.

Notes on the Components of the formulae

F_i Pending the establishment of more exact values (see footnote to Article 4.17 (*b*) (ii) p. 60, a value of not less than 2·0 can be tentatively assumed at all values of H for conduits under roads and verges, carriageways and access roads,** railways, and airfields. The value of 1·0 may be used where only light agricultural vehicles will traverse the fill, provided the cover is not less than 2 ft.

C_t Is an influence value for the vertical component of the imposed load caused by a single point surcharge, depending upon its position relative to the conduit in plan, the depth of cover H, the overall

** Now amended. See Table D.

width of the conduit B_c and an arbitrary pipe length of 3 ft. It is computed by the method given in Chapter 4, Article 4.17 (*b*) (iii). If more than one surcharge is active simultaneously, C_t and the imposed load must be computed for each load of the group, in the position of the group which imposes the greatest load on the conduit. W_{csu} is then the sum of the simultaneously imposed loads.

H — Is the depth of cover above the top of the conduit. Estimates of W_{csu} should be made for both the minimum and maximum values of H, *excluding* future raising of the ground level above the immediate value of H_{max}.

P — The actual static value of the wheel load or other concentrated surcharge, in pounds, should be used.

F_L — The values given in Charts C6 (*a*) and (*b*) are only strictly applicable to rigid pipes and a single wheel load, or to equal wheel loads at 3 ft spacings on axles which are not less than 6 ft apart. For flexible pipes its value may be tentatively assumed as 1·0.

W_L — The values given in Charts C7 (*a*) and (*b*) are the corrected static loads for the various specified bedding classes. They are directly applicable only to rigid pipes and two equal wheel loads each of 7,000 lb spaced 3 ft apart and a wheel track width of 6 ft or more. They may be used for other values of P and the same wheel spacing by multiplying the graph values of W_L by $P/7{,}000$. Chart C7 (*a*) is not applicable to pipes with arch support or to Bedding angles $> 120°$. Chart C7 (*b*) is applicable to rigid pipes with arch support and Class B bedding and it may be used for flexible pipes.

W_o — The values given in Charts C8 (*a*) and (*b*) are the corrected static values for B.S. 153 type HB road loading with $P = 20{,}000$ lb per wheel and are directly applicable only to that specific arrangement of wheels acting on rigid pipes under roads, verges, and heavy vehicle approaches, in England and Wales. They may be used for lower values of P and the same wheel spacings, by multiplying the graph values of W_o by $P/20{,}000$. Chart 8 (*a*) is not applicable to pipes with arch support or to bedding angles $> 120°$. Chart 8 (*b*) is applicable to pipes with arch support and Class B bedding and it may be used for flexible pipes.

C_t, W_L and W_o all depend on H, B_c, and a pipe length of 3 ft or more. If the length of a single pipe is less than 3 ft, as possibly with some clayware pipes to B.S. 65 and 540: (1966), the error in using the chart values will not exceed about 7 per cent for sizes up to 8 in. and can usually be ignored. A correction is not necessary for short make-up lengths of any size.

8.1 (c) Loads Imposed by Uniformly Distributed Surcharges of Limited Extent for all Laying Conditions

Pipe type	*Load*	*Uniformly distributed surcharge load* (W_{sus} or W_{sust})
Any	Static	$C_{sus}U_{sus}B_c$
	Mobile (on tracks)	$C_{sus}U_{sust}B_cF_i$

These formulae are applicable to all temporary or permanent, static or mobile surcharges, which are uniformly distributed over a specific area $L \times B$ of known position, in which the greatest dimension is longer than $H/2$, when applied at or below ground level but above the top of the pipe, and not otherwise carried at a lower level by piles or piers penetrating below the bottom of the pipe, (e.g. existing or proposed column or wall footings, pylon bases, gantry stanchion footings, tracked vehicles including tanks, and heavy mobile construction equipment) and to rigid or flexible conduits of any shape, installed in any laying conditions, and in any sort of soil. Permanent static surcharges are preferably carried independently of the conduit.

They yield only the vertical component of the surcharge load and the position and intensity of a static surcharge must be foreseen at the design stage. For an isolated static load the effect on the conduit is purely local.

During construction the use of tracked vehicles is common and the intensity of the transient loads imposed on the fill must be known in estimating the depths of cover at which such vehicles can be allowed to work over or in the vicinity of the conduit without imposing excessive loads thereon. The maximum load is imposed when the vehicles are crossing or travelling directly over the conduit (see App. A, Exs. 15 and 16).

Notes on the components of the formulae

C_{sus} — Is an influence value which depends on the length L and the breadth B of the loaded area, on its position in plan relative to the conduit, and on the depth H of the fill above the top of the conduit. It is the algebraic sum of the influence values I_σ, obtained as described above in Chapter 4, Article 4.18, from Chart C5.

U_{sus} or U_{sust} — Is the uniform surcharge in lb/ft², i.e. usually the total surcharge or load divided by the area $L \times B$. For tracked vehicles there is no precise guide to its value (U_{sust}) which may vary somewhat in different soil conditions; but the total weight of the vehicle divided by the total contact area of the tracks is suggested as an approximation, trailers being separately considered from tractors. For wheeled trailers the loads imposed by the wheel system should be computed by using the concentrated load formula, if the maximum dimension of the area of contact of a wheel is less than $H/2$, i.e. using C_t instead of C_{sus}.

B_c — Is the overall width of the conduit as for trench loading.

F_i — Is the impact factor for mobile loads. Its value is a matter of judgement, depending upon the degree of unevenness of the surface. If in doubt a value of not less than 2·0 should be assumed.

8.1 (d) The Total Effective External Design Load (W_e)

The components of the total load should be summed, using Table P in accordance with Article 4.20 Formulae (4.57) to (4.61) (see also Article 8.3 (*a*) and (*b*).

8.2 Selection of the Safe Crushing Test Strength and Bedding Class for Rigid Pipes

8.2 (a) Rigid Circular Non-pressure Pipes

Min. safe crushing test load	*Laying conditions*
1. $W_T = W_e F_s / C_{BS} F_m$ (Form. 5.3)	Trench and sub-trench, heading and thrust bore conditions.
2. $W_T = W_e F_s / C_{BS} F_b$ (Form. 5.4)	Positive projection, very wide or induced trench conditions.

Formula (1) is only applicable when a conduit consists of a circular pipe in trench, heading, or thrust bore, or true negative projection conditions which, at any given depth of cover, is uniformly loaded over its width and length, and which is similarly uniformly supported by one of the specified beddings illustrated in Chart C9, or is in a thrust bore. The bedding factor F_m should be used for all usual widths of the trench or sub-trench in trench or negative projection conditions.

Formula (2) and the factor F_p should only be used for rigid pipes in very wide trenches $B_d \gtrsim \{B_c + 2(H + mB_c)/\mu\}$ conditions, or positive projection or induced trench conditions under embankments or valley fills, and when the pipe is supported by one of the specified beddings illustrated in Chart C10.

Notes on the components of the formulae

W_T For all unreinforced rigid pipes, this is the British Standard or other guaranteed minimum two- or three-edge ultimate crushing test load. For steel reinforced concrete pipes it is the British Standard or other guaranteed minimum crushing test load required to produce a crack having a width not exceeding 0·01 in. measured at close intervals over a length of 1 ft or more, or 80 per cent of the ultimate crushing test load, whichever is the less.

W_e Is the total effective external design load computed as in Article 8.3 (*a*) or (*b*) for the respective laying conditions.

F_s Is the factor of safety against crushing. In the absence of special circumstances the values given in Table G may be used. The actual value chosen depends on local circumstances and is essentially a matter of good judgement with respect to pipe quality and site workmanship, especially in the construction of the bedding and back fill.

C_{BS} Is the correction factor applying to the British Standard 6 in. wide rubber pad strip 2-edge bearings. It may be assumed as 0·90 approximately for pipe sizes up to 33 in. internal diameter, or 0·95 approximately for larger sizes. For the British Standard three-edge bearings its value is 1·0.

F_m Is the bedding factor applicable to narrow or wide trench, heading or thrust bore, and negative projection conditions. It assumes that no active lateral soil pressure is imposed on the pipe. Its value depends on the type of bedding and is independent of H. The values and construction details usually recommended are given in Chart C9. The value of F_m for thrust bore conditions requires experimental investigation; it may be assumed tentatively as 1·5.

F_p Is the bedding factor applicable only to pipes in the positive projection or induced trench conditions under embankments or fills or in very wide trenches (see Fig. 5.3 and Chart C.10). It assumes that active lateral soil pressure is exerted on the pipe by the fill over the exposed vertical projection height

mB_c, and it is therefore somewhat larger than F_m for the same bedding. Its value depends on the class of bedding, on the exposure ratio 'm', and on the ratio H/B_c. Illustrations showing the requirements for the various standard classes of bedding and the site conditions which are generally suitable for use in wide trenches or under embankments are given in Chart C.10.

Spangler[38] gives its value as

Equation 8.1

$$F_p = 1{\cdot}431/(N - xq)$$
$$\text{or } 1{\cdot}431/(N' - x'q)$$

where N and N' depend on the distribution of the load and reaction on the pipe periphery, i.e. on the class of bedding; x and x' depend on the vertical projection of the pipe on which the active lateral pressure of the fill material acts, i.e. on mB_c, and q is the ratio of the lateral soil load to the vertical fill load per unit length of pipe. N and x are applicable when the pipe cracks first at the bottom, (i.e. with Classes B, C, and D beddings.) N' and x' are applicable when the pipe cracks first at the top, (i.e. with Class A concrete beddings.) The value of N, N', x, x', and mB_c are given in Chart C10.

Equation 8.2

$$q = \left(H + \frac{mB_c}{2}\right) KmB_c\gamma_s/C_c\gamma_s B_c^2$$
$$= \frac{mK}{C_c}\left(\frac{H}{B_c} + \frac{m}{2}\right)$$

If mB_c varies in the length of conduit under consideration, its minimum value should be used in design.

8.2 (b) Rigid Circular Pressure Pipes

For concrete pipes reinforced with steel:

$$W_T = W'_e F_{se}/C_{BS}F_m(1 - p_i F_{si}/p_{ult}) \text{ (see Formula (5.5))}$$

For cast and spun iron pipes and for asbestos cement pipes:

$$W_T = W'_e F_{se}/C_{BS}F_m(1 - p_i F_{si}/p_{ult})^{\frac{1}{2}} \text{ (see Formula (5.6))}$$

For prestressed concrete cylinder type pipes:

$$W_T = W'_e F_{se}/C_{BS}F_m(1 - p_i F_{si}/p_{oc})^{\frac{1}{2}} \text{ (see Formula (5.7))}$$

For prestressed concrete non-cylinder type pipes the corresponding formula has not yet been established (see Article 5.3 (*b*)).

These formulae are applicable to rigid circular pipes acting as water or gas mains, or as rising (pumping) mains, or inverted syphons, or as surcharged pipelines in sewerage systems, or otherwise when subjected to internal pressure, when laid in 'wide' or 'narrow' trench conditions, or in any conditions under embankments or fills. They assume that the pipes are uniformly loaded externally over their length and width and that they are similarly uniformly supported by one of the specified beddings illustrated in Charts C9 and C10. The specified bedding factor F_m should be used for all usual 'wide' or 'narrow' trench heading and thrust bore and negative protection conditions. The bedding factor F_p may be substituted for F_m in 'very wide' trench or positive projection or induced trench conditions under embankments.

The required value of W_T in a pressure pipe will always be greater than that in a non-pressure pipe of the same size, when carrying the same external load. Pressure pipes under railway embankments are usually required to be placed inside a protective sleeve which relieves them of external load other than partial vacuum. The sleeve however should be designed to carry the full external load excluding partial vacuum.

Notes on the components of the formulae

W_T — For steel reinforced concrete pipes is the minimum $\frac{1}{100}$ in. crack 2 or 3-edge test load *as for non-pressure pipes.* For prestressed concrete cylinder type pipes, it is 90 per cent of the minimum $\frac{1}{1000}$ in. crack 3-edge test load. For prestressed concrete non-cylinder type pipes, it is not yet established and for these pipes its value should be declared and guaranteed by the makers.

For cast and spun iron and asbestos-cement pressure pipes, it is the minimum ultimate 3-edge crushing test load.

N.B. All the above tests are made at atmospheric internal pressure.

W'_e — Is the effective total external load applied simultaneously with the internal pressure p_i. It is usually assumed that the maximum external surcharge loads and the maximum internal test pressure will not be imposed simultaneously. Consequently the required value of W_T will be the larger value determined by Article 4.20:

$W'_e = W_e$, when $p_i =$ the normal working pressure p_w

or $W'_e = W_c + W_w(+W_{us} + W_{sus}$ if present) when $p_i =$ the maximum field test pressure p_t.

F_{si} Is the factor of safety against bursting. Suggested values are:

4·0 for asbestos cement pipes on the minimum bursting pressure.

2·0 to 2·5 for cast and spun iron pipe on the minimum bursting pressure.

2·0 for steel reinforced concrete pipes on the minimum bursting pressure.

1·0 for prestressed concrete cylinder type pipes on the zero compression pressure.

For prestressed concrete non-cylinder type pipes refer to makers.

F_{se} Is the factor of safety against crushing. Suggested values are:

2·5 for asbestos cement pipes on the minimum ultimate crushing test load.

2·0 to 2·5 for cast and spun iron pipes on the minimum ultimate crushing test load.

1·6 for steel reinforced concrete pipes on the minimum $\frac{1}{100}$ in crack load.

1·0 for prestressed concrete cylinder type pipes on 90 per cent of the minimum $\frac{1}{1000}$ in crack load.

For prestressed concrete non-cylinder type pipes, refer to makers.

N.B. All the above factors assume first class sitework.

p_i Is the effective internal fluid pressure p_w or p_t (see W'_e).

p_{ult} Is the minimum bursting pressure of reinforced concrete, cast or spun iron and asbestos-cement pressure pipes, all when under zero external load.

p_{oc} Is the pressure required to produce zero compression stress in prestressed concrete pipes.

C_{BS}, F_m, and F_p All have the same values as above for non-pressure pipes.

8.3 The Detailed Procedure in Computing Loads and Corresponding Rigid Pipe Strengths

8.3 (a) Trench Loads

The detailed computing procedure for trench loading is as follows (see also App. A, Ex. 1 to 6):

Step 1. Select a suitable pipe material for the physical and chemical conditions in the local soil and in the effluent (see Table Q) and for the pressure conditions. Determine the inside diameter (D)

of the conduit for the required flow capacity, and the corresponding external diameter B_c (see Fig. 4.3). Refer to the pipe makers, or their catalogues, for B_c, or use Table N (with discretion), for rigid pipes. Determine a convenient maximum effective trench width (B_d) or heading width (B_t) (see Fig. 4.2 and Table N).

Determine the maximum (H_{max}) and minimum (H_{min}) cover depths in the length of pipeline under consideration, and the depths of submergence H_s where permanent, allowing with H_{max} for any anticipated raising of the ground level, thus ensuring that the maximum present or future value of W_c is considered. If there is a large difference in cover it may be advisable to divide the length of the pipeline into two or more sections and use either a different class of bedding, or pipes of a different strength class, in each section.

Step 2. Determine the maximum saturated density (γ_s) of the soil fill, noting that γ_s is the weight of unit volume of the soil when all the voids are filled with water, and the minimum value of the frictional characteristics ($K\mu$ or $K\mu'$) for the worst soil conditions in the section of pipeline under consideration, and note that the submerged density is $(\gamma_s - \gamma_w)$. In the absence of soil tests the values given in Table O may be assumed.

For a free draining coarse granular material (e.g. some gravels use the drained density γ_d and note that the submerged density will then be $\gamma_d - \gamma_w(1 - n)$.

If imported fill is to be used, the values of γ_s or γ_d and $K\mu'$ appropriate to that material should be used but $K\mu$ for the undisturbed soil in the trench sides if less than $K\mu'$ for the imported soil.

Step 3. With the values H_{max}/B_c, H_{min}/B_c enter the Chart C3 and obtain $C_{c\ max}$ and $C_{c\ min}$ for $r_{sd}p = 1{\cdot}0$ or $0{\cdot}5$ (see Table M) for each section of the pipeline. With the values of H_{max}/B_d and H_{min}/B_d enter the Chart C1 having the nearest lower values of $K\mu'$ to that determined in Step 2 to obtain the values of $C_{d\ max}$ and $C_{d\ min}$ for each section of the pipeline.

Compute the maximum and minimum values of $W'_c = C_c\gamma_s B_c^2$ and $W_c = C_d\gamma_s B_d^2$.

N.B. Alternatively the Load Charts given in N.B.S., S.R.37[13] may be used to save time.

Step 4. Select the lesser value of $W'_{c\ max}$ and $W_{c\ max}$ and the lesser value of $W'_{c\ min}$ and $W_{c\ min}$. These are then the maximum and minimum effective design values of the fill load, W'_c or W_c. Correct these values for permanent submergence if safe and necessary using Formulae (4.31) or (4.33).

Compute the water load $W_w = (0{\cdot}75)\ (\pi/4)\ (D^2)\ (62{\cdot}4)$ lb/lin. ft for rigid pipes or $W_w = (0{\cdot}5)\ (\pi/4)\ (D^2)\ (62{\cdot}4)$ lb/lin. ft for flexible pipes (see Table M).

N.B. $W_w = 0$ if the pipeline is *permanently* submerged.

Step 5. Determine an appropriate value for the uniformly distributed surcharge (U_s) or, in the absence of specific requirements, assume a value not less than 500 lb/ft².* If, in Step 4, W'_c has proved less than W_c for either H_{min} or H_{max} in any section of the pipeline, compute the corresponding value of $H' = H + U_s/\gamma_s$ in feet. Enter the Chart C3 with the value of H'/B_c, to obtain a new value, C'_c. Multiply this value by $\gamma_s B_c^2$ and subtract from it the fill load W'_c obtained in Step 4. This is then the effective design value of the uniform surcharge load W'_{us} for that section.

If, however, in Step 4, W_c is less than W'_c for either H_{min} or H_{max} in any section, enter Chart C2 for the same value of $K\mu'$ as for the fill load, and, with the appropriate value of H/B_d, obtain the value of C_{us}. Multiply this value by $B_d U_s$ to obtain the effective design value of the uniform surcharge load, W_{us} for that section of the pipeline.

N.B. For Flexible Pipes Steps 1 to 5 are similar using

$$W_c = C_d \gamma_s B_c B_d \text{ and } W_{us} = C_{us} B_c U_s$$

in place of the rigid pipe 'narrow' trench formulae, and

$$W'_c = H\gamma_s B_c \text{ and } W_{us} = B_c U_s$$

in place of the rigid pipe 'wide' trench formula.

Step 6. If the conduit is under a road, or verge, in England and Wales, enter the Chart C8 (*b*)* with the values of B_c in inches and H_{min} and H_{max} in feet for each section of the pipeline, and obtain the values $W_{o\ max}$ and $W_{o\ min}$ for class B bedding. Multiply these values by an appropriate impact factor (F_i) (but see the footnote to Article 4.17 (*b*) (ii) p. 63) to obtain the effective design values of the concentrated surcharge loads $W_{csu\ max}$ and $W_{csu\ min}$.

If the conduit is not under a road, enter the Chart C7 (*b*) with the values of B_c in inches and H_{min} and H_{max} in feet, for each section of the pipeline, and obtain values of $W_{L\ max}$ and $W_{L\ min}$ for class B bedding and wheel loads of 7,000 lb each. Multiply these values by an appropriate impact factor F_i (not less than 2·0 at present under carriageways and access roads,* (but see the footnote to Article 4.17 (*b*) (ii), p. 63), or 1·0 under fields and gardens to obtain the effective design values of the concentrated surcharge

* But see Table D.

loads $W_{csu\ max}$ and $W_{csu\ min}$. If the track width is less than 6 ft W_{csu} should then be computed for the actual wheel spacing and value of P as in Article 4.17 (*b*) (iii) including all wheels on the rear axles of both vehicles, and applying the Correction Factor F_L as in Article 4.17 (*c*).

N.B. If after computing the required value of the safe crushing test strength of the pipe, W_T (see below), it is found preferable to adopt the Class A or A_{RC} bedding in any section of the pipeline, this step must be repeated using the appropriate charts C7 (*a*) or C8 (*a*) to obtain the revised values of $W_{csu\ max}$ and $W_{csu\ min}$.

Step 7. If static or mobile uniformly distributed surcharges of small extent are to be provided for, determine the dimensions $L \times B$ in feet and the intensity of the surcharge, U_{sus} lb/ft^2 for static surcharges, and U_{sust} for mobile surcharges. For a static surcharge construct a family of rectangles embracing the surcharge and with sides parallel to its sides, each rectangle having one common corner lying over the centre line of the conduit at the point nearest to the centre of gravity of the surcharge, see Fig. 4.30 (*b*) and determine the lengths of the sides of each rectangle. For a mobile surcharge proceed as in App. A, Ex. 16. With the known cover depth H_{min} obtain $m = L/H$ and $n = B/H$, and from Fadum's graph (Chart C5) or Table C obtain the values of I_σ for each rectangle. The value of C_{sus} is then the algebraic sum of the influence values (Article 4.18) which yields the resultant value for the surcharged area. Select a value for the impact factor and determine the value of $W_{sust\ max} = C_{sus}U_{sust}B_cF_i$ lb/lin. ft for a mobile surcharge. Determine the value of $W_{sus\ max} = C_{sus}U_{sus}B_c$ for a static surcharge. Repeat the procedure for H_{max} to obtain $W_{sust\ min}$ or $W_{sus\ min}$.

Step 8. Enter all the respective effective design values obtained above as in Table P for each section of the pipeline, and sum each column. The largest of these sums for each section of the pipeline is then the maximum total effective design load, W_e, in lb/lin. ft of pipe length for that section.

N.B. If it is finally decided to use a concrete arch support all the steps 1 to 8 must be repeated using the reduced values of H and the increased values of B_c imposed by the arch dimensions. Add the weight of the concrete arch to the fill load and use Charts C7 (*b*) or C8 (*b*) for concentrated surcharges.

Step 9. For pressure pipes determine the effective design values of the internal fluid pressure, i.e. p_w and p_t. For rigid or flexible pressure pipes or for flexible non-pressure pipes determine the effective design value of the external fluid pressure p_e (see Table E).

8.3 (*b*) *Embankment Loads* (see also App. A, Exs. 7, 8, 9, 14, 15, 16)

Step 1, is the same as for trenches.

Step 2. Select one or more alternative methods of installation and determine the maximum fill density γ_s or γ_d.

Step 3. Select appropriate values of r_{sd} and p or p' and with the values of H_{max}/B_c, H_{min}/B_c, or H_{max}/B_d, or H_{min}/B_d enter Charts C3 and C4 and obtain $C_{c\ max}$ and $C_{c\ min}$, or $C_{n\ max}$ and $C_{n\ min}$ as appropriate, and compute the max. and min. values of $W'_c = C_c\gamma_s B_c^2$ or $W_c = C_n\gamma_s B_d^2$ or $W_c = C_n\gamma_s B_c^2$.

Step 4. Choose the effective value of W_c or W'_c from Step 3 and correct this value for submergence if necessary *and safe*. Compute the water load as for trenches.

Step 5. Determine the value of U_s.* Compute $H' = H + U_s/\gamma_s$ and from the appropriate chart obtain C'_c or C'_n. Multiply this value by $\gamma_s B_c^2$ or $\gamma_s B_d^2$ as appropriate, and subtract the fill load obtained in Step 4 to obtain W'_{us} or W_{us}.

Step 6. Proceed as for trench loading to obtain $W_{csu\ max}$ and $W_{csu\ min}$.

Step 7. Proceed as for trench loading to obtain W_{sus} or W_{sust} or both.

Step 8. Proceed as for trench loading to obtain $W_{e\ max}$.

Step 9. Proceed as for trench loading to obtain p_i and $p_{e\ max}$.

Step 10. To determine the safe depth of fill at which to operate construction equipment proceed as in the worked examples App. A, Ex. 9, 15 or 16.

8.3 (*c*) *Computation of Pipe Strength*

8.3 (*c*) (i) *For Rigid Non-pressure Pipes* (see also App. A, Exs. 1–4, 7, 8)

Step 1. Select a suitable class of bedding and its bedding factor F_m or F_p.

Step 2. Select appropriate values of C_{BS} and F_s.

Step 3. With the value of W_e obtained as above compute

$$W_T = W_e F_s / C_{BS} F_m \text{ lb/lin. ft}$$

$$\text{or } W_T = W_e F_s / C_{BS} F_p$$

as appropriate.

8.3 (*c*) (ii) *For Rigid Pressure Pipes* (see also App. A, Exs. 5 and 6)

Step 1. Choose the type of pipe to be used and obtain its minimum bursting pressure p_{ult} (or p_{oc}) under zero external load.

*But see Table D.

Step 2. Select a suitable class of bedding and its bedding factor F_m or F_p.

Step 3. Select appropriate values of C_{BS}, F_{si} and F_{se}.

Step 4. Determine the values of $W'_e = W_e$ when $p_i = p_w$;
and of $W'_e = W_w + W_c + W_{us} + W_{sus}$ when of $p_i = p_t$.

Step 5. Compute the values of W_T from the appropriate formula for the type of pipe 5.5, 5.6 or 5.7 and for the two values of W'_e in conjunction with p_i from Step 4 (see also App. A, Exs. 5 and 6).

Step 6. Choose the higher value of W_T obtained in Step 5.

Step 7. Choose the appropriate British Standard class of pipe having a value of the safe crushing test load $\geqslant$ the value of W_T obtained in Step 6 (see Table H). If the required value of W_T is not within the standardised range, refer to the makers.

8.4 Computation of the Wall Thickness of Flexible Non-pressure or Pressure Pipes

To obtain the required wall thickness of corrugated or smooth walled steel pipes, proceed as in Articles 6.4 and 6.5 of Chapter 6 and as in the worked examples 10 to 14 in Appendix A.

N.B. (i) This method is not applicable to pitch-fibre pipes.

(ii) It might be applied to ductile iron and uPVC pipes when the values of yield strength in tension and compression and Young's Modulus (after 50 years in service) can be established or reliably estimated.

Section E

Secondary Loads and Forces
Flexible Joints

Chapter 9
Secondary Loads and Forces

9.1 The Nature of the Secondary Loads

In the previous chapters the theory of pipe design has been mainly concerned with the vertical loads imposed on a pipeline by the trench or embankment fills and by surface surcharges, and with the ability of the pipes to sustain such loading. Observations of many fractures of rigid pipes in service, however, have revealed that only under rather exceptional conditions is an underground pipeline subjected to simple vertical loading. Usually, other forces are active, each contributing its own particular component to the total principle tensile stress on the pipe materials.[81] Some of the additional sources of stress which commonly occur are:

(i) Volume changes in clay soils due to wetting or drying, or drying by tree roots; frost heave in some wet fine grained soils;
(ii) Tree root growth pressures;
(iii) Impact forces and pressures caused by compacting the fill;
(iv) Non-uniformity of the foundation;
(v) Non-uniformity of the bedding;
(vi) Differential subsidence, especially in mining areas;
(vii) Settlement of buildings, or of the ground under embankments or fills;
(viii) Thermal and moisture changes in the material of the pipes, joints, and concrete beddings;
(ix) Restraints caused by bends, branches, valves, manholes, etc;
(x) Vibration caused by heavy traffic or machinery.

These forces can be large, highly variable, and localised. They do not lend themselves to measurement and they cannot be estimated quantitatively with any confidence. The types of fracture they cause in rigid pipes are shown in Table R. Rigid pipes are usually weak in tension (or shear). With this type of pipe it will usually be better to eliminate the secondary forces wherever possible, rather than to attempt to resist them. If the resistance to axial deformation of the pipeline, as distinct from that of individual pipes, can be reduced or eliminated, the corresponding stresses will also be reduced or eliminated. This desirable relaxation can be achieved very largely by the use of joints which permit some freedom of movement to individual pipes both axially and transversely. In effect these joints are comparable with the universal joints and splines used in the transmission system of a motor car but without the necessity to transmit torque.

The differential soil movements caused by seasonal moisture changes in clay soils and the growth of tree roots,[102] both tend to bend a pipeline slightly in any plane containing the axis, or to displace the pipes axially. Variations of hardness in the soil foundation of the pipeline induce differential vertical movements of, and consequent axial bending stresses in, the pipes. Flexible joints permit the pipes to yield to all these forces and so greatly reduce the axial tensile stress and lessen the risk of rupture. They will not however protect the pipes against breakage by local concentrations of vertical load or reaction, such as occur at a hard spot in the bedding, or by pressure from a swelling tree root.

The large extensions and contractions and the more or less severe axial bending of the pipeline caused by gross progressive ground subsidence, such as occurs in mining areas, are sufficient to fracture any rigid pipeline which is not flexibly and telescopically jointed. These movements can also cause serious trouble with flexible pipelines which are not provided with sliding (i.e., telescopic) joints (Plate 7). They are of such importance that parts of a note on them originally prepared for the I.C.E. Pamphlet on Mining Subsidence[83a] is reproduced as Appendix B for convenient reference.

Another source of secondary stress is the settlement of a manhole or other structure to which the pipeline is connected, relative to that of the adjoining pipe, or vice versa. In addition to flexible joints, the relief of the bending and shear forces involved under these conditions may require special construction, especially for the smaller sizes of pipe (see Fig. 11.2). Similar conditions occur when pipes are in contact with or in close proximity to other pipes or structures which they cross either above or below.

If any rigid pipe is rigidly jointed with a material which has a higher rate of thermal or moisture expansion that than of the pipe material itself, the bursting of the pipe socket is highly probable, (e.g. this is a very frequent cause of fracture of clayware pipes jointed with cement mortar) (Plate 6). A similar type of fracture may occur with flexible joints of the radially-compressed rubber ring type described below, if the radial pressure of the sealing ring is excessive.

The axial pulls and thrusts in either rigid or flexible pipes caused by shrinkage or swelling of the soil, or of the pipe materials, or of the concrete bedding, if any, due to moisture or thermal changes, can be limited in their effect to the length of a single pipe by the use of telescopic joints.

Since some of these causes of stress may be present in any pipeline, anywhere, it is considered that the use of flexible telescopic joints should be used, even in *normal* practice, and that the more rigid jointing methods which have been commonly used hitherto should be abandoned.

9.2 The Cause and Effects of Axial Bending in a Pipeline

9.2 (a) General Notes It must be emphasised that the loads computed as in Chapters 4 and 8 take no account of the effects of irregular bedding or differential movements of the soil, either of which may produce beam action (axial bending) in the pipeline. When the pipes are acting as a beam, or a cantilever, the crushing load at the points of support will depend on the span of the beam and the length of the support, and it will be several times larger than the uniform load estimated by the Marston formula, especially if rigid pipes are so rigidly jointed as to become continuous beams (see Fig. 9.1). Such differential movements frequently occur and are the cause of many failures, e.g.:

1. Where a building, manhole, or other structure, settles relatively to the soil in its vicinity in which the pipe is embedded, especially if the pipe is rigidly connected to the structure.
2. Where the pipe bed settles unevenly or is undermined for example by the erosion of the soil below it into a water course or leaky sewer.
3. Where the soil is susceptible to seasonal rise and fall with changes in its moisture content, when the pipe is near the surface, e.g. in most clay soils. This effect is most common in house drainage pipes or where the soil is subject to frost heave.
4. In the vicinity of tidal water, where there may be a considerable daily ground movement synchronised with the tide.
5. Where general differential ground movement occurs as in mining subsidence.

The disruptive effect on the pipeline of any of these conditions is reduced or eliminated by the use either of flexible pipes or of short-length rigid pipes with flexible telescopic joints.

The strength of rigid pipes as beams requires careful consideration since the complete prevention of axial deformation can rarely be achieved. This is especially true of small diameter pipes laid with shallow cover. It cannot be too strongly emphasised that axial deformation of a pipeline is likely to occur in most soils, especially if the pipeline is near the surface. If the pipeline cannot yield and adapt itself to these deformations it will break. It is not surprising therefore that the rigid jointing of rigid pipes, such as has been common practice with concrete and ceramic pipes hitherto, is a very frequent cause of failure. When the pipe bed has an uneven bearing capacity the condition shown in Fig. 9.1 is likely to arise. It will be seen that, in addition to axial bending tensile stress varying as the square of the span the crushing load at the support is increased in the ratio of the span length to the support length. The ring stresses under these conditions then

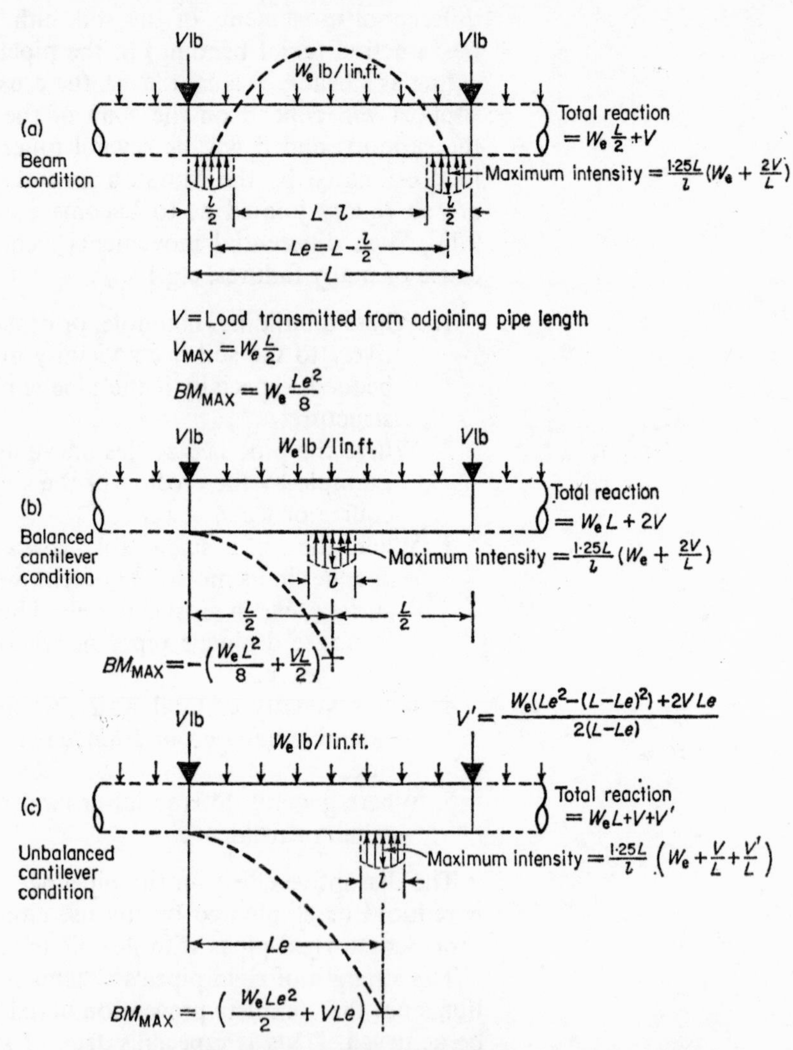

Fig. 9.1 Effect of non uniform bedding and position of hard spots on ring bending loads and longitudinal bending moments in flexibly jointed pipes

approximate to the self load stresses which would be produced with the same type of bedding in a pipe weighing as much as the sum of the actual self load plus the sum of all the distributed loads on the span plus the shear forces at the ends of the pipe. These potentially dangerous conditions can be largely, if not entirely, eliminated by:

1. The use of flexible pipes, e.g. steel, pitch fibre, plastics.
2. The use of suitable flexible joints for rigid pipes.
3. The adoption of practical precautions in the preparation of the pipe bed and the laying of the pipe, which will ensure uniformity of support of the belly of the pipe.
4. The longitudinal reinforcement of concrete pipes combined with 2 and 3.

Flexible pipes, or very short length rigid pipes with flexible joints, should normally be used in made or very uneven ground in which large differential movements are to be expected.

9.2 (*b*) *Effects of Soil Swelling or Shrinkage and of Frost Heave*

Most clays are subject to seasonal swelling and shrinkage with changes in their moisture content to a depth of about 5 ft below the surface or several times that value in the vicinity of trees. Waterlogged fine sands, silts, chalk, or flyash are subject to frost heave at depths down to about two feet in the British Isles.

Pipelines laid in such soils within these depths are likely to be deformed axially in any direction by either of these soil movements.

9.2 (*c*) *Effects of Soil Disturbance*

In some fine grained soils the sinking of a new trench alongside an existing pipeline, without appropriate precautions, may give rise to unbalanced lateral loading on the old pipe and so to deformation of its axis more or less in the horizontal plane. This effect has also been experienced behind a river wall which settled and moved outwards as a result of dredging the river, so causing disturbance and fracture of a large water main laid behind the wall.

9.2 (*d*) *Effects of Groundwater Movement*

In some fine grained soils, particularly on steep gradients, the pipe bed may become a subsoil drainage channel with consequent erosion of fine grained soil and increased loads locally on the pipe. The remedy is to prevent erosion if possible, by the placing of obstacles to the water movement at frequent intervals, such as puddled clay water stops across the trench, and by reducing the water velocity, by any suitable means, as required by site conditions.

Heavy pumping in excavations in fine grained non-cohesive soils is dangerous and may produce similar effects on existing pipelines.

9.2 (*e*) *Effects of Sudden Ground Subsidence*

In salt mining areas, and possibly in areas overlying calcareous bedrock which is subject to caving, destructive shear forces sometimes occur owing to sudden local subsidence partaking of the nature of a geological fault. The downthrow may be large enough to rupture the pipe or to pull or break its joints. The occurrence is fortuitous and there are no reasonable precautions against it known to the author. Where there is prior warning of the caving, welded steel pipes *may* survive but such areas should be avoided wherever possible.

9.2 (*f*) *Effects of Lateral Spread of, and Subsidence under, Embankments*

Pipelines crossing a new embankment may have their axes bent differentially by the variable compression of the foundation soil caused by the varying depth and weight of the fill. Simultaneously they may be stretched outward by the frictional drag of the fill which tends to spread laterally as it consolidates. These movements may be quite large under a high fill[98] and they must be allowed for in the design and construction of the conduit, (see Fig. 11.6). Flexible pipes are subjected to the same forces as rigid pipes but are better able to resist them if their tensile strength is adequate and proper attention is given to the design and execution of their joints. Frictional drag on corrugated steel pipes is likely to be greater than that on smooth pipes and their joints must be strong enough to resist the tensile forces induced in the pipeline.

9.2 (*g*) *Effects of the Subsidence of Structures*

Pipelines passing into or beneath walls or other structures require special consideration if they are to remain watertight during and after any differential settlement which may occur. Some possible solutions to this problem are given in Figs. 11.2 which may serve to illustrate the essential principles and advantages of a double hinge in the pipeline, and the elimination of the potentially large shear forces imposed on the pipes.

9.2 (*h*) *Effects of Mining Subsidence*[83,83a]

In modern coal mining, surface subsidence usually takes the form of a wave-like depression advancing as the coal is extracted and leaving a 'standing wave' at the sides of the working. A pipeline within the subsidence zone, which is more or less parallel to the direction of coal working is then usually stretched and bent when the wave reaches it and is restored more or less to its original length as the wave recedes. In sloping ground however the movements may be reversed. In either event it is unlikely that the movement will be the same at every joint in the pipe line.

The advice of the National Coal Board, or other appropriate mining authority should always be obtained regarding the protective measures required for pipelines in their areas.

9.3 Effects of Temperature and Moisture Changes

9.3 (a) Axial Expansion and Contraction

If the temperature of the pipe contents varies considerably with the season, or otherwise, axial forces large enough to fracture a rigidly jointed pipeline transversely may and do occur, or the joints may be drawn or broken. Gas pipes and pipes carrying water extracted from a river are more liable to this kind of damage than public sewers or pipes conveying well water. Some industrial drains in which the effluent temperature varies are similarly vulnerable.

Clayware pipes surrounded by, or bedded and rigidly jointed in, concrete may be fractured by a sharp change in temperature after laying and before back filling, because the thermal movement of the concrete is about double that of the pipes.

The axial pulls and thrusts in either rigid or flexible pipes caused by shrinkage or swelling of the pipe materials (Plate 8), or of concrete beddings, due to moisture or thermal changes, can be limited in their effect to the length of a single pipe by the use of telescopic joints and corresponding relief gaps in the site concrete (see Fig. 11.3).

9.3 (b) Differential Heating and Cooling or Moisture Change

Non-uniform heating of the periphery of a pipeline by the sun, or cooling by cold winds, may cause longitudinal fractures in some pipes (Plates 9 and 10). Differential moisture changes can produce similar effects. These fractures have occurred in unreinforced concrete pipes laid alongside a trench and waiting to be laid, and sometimes even in stock yards. There is only one crack and it always faces the sun, or the direction of the cold wind, or it occurs in the wetter zone, e.g. where the pipe is laid on its side in damp ground.

Cold water should not be introduced suddenly into exposed clayware or concrete pipes which have been heated on top by the sun, otherwise the pipes may be cracked.

It is always advisable, whilst awaiting test, to protect such pipelines against uneven heating or cooling, using any available insulating materials such as straw, cement bags, building paper, or placing a sun (and wind) shade over the pipeline and, if there is considerable difference in temperature, to introduce test water slowly, especially in small diameter pipes.

9.3 (c) Effects of Differential Volume Change and Stress Relaxation in the Jointing Material

A rigid jointing material having a thermal or moisture expansion differing considerably from those of the pipe material is very likely to expand and cause fracture of the pipe sockets or collars, or shrink and cause leakage. Thus the use of cement mortar or certain acid resisting compounds which set and harden, is known to be a frequent cause of fracture of ceramic pipes (Plate 8). The type of fracture has occurred above ground, or before backfilling, with a rise in temperature (exposure to the sun) or, with mortar, after backfilling, due to differential moisture expansion. The richer the mortar the greater the risk of fracture from either cause.

The overheating of bituminous jointing compounds may cause such hot poured material to crack after cooling, and the joint to leak in consequence.

Rubber and plastics materials are subject to stress relaxation[80] under sustained pressure and so may cause leakage in a joint after a time, if they are not correctly designed allowing for the relaxation.

Run or caulked lead joints are subject to creep and are frequently loosened by vibration sufficiently to permit serious leakage in pressure pipes especially if subjected to seasonal draw.

Cold bituminous mastic joints in non-pressure pipelines may suffer somewhat similarly.

9.4 Axial Thrust in Pressure Mains

The axial end thrust in a pressure pipe acquires increased significance where flexible telescopic or lead joints are used. The thrust blocks provided at bends, junctions and control valves should therefore be of ample strength and generous size to minimise their possible movement under sustained loading. The modulus of soil reaction (*k*), previously referred to in Article 6.4 (*c*) in connection with the deflection of flexible pipes, is of interest in this respect as giving an approximate value of the movement to be expected for a given intensity of pressure of the block on the soil, after allowing for sliding friction between the block and the surrounding soil. The use of 'positive' flexible joints which have little or no draw but provide positive resistance to axial tension (e.g. the victaulic joint) enables the pipeline itself to carry the axial tension and largely eliminates the need for anchorages. Provision for thermal expansion and contraction may need consideration however.

9.5 Effects of Vibration

Vibration of the soil by machinery or traffic may destroy or even reverse the upward acting frictional forces in a trench, or accentuate differential settlement of the pipe bed and, in either event, increase the load on the pipe. It tends to aggravate other potential sources of stress but its effects are considerably alleviated by the pliability and damping effects of rubber joint rings. Unlike lead or cement mortar, rubber rings, particularly of the sliding type, do not tend to loosen and become displaced by vibration.

Chapter 10
The Design and Use of Flexible Telescopic Joints

10.1 General Notes on Joints

10.1 (a) Functional Requirements

The essential functional requirements of flexible telescopic joints are as follows:

1. They should be so designed as to permit angular and axial movements of the pipes, or pipeline, large enough to tolerate the maximum displacement to which the pipeline may be subjected, without damage to the pipes or loss of watertightness under the maximum internal or external water pressure to which the pipeline may be subjected in service, or on test.
2. They should retain their efficiency throughout the useful lifetime of the pipeline and hence be made with materials whose physical and chemical properties do not change considerably adversely with the passage of time in the internal or external environments to which they may be subjected.
3. They should retain their efficiency under any vibration to which the pipeline may be subjected.
4. They should be easily and quickly made, or unmade, in the limited working space available in a trench, with the the minimum of individual adjustment.

10.1 (b) Existing Varieties of Joints

Many varieties of flexible joints have been suggested from time to time but those which have survived practical trials and are in current use, or course of development, are classified in Tables Ta and Tb and some of them are illustrated in Figs. 10.1 to 10.10. With the exception of the *Oanco* joint all of them utilise a rubber type of sealing ring.

These flexible joints fall into two main classes, viz. those which permit draw (Table Ta) and those which do not (Table Tb). The latter are known as 'positive' joints since they will transmit axial tensile forces throughout the length of a pressure pipeline. This property is sometimes useful as it may obviate the need for end anchorages in straight runs, and thrust blocks at bends, etc., in pressure pipelines.

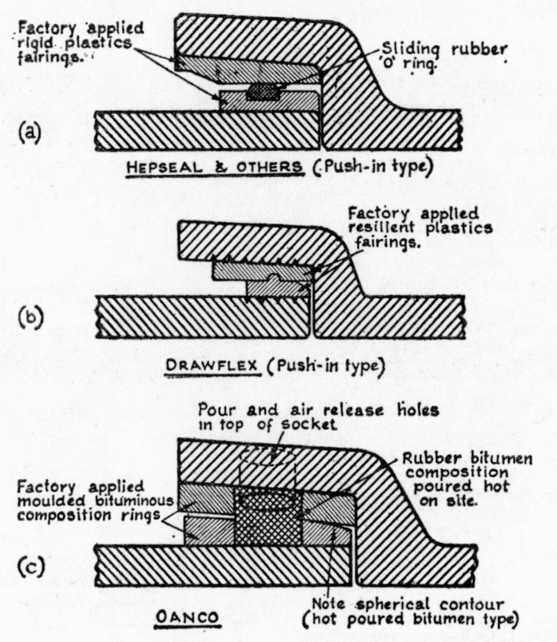

Fig. 10.1 Flexible joints in current use for clayware pipes

Crown Copyright—by permission

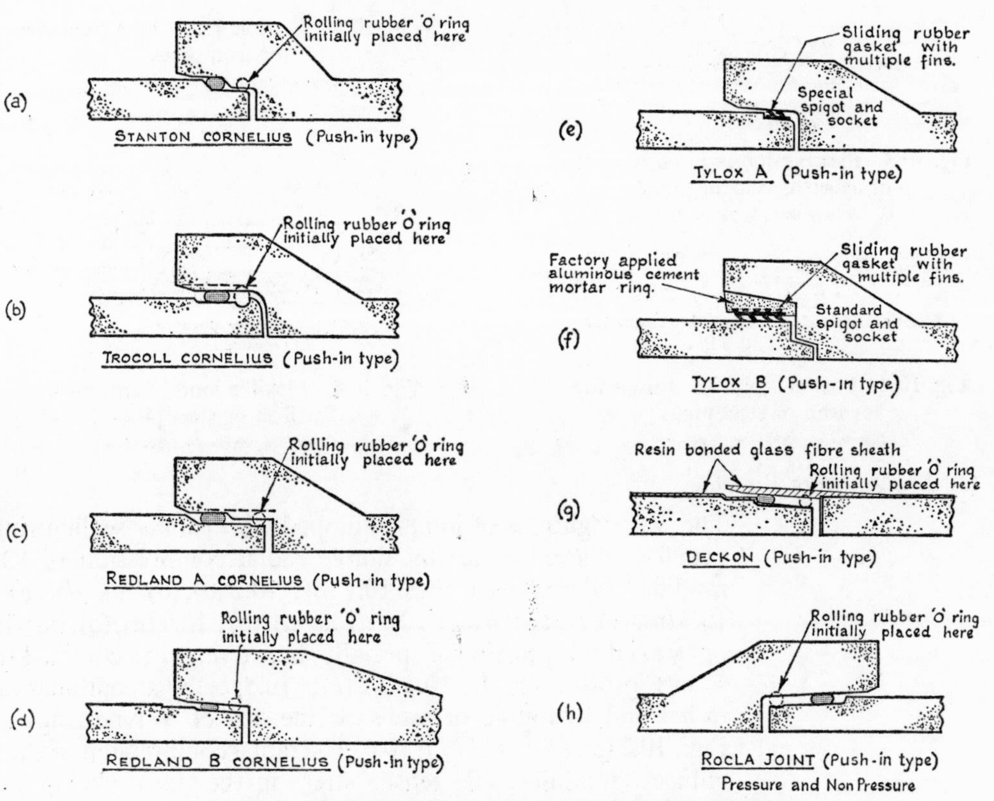

Fig. 10.2 Flexible joints in current use for concrete pipes

Crown Copyright—by permission

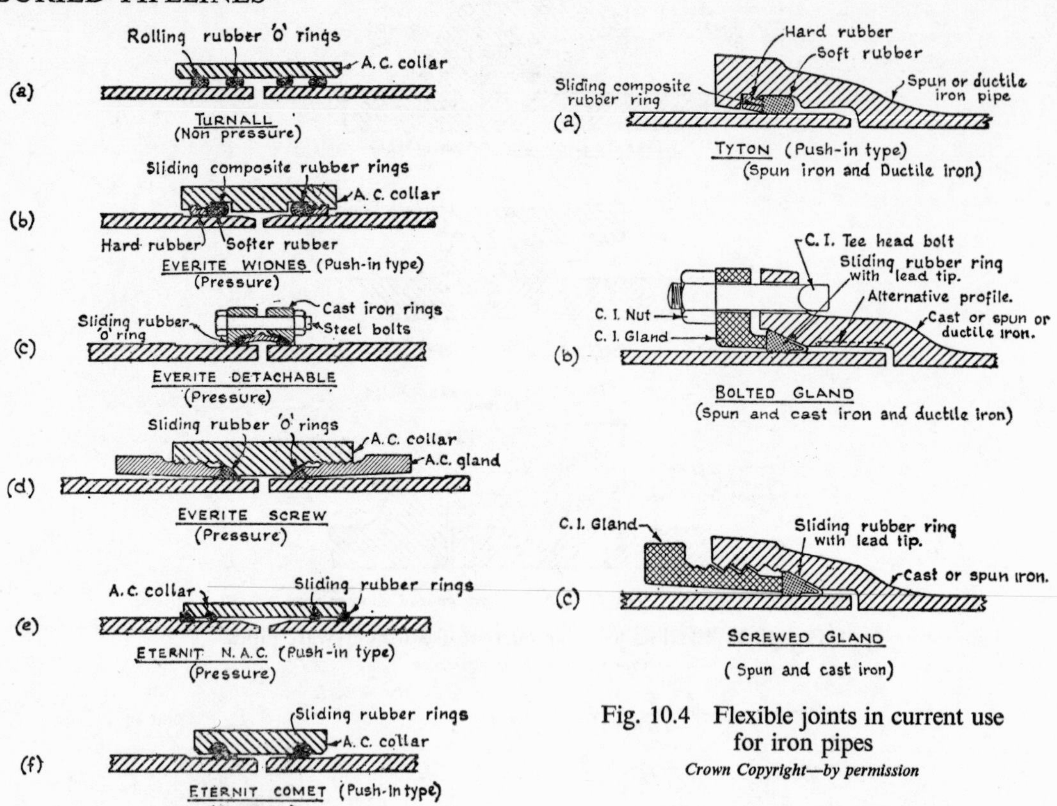

Fig. 10.3 Flexible joints in current use for asbestos cement pipes
Crown Copyright—by permission

Fig. 10.4 Flexible joints in current use for iron pipes
Crown Copyright—by permission

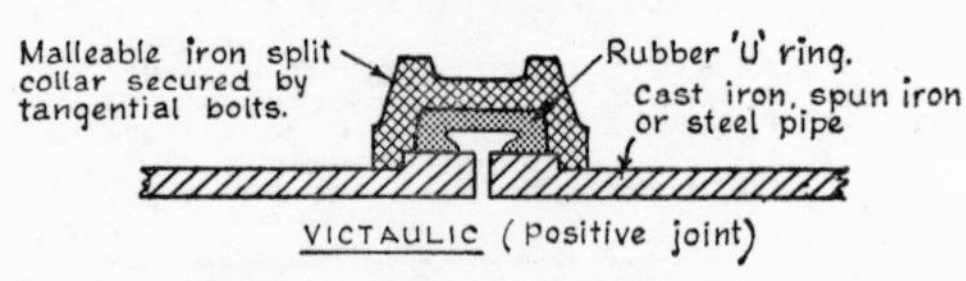

Fig. 10.6 Flexible joint in current use for iron or steel pipes
Crown Copyright—by permission

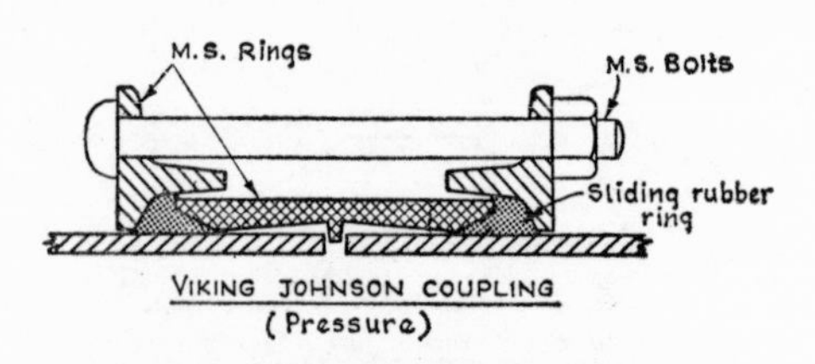

Fig. 10.5 Flexible joint in current use for iron or steel pipes
Crown Copyright—by permission

10.1 (c) Sealing Ring Pressure

The watertightness of joints equipped with rubber sealing rings is usually achieved either by simple radial compression of 'O' or modified 'O' rings as in Figs. 10.1 (*a*); 10.2 (*a*), (*b*), (*c*), (*d*), (*g*), (*h*); 10.3 (*a*), (*b*), (*e*); 10.4 (*a*); 10.7; 10.8 (*a*), (*b*); 10.9 (*a*), (*b*), (*c*); 10.10 or by axial compression of specially shaped rings in a confined space as in Figs. 10.3 (*c*), (*d*); 10.4 (*b*), (*c*); 10.5, or by a combination of radial and hydraulic pressure on the lips of V type rings as in Figs. 10.2 (*e*), (*f*); 10.3 (*f*); 10.6. The radial compression of the ring induces circumferential tensile stress in the pipe socket or collar,

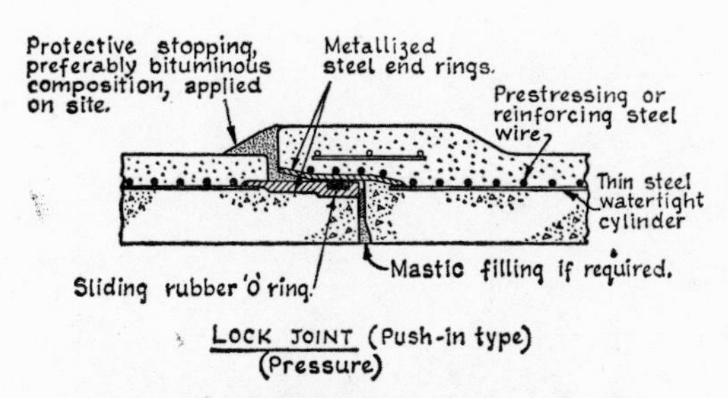

Fig. 10.7 Flexible joint in current use for prestressed concrete pipes
Crown Copyright—by permission

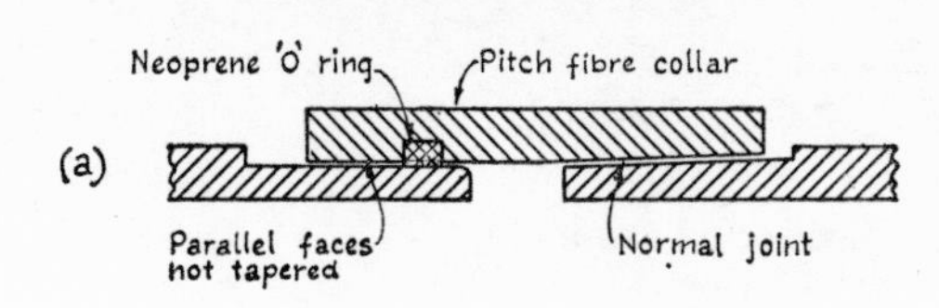

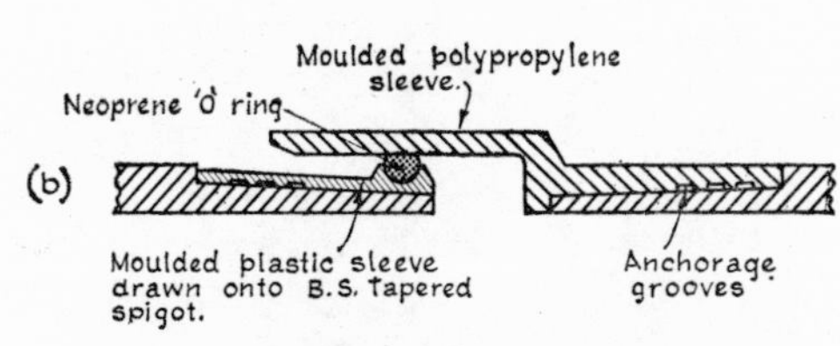

Fig. 10.8 Flexible joints in current use for pitch fibre pipes
Crown Copyright—by permission

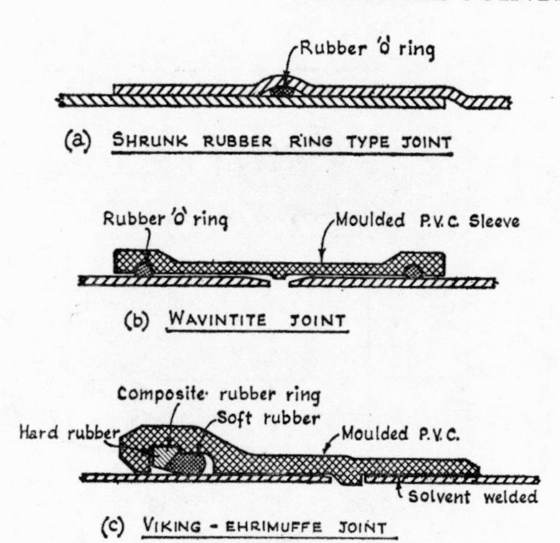

Fig. 10.9 Flexible joints in current use for plastic pipes
Crown Copyright—by permission

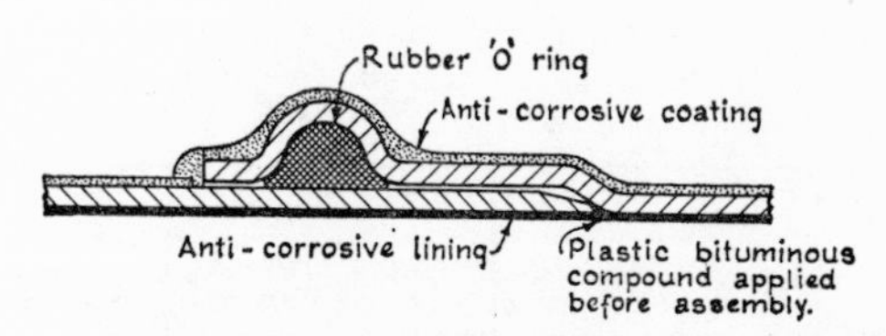

Fig. 10.10 Flexible joint in current use for steel pipes
Crown Copyright—by permission

the intensity of which depends upon the hardness of the rubber and the degree of compression imposed upon the ring. This stress is additional to the stresses imposed by the external and internal loads imposed on the pipe when in service or under test, and, if excessive in relation to the strength of the socket or collar, could be the cause of rupture. The graphs in Fig. 10.11 show the very large rise in the radial loading imposed by these rings as the degree of compression increases, (see also Refs. 80 and 80 *a*).

Clearly then the compression of the rings should be no more than is required to produce a watertight seal under the test pressure for which the joint is designed, after allowing for pressure losses caused by shrinkage and stress relaxation* of the rubber,

* *Stress relaxation* is the gradual decrease with time of the stress imposed on the rubber ring by its sustained compressive strain. It is several times higher for the harder carbon-black filled rubbers (hardness about 70° B.S. or more) than for the softer 'gum' rubbers containing little or no filler (hardness < 70° B.S).[80]

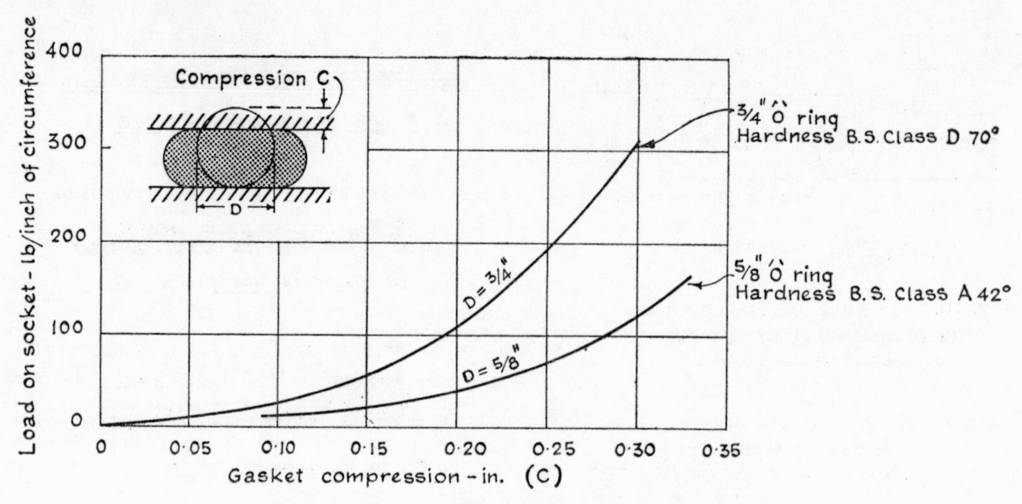

Fig. 10.11 (*a*) Radial load on pipe socket caused by compression of rubber 'O' ring

Crown Copyright—by permission

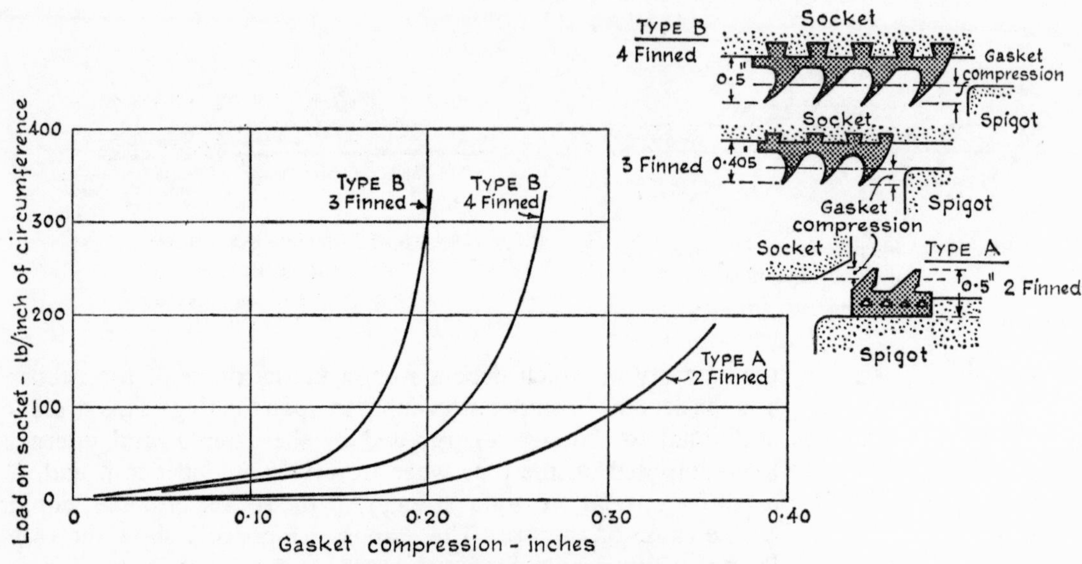

Fig. 10.11 (*b*) Radial load on pipe socket caused by compression of Tylox rubber gaskets

Crown Copyright—by permission

which varies considerably with the composition and hardness of the rubber and which may amount to one half to two thirds of the initial pressure over a period of many years under stress. If water pressure tends to cause movement of the sealing ring relative to the pipes it would obviously be preferable to prevent blowouts by means of a stop formed in the pipe rather than by additional compression of the ring. In pressure pipe joints, blowout is usually prevented by anchoring the rubber ring in a groove in the spigot or socket or collar, as in Figs. 10.3 (*b*); 10.4 (*a*); 10.7; 10.9 (*a*), (*b*), (*c*); 10.10, or by some other similar device as in Figs. 10.3 (*c*), (*d*); 10.4 (*b*), (*c*); 10.5; 10.6.

Variations in the ring compression around the circumference of the pipe which might affect watertightness are better reduced or eliminated by reduction of the manufacturing tolerances on the rubber section and hardness, and on the tolerances controlling the width of the annular gap between the spigot and socket, or collar, than by additional compression of the ring. Similarly the curved surfaces of the pipes in contact with the sealing ring should be smooth and free from humps and hollows. In rolling or sliding ring joints these surfaces should be as nearly parallel to the pipe axis as possible so that variations in compression do not occur with changes in draw or ring position. In design, these variations may be covered by an empirical factor (see Article 10.2 (*d*).

Care is needed to make the sockets of spun concrete pipes impervious otherwise leakage may occur through the pipe ends regardless of ring pressure.

10.1 (d) Control of Variation in the Width of the Annular Gap

In all compressed rubber ring type joints a critical factor is the variation in the width of the gap between the spigot and the socket or collar. Where both surfaces can be machined, as in asbestos-cement or pitch-fibre pipes, very small tolerances are practicable, a low compression variation factor* is required and a good joint is obtained without difficulty. Cast iron and concrete pipes, being moulded, occupy an intermediate position. In clayware pipes large variations in gap width may occur, making the jointing of these pipes far more difficult. It will be readily appreciated that if an elliptical spigot is mated with a circular socket or vice versa, a simple 'O'-section joint ring will have tight and slack zones at the extremities of the major and minor axes of the ellipse respectively. The resulting variations in radial pressure may then induce over-stress, or even rupture of the socket, at the one extreme, or lack of watertightness at the other. In 1956 it was suggested[5] that it might prove feasible to form a sufficiently uniform seating for the ring by coating the spigots and sockets of clayware pipes with a die cast compound in the pipe factory. This has been achieved in the two American type joints now in current use here (see Figs. 10.1 (*a*) and (*b*). In both these joints variation in the annular gap width is virtually eliminated by the use of die cast plastic fairings to both spigots and sockets. Another promising new joint utilises a flexible PVC sleeve, in

* See Article 10.2 (*d*).

place of the usual socket, with rubber rings sealing the spigots of the adjoining double spigot pipes.

For concrete and cast iron pipes, provided the tolerances on the diameters of the spigot and socket, and on the roughness of the contact surfaces, are sufficiently small, the resilience of a rubber 'O' ring is usually sufficient in itself to cope with the relatively small variations in the gap width for non-pressure pipes, as exemplified in the original and still widely used Cornelius type joint. Fig. 10.2 (*a*), (*b*), (*c*), (*d*), (*g*), (*h*). For these pipes, however, a relatively high compression variation factor* is usually required.

10.1 (*e*) *Dimensions, Hardness and Quality of Rubber Sealing Rings*

Another critical factor in joints using 'O' type sealing rings is the extent of the variation in the cross section and hardness of the rings. Evidently any considerable departure from the nominal diameter and hardness of the ring assumed in designing the joint will adversely affect the radial pressure exerted by the ring either in different joints or in random places in the same joint. To avoid loss of watertightness on the one hand, and excessive hoop stress in the sockets or collars of the pipe on the other, it is important therefore that the diameter and hardness of these rings should be as consistently uniform as is practicable. It is understood from the makers of the rings that with respect to dimensional uniformity, moulding the rubber, either as a complete jointless ring or as a cord, is preferable to extruding it. The splicing of cord rings needs careful attention. In view of the cost of moulds, standardisation of the external diameters of pipes and of the annular gap dimensions would be desirable but this does not appear to be economically practicable at the present time.

The physical characteristics of the rubber, including its hardness, are specified in B.S. 2494 for natural and synthetic rubber and in B.S. 3514 for oil resisting synthetic rubber. Neither of these standards covers dimensional or hardness tolerances, stress relaxation, or ring sizes; B.S. 2494 was intended to cover 'O' rings but compliance with it has hitherto been required for all joint rings in pipelines requiring loan sanction in England and Wales. The Standard needs modification and extension however, to cover some of the multilipped and modified 'O' type rings now being used. It is understood that this matter and the question of tolerances have been brought to the attention of the British Standards Institution.†

The degree of hardness required for any particular joint design evidently depends on the geometry of the joint and the design test pressure. Broadly speaking for equal test pressures hardness

* See Article 10.2 (*d*).

† B.S. 2494 is being revised and the new (1968 ?) edition should be consulted.

will be inversely proportional to the degree of compression of the ring imposed by the width of the annular gap in the joint; i.e. for Cornelius-type rolling ring joints with a given gap width, the harder the rubber the less its uncompressed diameter must be (see Article 10.2 (*d*) (i)). Another consideration is the possible differential compression of the ring at crown and invert caused by any tendency to differential settlement of the jointed pipes, or the transmission of loads from one pipe to another when the bedding is not as completely uniform as it should be. Under these conditions, too soft a rubber might not maintain a watertight seal in some joints. On balance, however, and because of its lower stress relaxation, it would seem preferable to utilise the softer and larger diameter ring. In current practice the hardness does not usually exceed 60 ± 5 B.S. degrees for Cornelius-type rolling rings for sewer pipes.

Conservation of the physical properties of the rubber, whether natural or artificial, over a very long period is essential. Whilst rubber technologists hesitate to offer unqualified opinions on a life in excess of about 30 years, there are remarkably few reports of failure in this respect where rubber of the proper quality and type, has been used in this country and the U.S.A.[80b] over the past 40 years. Some examples of natural rubber deterioration have been reported from overseas countries recently but it is not yet clear whether the cause lay with the rubber composition or with peculiar local conditions.[80a] In view of the high cost and difficulty of replacing faulty rings it would evidently be false economy to use anything but best quality material initially. Materials inferior to the B.S. quality should therefore be rigorously rejected regardless of any saving in initial costs. In this respect it is worth noting that the underground conditions of darkness, wetness, and scarcity of ozone both inside and outside the pipes, favour the preservation of both natural and artificial rubbers. So far as the author is aware no better alternative material is available for the purpose at present. The use of synthetic materials is too recent to enable an opinion on their durability to be expressed with equal confidence.

Natural rubber however is not suitable for use above ground when exposed to the air or sunlight, or below ground in the presence of oils, fats or solvents (see Table Q). In gas mains B.S. 2494 natural rubber rings may require protection against direct contact with the gas. (See amendment No. 1 to B.S. 2494 dated May 16, 1960).*

10.1 (f) Provision for 'Slew' and 'Draw'

The angle through which a pipe can be deflected relative to its neighbour ('slew'), without locking or leakage, depends on the

* B.S. 2494 is being revised and the new (1968 ?) edition should be consulted.

average or nominal width of the annular joint gap and of the depth of the socket, or the half length of the collar. As a practical necessity, this angle decreases as the pipe diameter increases, since for a given angle of S°, the displacement of the spigot on the convex side of the curve will be $B_c \frac{\pi}{180} S$ approximately, where B_c is the outside diameter of the pipe. Chamfering or rounding the end of the spigot as in Fig. 10.2 (*b*), (*e*) or (*h*), permits greater angular movement than a square end before the locking point is reached.

The axial distance through which a pipe can withdraw from its neighbour without loss of watertightness ('draw') depends upon the length of the socket or half length of the collar, and on whether or not the sealing ring rolls, or slides during the movement. Theoretically the length of draw required for a given displacement of the pipeline, or for a given change in temperature, depends on the length of the individual pipe, and is independent of pipe diameter. The depth of sockets or the length of collars should therefore vary with pipe length rather than with pipe diameter. However, an adequate angle of slew for large pipes must be provided, and as the socket is deepened, the angle of slew for a given size of pipe will be decreased unless the internal diameter of the socket is increased (i.e. the annular gap is widened). Thus, where the same joint design is to be used for pipes of various standard lengths it should be adequate for the longest of such lengths.

10.1 (*g*) *Ease of Assembly and Site Work*

Builders and contractors are mainly interested in the ease and speed of jointing and, even at increased cost, they are likely to prefer a joint which is 'right first time'. By and large, this means either that loose sealing rings must be anchored, or automatically held in the correct position, or that there are no loose rings, and, in either design, that the tolerances on the dimensions of the pipes and rings are small enough to ensure watertightness in every joint and no subsequent burst sockets. Practice indicates however that certain precautions during the handling, laying and jointing of the pipes, and in the backfilling, are necessary to ensure the efficient performance of flexible telescopic joints (see Article 11.1 (*k*) and Ref. 31 *a*) and that site personnel should be trained to observe them rigorously. Failures on test, and loss of time in making good, are thereby avoided; backfilling can proceed without delay; and it becomes possible to work to a time limit with greater confidence. In emergency, some sliding ring type joints can be made under water, but this practice should be avoided whenever possible.

10.2 Joint Design

10.2 (a) Basic Requirements

It would appear that the 'push in' or sliding ring types of joint such as some of those shown in Figs. 10.1–10.10 most nearly meet the requirements for watertightness coupled with rapid assembly outlined above, provided they have adequate slew and draw potentials relative to their individual pipe length.

The unsatisfactory performance of run lead joints under the vibration caused by modern road traffic has already encouraged the use of rubber sealing rings in water and gas mains. The need for this type of joint has been further emphasised by investigations in London[82] and elsewhere[79,85] which have confirmed the seasonal incidence of fractures and pulled rigid or semi-rigid joints in water and gas mains, caused partially or wholly by thermal changes.

Much valuable quantitative work on movement caused by mining subsidence has been recently carried out by the National Coal Board[83] and their conclusions are of great importance. They speak of 'strain' in the sense of draw and indicate that this is the major disturbance, the magnitude of which depends on the depth of the working coal seam below the surface rather than on the nature of the intervening strata, unless the latter is disturbed by a discontinuity such as fault. As previously noted some strains can be positive or negative, i.e. compression or extension, and may reach 10 mm per metre, i.e. ± 1 per cent, or even more, of the effective length. On the other hand slew due to subsidence is very slight (i.e. the radii of curvature of a pipeline during and after subsidence are very large), far less in fact than the movements experienced in any area with differential settlements and shrinkable soils. These findings also indicate the necessity for telescopic joints in pitch fibre, steel, plastics or other flexible pipelines, especially at all anchorages such as valves, branches and bends. Similarly a more generous provision for draw, or shorter individual pipe lengths, in rigid pipelines, than those provided by current practice, may be needed for pipes other than clayware.

There is ample evidence, however, that small diameter pipes, when rigidly connected to manholes or other structures are fractured, by excessive slew, axial bending, or shear, which suggests that close attention is needed to the relationship between slew, draw, pipe length and pipe diameter in flexible joints used under *normal* soil conditions.

10.2 (b) Proposed Minimum Performance Requirements for Draw

As yet no minimum performance requirements for flexible joints have been specified. As a result of the N.C.B. findings, the

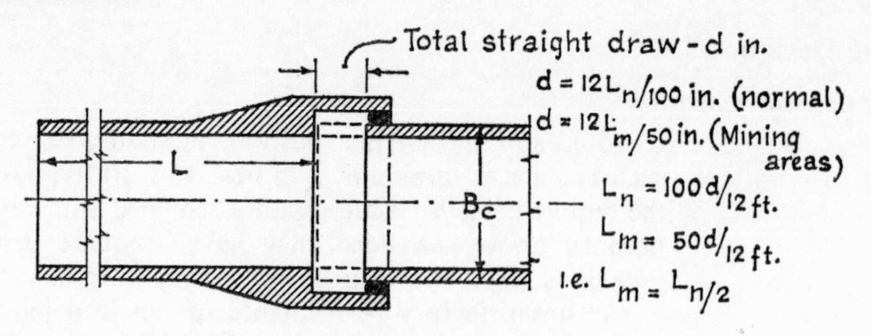

(a) MINIMUM DRAW FOR SOCKETED PIPE JOINT (Rolling ring type).

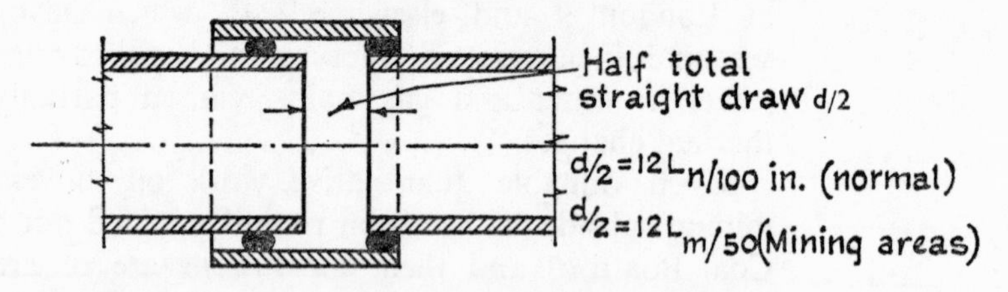

(b) MINIMUM STRAIGHT DRAW FOR DOUBLE-ENDED SLEEVE JOINT (Rolling ring type).

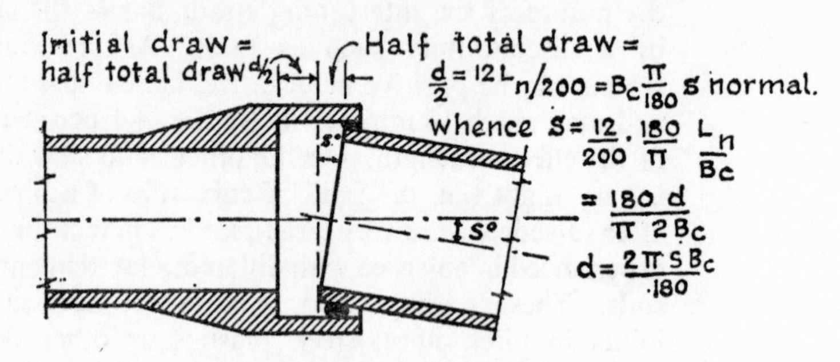

(c) MAXIMUM SLEW FOR SOCKETED PIPE JOINT (Rolling ring type)

Fig. 10.12 (*a*), (*b*), (*c*) Joint geometry

Crown Copyright—by permission

following tentative suggestions regarding draw might be considered:

1. That for use in *normal* conditions a joint should have a minimum draw of $\pm 0{\cdot}5$ per cent of the individual pipe length, i.e. a total working draw of at least 1 per cent of the pipe length.
2. That for use *in mining areas* a joint should have a minimum draw of ± 1 per cent of the individual pipe length,

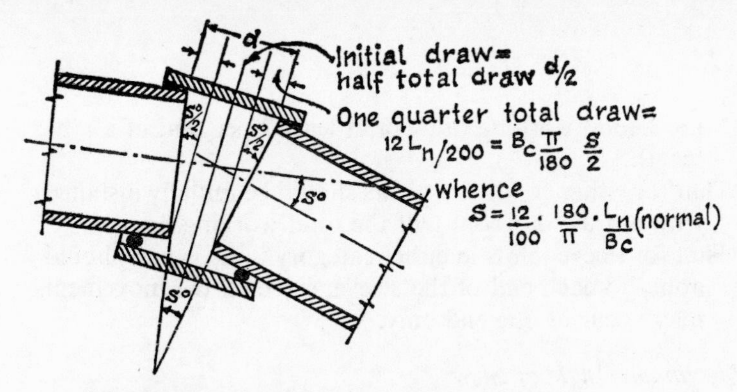

(d) MAXIMUM SLEW FOR DOUBLE ENDED SLEEVE JOINT. (Fixed ring (sliding) type)

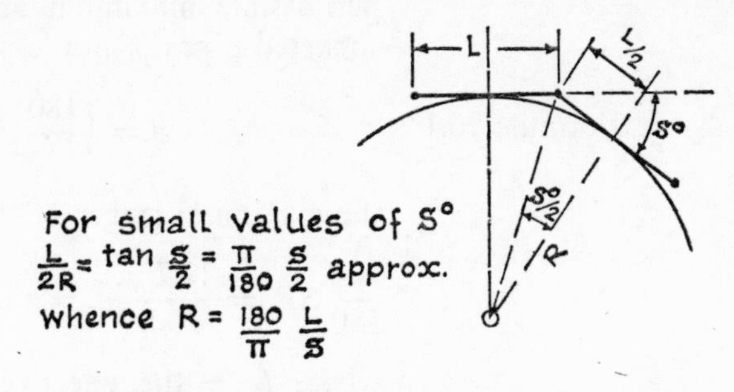

For small values of S°

$\frac{L}{2R} = \tan\frac{S}{2} = \frac{\pi}{180}\frac{S}{2}$ approx.

whence $R = \frac{180}{\pi}\frac{L}{S}$

(e) RADIUS OF CURVATURE OF A PIPELINE WHEN THE JOINTS ARE SLEWED THROUGH THE ANGLE S°

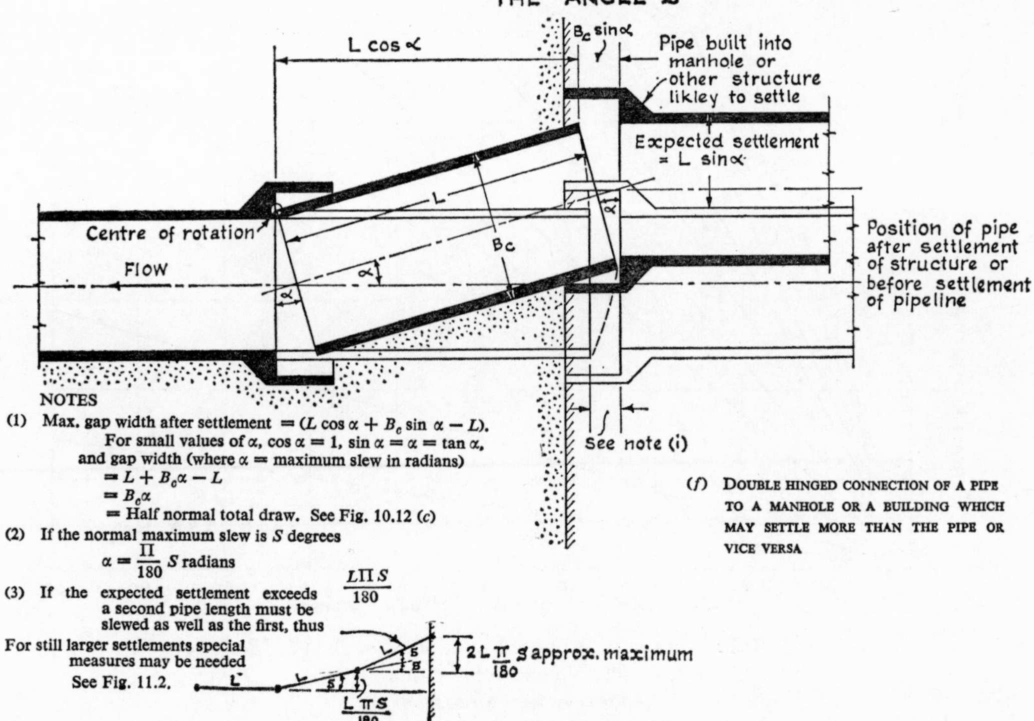

NOTES

(1) Max. gap width after settlement $= (L \cos\alpha + B_c \sin\alpha - L)$.
For small values of α, $\cos\alpha = 1$, $\sin\alpha = \alpha = \tan\alpha$, and gap width (where α = maximum slew in radians)
$= L + B_c\alpha - L$
$= B_c\alpha$
= Half normal total draw. See Fig. 10.12 (c)

(2) If the normal maximum slew is S degrees
$\alpha = \frac{\Pi}{180} S$ radians

(3) If the expected settlement exceeds $\frac{L\Pi S}{180}$ a second pipe length must be slewed as well as the first, thus

For still larger settlements special measures may be needed See Fig. 11.2.

(f) DOUBLE HINGED CONNECTION OF A PIPE TO A MANHOLE OR A BUILDING WHICH MAY SETTLE MORE THAN THE PIPE OR VICE VERSA

Fig. 10.12 (d), (e), (f) Joint geometry

i.e. a total working draw of at least 2 per cent of a pipe length.

3. That for either condition, joints should be initially installed with a draw of about half the total working draw.
4. That for sleeve joints in either category these values should apply to each end of the sleeve, since all the movement may occur at one end only.

10.2 (c) Proposed Formula for Maximum Angle of Slew

The adoption of these suggestions for minimum draw would lead logically from the geometry of the joints (see Fig. 10.12) to a permissible maximum angle of slew *for normal conditions* and socketed pipes of not less than:

Formula 10.1
$$S = \left(\frac{180}{\pi}\right)\left(\frac{12L_n}{200\,B_c}\right) = \frac{3{\cdot}45\,L_n}{B_c}\ \text{degrees}$$

the maximum length of the slew arc is then

$$\frac{\pi}{180}\,SB_c = \frac{12\,L_n}{200}\ \text{in.} = \text{a normal draw of } 0{\cdot}5 \text{ per cent of pipe length}$$

where L_n = the effective length of an individual pipe in feet. B_c = the external diameter of the pipe spigot in inches (see Fig. 10.12). These relationships are illustrated by the graph Fig. 10.13.

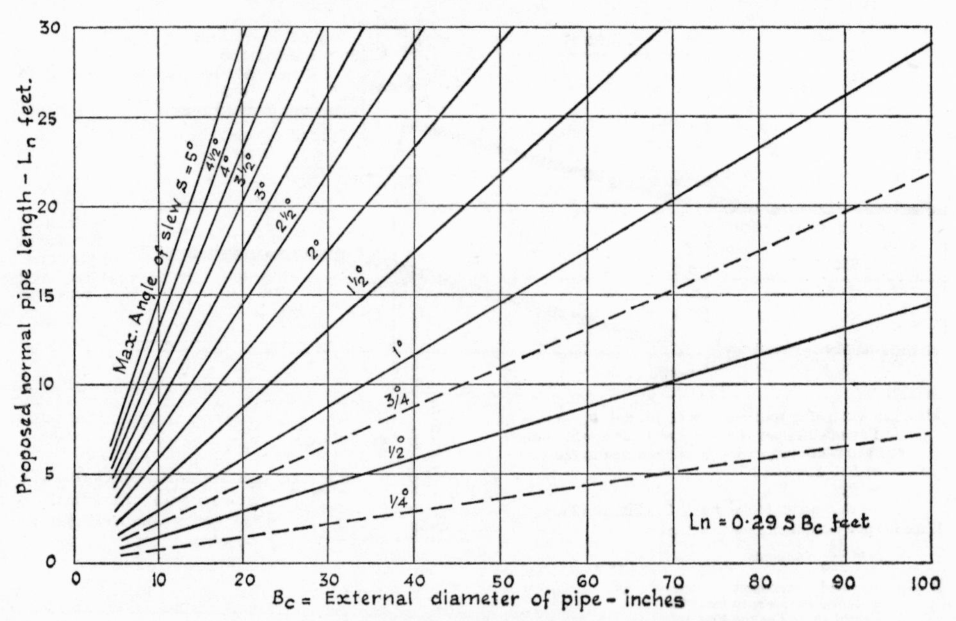

Fig. 10.13 Graph of proposed relation $S^\circ = 3{\cdot}45\ Ln/Bc$ for flexible-telescopic joints

Crown Copyright—by permission

The corresponding formula for sleeve joints which will have twice the angle of slew of socketed joints is:

Formula 10.2
$$S = \left(\frac{180}{\pi}\right)\left(\frac{12\,L_n}{100\,B_c}\right) = \frac{6{\cdot}9\,L_n}{B_c}$$

Slew is thus made to depend on both pipe length and diameter.

If provision is to be made for simultaneous maximum draw and maximum slew, the length of the socket, or half length of the collar must be somewhat greater than 3 × normal draw

$$\text{i.e.} > \frac{36\,L_n}{200}\ \text{in.}$$

Since mining subsidence can be assumed to require little greater slew than normal conditions, sockets or collars designed for normal conditions could be expected to satisfy requirements in mining areas provided that the length of individual pipes is half that appropriate for normal conditions.

For any given value of B_c Formula 10.1 or 10.2 enables either the slew or the pipe length to be chosen as the basis of joint design, from which the other dimensions of the joint, such as depth of socket, length of collar, and width of annular gap follow logically It should be noted that the slew formula (10.1) yields

Equation 10.3
$$\frac{L_n}{S} = \frac{B_c}{3{\cdot}45}$$

which is constant for a given value of B_c.

The minimum radius of curvature (R_n) of a fully slewed socketed pipeline is given by

$$\frac{L_n}{R_n} = \tan S = \frac{\pi}{180}\,S \text{ approx (see Fig. 10.12)}$$

Equation 10.4
$$\text{whence } R_n = \left(\frac{180}{\pi}\right)\frac{L_n}{S} = \frac{200\,B_c}{12} \text{ from Equation 10.1,}$$

which is also constant for a given value of B_c.

Then in mining areas, if *normal* socket values are used and L_n is halved, R is also halved, i.e. $R_m = 100B_c/12$. If collars are used, and the total slew is twice that for sockets, the value of R_n or R_m would theoretically be halved. This value is of interest in connection with the adverse conditions envisaged in Fig. 11.2.

10.2 (*d*) *Design of Rubber 'O' Rings*

Evidently both the initial contact pressure and the radial load imposed by the ring on the pipe socket or collar will depend on the fluid test pressure to be resisted, the diameter and elastic modulus (or hardness) and stress relaxation of the ring, the degree and variation of the compression imposed on it by the gap dimensions of the joint, the relative smoothness of the contact surfaces, and on whether or not the rubber is restrained laterally.

The rigorous theoretical estimate of this load and pressure is difficult and probably unnecessarily tedious in view of the relatively wide variation in practical conditions. A much simpler empirical estimate can be made as follows, however, and it may be quite adequate for the design of *rolling* 'O' rings in the absence of more exact analysis. It is not applicable to 'O' rings restrained in a groove or to rings of other shapes.

Essential principles of rolling 'O' ring design. The essential principles to be observed in the selection of the size and hardness of a rolling rubber ring are:

1. That the effective long-term contact pressure between the ring and the pipe surfaces must never be less than the maximum fluid pressure to be resisted.
2. That the frictional resistance of the ring to sliding on the pipe surfaces must be greater than the force exerted by the maximum fluid pressure on the exposed peripheral projected area of the ring.

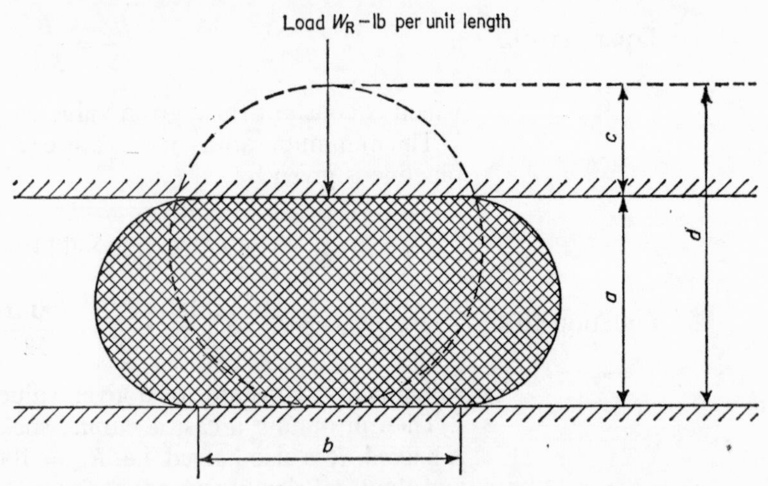

Fig. 10.14 Dimensions of cross section of laterally unrestrained rubber 'O' ring compressed between parallel surfaces

Theory. In a Cornelius type joint in which the rubber sealing ring is unrestrained laterally, as in Fig. 10.14 let:

a be the nominal width of the gap between spigot and socket—in.

b be the width of the contact area of the compressed ring on the pipe surfaces—in.
c be the reduction in ring diameter after compression—in.
d be the uncompressed diameter of the ring section—in.
D be the internal diameter of the socket or collar—in.
E be the initial Young's modulus of the rubber—lb/in.^{2}
F_r be the factor to allow for stress relaxation of the rubber, $= 100/(100 - r)$—where r is the per cent loss of stress with time.[80]
F_c be the compression variation factor to cover practical variations in 'a' and surface irregularities in the pipes.
h be the hardness of the rubber in B.S. degrees.
p be the maximum fluid test pressure to be resisted—lb/in.^{2}
p' be the required initial mean contact pressure—lb/in.^{2}
W_R be the initial radial load per unit length of circumference imposed on the socket, or collar, by the compression of the ring—lb/in.

By testing a number of rings of various sectional diameter and hardness between flat rigid plates and so obtaining the empirical relationships between W_R, c, d, and E for each specimen, and computing b, Lindley[80a] recently found that when the dimensionless values of W_R/Ed are plotted against c/d and b/d respectively, all the values lie within narrow limits, on master curves, as reproduced in Chart C14 (thus, as regards W_R/Ed confirming previous approximate observations at B.R.S.), also that as there is no change in the rubber volume after compression,

Equation 10.5 $$b = 2{\cdot}4\,c \text{ in. approx.},$$

and that the relation between 'h' and 'E' is:

$h =$	30	35	40	45	50	55	60	65	70	75	BS, IRH, or Shore A degrees
$E =$	130	168	213	256	310	460	630	830	1040	1340	lb/in.^{2}

This relationship is plotted in Chart C13.

Then to satisfy the *first* principle, i.e. to resist a maximum long term fluid test pressure of p lb/in.^{2}, the initial mean contact pressure must incorporate the factors F_r and F_c,

Equation 10.6 $$\text{then } p' = F_r F_c p \text{ lb/in.}^2$$

The mean radial force initially exerted on the socket, or collar, is then

Equation 10.7 $W_R = bp' = bF_rF_cp$ lb/in. of pipe circumference. This force is subject to local variations plus or minus.

The tangential tension imposed on the socket, or collar, is

Equation 10.8 $$T = W_R D/2 \text{ lb.}$$

To satisfy the *second* principle after relaxation of the contact pressure, if μ_r is the coefficient of static friction of rubber on the surfaces of the pipes,[80]

$$2bF_c p\mu_r > ap$$

whence

Equation 10.9 $$\mu_{r\,min} = a/2bF_c = a/4{\cdot}8\,cF_c \text{ (see Equation (10.5))}$$

Design

Determination of c, d, h and $\mu_{r\,min}$. It is convenient to assume a value of c/d as the basis of design, and to assess reasonable values of F_r and F_c depending upon the rubber hardness and the particular quality of the pipes respectively.

Thus assuming a compression ratio $c/d = x$ and a gap width a, $(d - c) = a$, $d = a/(1 - x)$, and $c = ax/(1 - x)$.

From Equation 10.5, $b = 2{\cdot}4c = 2{\cdot}4ax/(1 - x)$.

From Chart C14, for $c/d = x$, obtain $W_R/Ed = y$.

Then $W_R = Eay/(1 - x)$.

But from Equation 10.7, $W_R = bp' = 2{\cdot}4axp'/(1 - x)$.

Then $E = W_R(1 - x)/ay = 2{\cdot}4axp'(1 - x)ay = \underline{2{\cdot}4xp'/y.}$

From Equation 10.9, $\mu_{r\,min} = a/4{\cdot}8cF_c = a(1 - x)/4{\cdot}8axF_c = \underline{(1 - x)/4{\cdot}8xF_c}$.

Thus as x varies the other quantities have the following values:

x	d/a	c/a	y	E/p'	$\mu_r F_c$	W_R/ap'
0·25*	1·33	0·33	0·17	3·52	0·625	0·8
0·3	1·43	0·43	0·24	3·00	0·486	1·03
0·4	1·66	0·66	0·52	1·85	0·312	1·60
0·5	2·00	1·00	1·22	0·99	0·208	2·40

Example. If $p = 30$ lb/in.², $F_c = 4{\cdot}0$, and $F_r = 2{\cdot}0$,
From Equation 10.6, $p' = (4)\ (2)(30) = 240$ lb/in.² and the essential quantities can be evaluated as follows:

c/d	E	$h°$††	d/a	c/a	$\mu_{r\,min}$†	W_R/a	T/aD	Ring Section
0·25*	845	70	1·33	0·33	0·156	192	96	Small and hard
0·3	720	65	1·43	0·43	0·121	248	124	Larger and softer
0·4	445	55	1·66	0·66	0·078	384	192	Larger and softer
0·5	238	45	2·00	1·00	0·052	576	288	Large and soft

* Minimum value specified by ASTM C443 - 62T

† For most pipe materials in common use, μ_r is not likely to have an actual value less than about 0·7.

†† To the nearest higher value of E.

Section F

Sitework and Maintenance of Pipelines—Failures, etc.

Chapter 11
Sitework and Maintenance

11.1 Construction in Trenches

11.1 (a) Introduction

The modern methods of underground pipe design described above demand complementary construction methods to ensure that the assumptions regarding loading and pipe strength made in design are realised as far as possible in practice. It is the purpose of this Chapter to indicate the methods of construction that contribute essentially to the structural strength and efficiency of a pipeline. It is also to be expected that careful attention to site operations during construction will reduce maintenance costs considerably and extend the useful life of a pipeline.

Much valuable guidance on the excavation, timbering, dewatering and back-filling of trenches in various kinds of soil and in rock, and on the construction of embankments, is given in the B.S. Code of Practice, 2003:1959, 'Earthworks',[88] to which reference should be made. Methods specially applicable to pipe laying, and their effect on the integrity and load-carrying capacity of rigid and flexible pipes, based on the investigation of many failures in the field, are described here.

Proper design of a pipeline is not in itself sufficient to ensure the survival of the line as an efficient watertight structure, because field experience has shown that the loads imposed on the pipes, the load-carrying capacity of the pipes, and the elimination of destructive secondary loads and forces all depend very greatly on the site operations. Inspection of all stages of the work, under the direction of a competent engineer, and reference of any matters of doubt to the designer are therefore always desirable and frequently essential.

For technical as well as economic reasons, the more quickly each stage of the work can be completed, consistent with sound workmanship, the better. Design, construction methods, and site organisation should, therefore, all be directed towards the avoidance of delays during construction.

The various methods of installing pipelines in trenches or headings and under embankments, and for which the primary external loading can be computed, are shown diagrammatically in Fig. 3.1.

11.1 (b) Trench Width

The permanent load on a pipe in a 'narrow' trench depends on the overall width of the trench as measured at the top of the pipe between the undisturbed trench faces. This 'effective' width (see Fig. 4.2) should, therefore, be kept as small as practicable and should never exceed the value specified by the designer, measured between soil faces, any side timbering being inside the limit, without his consent.

There is no such width limit in a 'wide' trench but the distinction between 'narrow' and 'wide' conditions rests with the designer. Above the top of the pipe the width may be increased as necessary or convenient (see Fig. 4.2) without adverse effect on the loading. It is therefore necessary to ensure that the dig lines as set out on the surface will permit the required effective trench width to be obtained after allowing, where necessary, for timbering or piling thicknesses, insets, and bottom berms, or, in 'V' trenches, for safe slopes and bottom berms.

11.1 (*c*) *Existing Underground Works and Services*

The location of existing underground pipes and cables within the limits of the proposed excavations must be established by preliminary investigations (e.g. by trial holes) and consultations with public utility authorities, and those which must remain in position during the excavation of the new work must be adequately supported by slings, or otherwise, before excavating below them. Adequate side support should be provided for existing pipes lying closely outside but more or less parallel to the new trench and above the proposed trench bottom; otherwise rigid pipes may have their bedding disturbed, and so weakened, and flexible pipes may suffer excessive deformation or collapse, depending on their depth of cover, when the trench is excavated.

Particular care should be taken to avoid deforming or displacing rigidly jointed rigid pipelines (e.g. clayware or concrete), since very small movements may crack them. Supporting timber or steel beams should be of ample section and length and should be carried on firm bearings that are unlikely to sink, placed well back from the edge of the trench. Chain slings of ample strength, evenly tensioned by means of folding wedges or turnbuckles, should be placed at frequent intervals along the pipeline, using strips of wooden packing to prevent point or line contacts between slings and pipes. Ropes are liable to stretch and so are unsuitable as rigid pipe slings. Clayware and concrete pipes that are exposed should also be protected from direct sunshine and cold winds since uneven heating or cooling is likely to crack them.

For the shoring of walls, underpinning of foundations and similar operations, reference may be made to the Civil Engineering Code of Practice No. 4 'Foundations'.[88]

*11.1 (d) Excavation and Timbering**

Methods of excavation and timbering naturally vary widely with local conditions and are dealt with in some detail in C.P. 2003 'Earthworks'[88] If road material is to be re-used, care should be taken to avoid mixing it with earthy spoil which would render it unsuitable for reinstatement.

In the excavation of the shallower portions of a road trench, special care should be taken to avoid damage to existing pipelines and cables, and consequent danger to personnel, especially as regards gas and water mains and electric cables. Not infrequently their position departs from the 'records', on which too much reliance should not be placed. Any leaks or fractures discovered, or any accidental damage which does occur, should be reported immediately to the authority concerned.

Any unusual article, unexpected pipes, or antiquities encountered should be treated with care and respect, and appropriate advice should be obtained before disturbing them in any way.

The resistance of a flexible pipe to vertical loading depends upon effective lateral support and that of a rigid pipe is increased thereby. Therefore, if timbering below the top of the new pipeline is necessary, it should be so arranged that it can be withdrawn as the bedding material alongside the pipes is placed. Walings should be positioned with this object in view and so as not to interfere with the various operations which have to be carried out below the pipe top level. Normally there should be no waling within the overall height of the pipes including their bedding, and the sheeting must be strong enough and its penetration below the trench bottom deep enough to provide safely for this condition. Pipes exceeding say 60 in. diameter may be strutted across the horizontal diameter, with walings inside as well as outside the pipes. This operation requires care and experience to avoid deforming or cracking the pipes. When timber must be left in place above the permanent water table, it should preferably be of rot-resisting hardwood or, if of softwood, be properly impregnated with creosote or other suitable wood preservative.

If, in bad ground, walings and struts are required beneath the pipes it will not be possible to withdraw the struts. Such struts may be plain timber below permanent water level, or treated timber or reinforced concrete above; they should be sunk well below the trench bottom so that they do not constitute hard spots in the foundation, and should preferably be so spaced as to lie approximately beneath a joint in the pipeline. Side timbers may be left in

* The word 'timbering' is used to mean any temporary support to excavations. The material used may be timber, steel or reinforced or prestressed concrete, or any combination thereof.

place below permanent water level if necessary. If timber or sheeting is withdrawn after the fill is placed, the load on the pipes will be increased considerably in 'narrow' trench conditions and the lateral support to the pipe and to Class B bedding will be impaired. The subsequent soil movement may cause uneven loading of the pipe, with consequent excessive displacement, deformation, or fracture. When such withdrawal is unavoidable, Class A_{RC} bedding (as Chart C9) (or a reinforced concrete surround) is to be preferred. Above the pipe top, struts should normally be so placed and spaced that the pipes can be lowered into the trench without disturbing them.

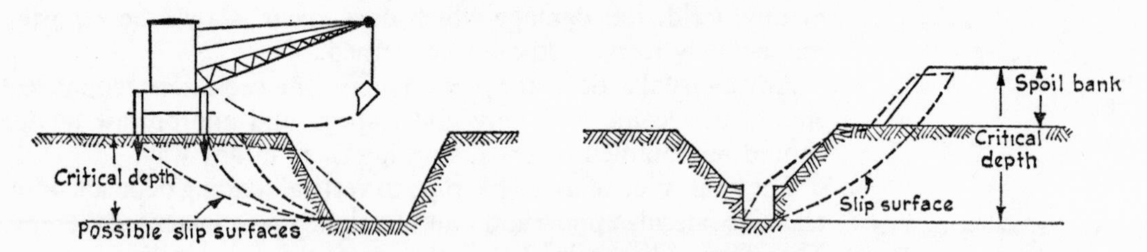

Fig. 11.1 Typical slip failures in clay soils
Crown Copyright—by permission

In untimbered 'V' trenches, slips of the type shown in Fig. 11.1 are likely to occur in soft clayey soils if the depth of the trench exceeds about 12 ft. Similar failures may occur at even shallower depths if the spoil bank is too near the edge of the trench, or in waterlogged silts, sands and clays. Whenever there is any doubt as to the stability of the trench sides the advice of a soil specialist should be obtained before proceeding with the excavation. The loads imposed by excavating machines and mobile cranes and their safe distance from the edge of the trench also require consideration in this respect.

11.1 (e) Temporary Disposal of Spoil

Spoil heaps in roads should not be so placed, or be so high, as to endanger existing structures or underground works, or the stability of the trench itself (or to interfere unnecessarily with traffic, or with traffic vision). In built-up areas it may sometimes be feasible to place spoil in side streets, but spoil which is not to be returned to the trench quickly, as fill, should be removed from the site without delay. The load imposed by spoil on a roadway should not normally be permitted to exceed 500 lb/ft^2. The loading of existing shallow pipelines, and the obstruction of surface water drainage or its diversion into the trench by the spoil, should be avoided.

To avoid excess lateral pressure, or slips, in many soils, and especially in clays, the toe of any spoil heap which is not to be immediately carted away should preferably not be nearer to the edge of a timbered trench than say the depth of the trench. Under difficult soil conditions, a soil specialist should be consulted regarding the stability of the soil in the trench walls, and the disposition and strength of the timbering.

Spoil containing clay which is to be used as fill should be protected against excessive wetting or drying. If it has been allowed to become appreciably wetter or drier than when excavated it should not be used as fill material in road trenches.

11.1 (f) Bottoming

The effective life, and overall cost, of a pipeline, whether rigid or flexible, depend very largely on the uniformity of its support. It is therefore necessary to eliminate from the foundation all hard spots which would cause hogging and local overloading of the pipes, and soft zones which would cause sagging and differential settlement (Plate 12). If either condition is uncorrected the pipes are likely to be subjected to loading as a beam; rigid pipes may then be fractured transversely or in shear, and flexible pipes may suffer serious deformation. For this reason beddings of Classes C and D (Chart C9) are unsuitable for non-uniform soils, in which Class B beddings are to be preferred even for metal pipes. In fine-grained soils (e.g. clay, silts, fine sands or mixtures of these), good drainage of the trench bottom is necessary and should be ensured, and ponding and flooding avoided at all stages of excavation. Artesian pressures should be investigated, and relieved if present, and, in sands or gravels, the water table should be lowered as necessary to prevent boils or blows. For advice on the dewatering of trenches see B.S. Code of Practice 2003. If sheeting is used as a water cut-off, adequate penetration below formation level should be ensured (see Civil Engineering Code of Practice No. 4, 'Foundations', Item 4.3306).[88]

Hand trimming of the trench bottom may be largely avoided by over-digging to a suitable depth by machine and replacing the excess depth with suitably selected free-draining bedding material (see Article 7.4 (*d*) (ii) and notes to Chart C9), carefully and evenly placed and compacted and finished to the required level and gradient. This procedure also ensures a more uniform foundation by eliminating large stones, boulders tree roots or other hard spots, and provides a clean, well drained bottom on which to lay the pipes. The excess depth will depend on the nature of the soil, the class of bedding to be used, and drainage conditions. In the absence of excess depth as above, extra depth should be provided under pipe sockets or collars where necessary to maintain the required minimum bedding thickness for Class B bedding.

Whether excavating by machine or by hand, it is essential to prevent disturbance and softening of wet fine-grained soils such as clays, silts and fine sands in the trench bottom; this is achieved, first, by proper drainage, and then by placing a blanket of an adequate thickness of coarse sand and gravel or other approved granular bedding material (see notes to Chart C9) on the virgin formation as soon as it is uncovered. All traffic on the trench bottom should be prohibited until the blanket is in place. In good ground this blanket may form part of the bedding, but in very soft ground it should be additional to the bedding. Extra soft spots should be hardened to the general condition of the bottom by tamping in granular material before continuing with the blanket.

Peat or boggy material below formation level should be removed whenever possible and replaced by sand or other approved stable filling material, unless it is overlain by a good bearing stratum thick enough to resist appreciable deformation.

In variable soils with alternating hard and soft zones, after testing with a bar or other instrument, the soft zones should be hardened as described above and the hard zones should be generously over-dug and the level restored with granular fill.

In mixed soils containing large boulders, stones or flints, care is needed to ensure that any such hard object which is near enough to the bottom to form a hard spot is removed and replaced by granular material. Likewise, rock bands occurring in the foundation should be generously overcut and replaced with bedding material.

In uniform coarse sands or gravels where it is intended to provide a Class B bed, whether or not the excavated material is suitable as bedding material, the bottom should be over-dug to the required depth of the underbed. If this is not done, and the pipes are laid on the undisturbed bottom, it is highly probable that the upper bedding (Stage II) will be compacted less than the bottom; the bedding may then be reduced to Class D and the pipes overloaded and broken after backfilling.

The positions of sharp transition from very hard to soft bottom should be clearly marked, so that whenever possible at least two flexible joints in the pipeline may be placed later at such junctions to form a double hinge and prevent excessive axial bending stresses in rigid pipes (cf. Fig. 11.2).

Most clays will take up water and expand (i.e. rise) when offloaded and sink again after the backfill is placed. Measurements of the rise must therefore be made and the sinking allowed for to avoid backfalls, especially in variable soils or with small gradients.

If the bottom is in clay overlying water-bearing sand or gravel, the water pressure under the clay should be measured and, if it is greater than the reduced overburden pressure, relieved by vertical sand drains; otherwise the trench bottom may heave and then sink excessively even if there is no actual 'blow'.[102] In difficult conditions of this kind it is advisable to consult a soil specialist at an early stage regarding the precautions needed.

11.1 (g) Loss of Ground

In fine sands or silts any flow of groundwater along the trench either during construction or after backfilling may impair the load-carrying capacity of the pipes by eroding the foundation or the bedding. Where groundwater is likely to make copiously, a sub-drain delivering to pump sumps clear of the trench, or to natural outlets, should be provided along one or both sides of the trench (see Chart C9) and the bottom should be sloped towards the drain(s). Where the flow of the groundwater is likely to transport fine soil particles, despite the sand filter provided by the granular bedding material, water stops of puddled clay extending up through the bedding and side fill should be placed across the trench at intervals along the line, say immediately down-stream of pump sumps. When no longer needed, temporary drains should be sealed permanently by grouting and sumps should be re-filled with well-compacted selected material as for the bedding.

The removal of groundwater or fine soil particles may well have adverse effects on existing structures in the vicinity of the trench, and special protective measures may be necessary, on which expert advice should be obtained.

11.1 (h) Construction of Bedding—STAGE I

The load-carrying capacity of a rigid pipe depends partly on the inherent crushing strength of the pipe itself and even more so on the manner in which it is supported by its bedding. Inadequacy in either component is likely to result in the failure of the pipe when loaded (see Plates 2 and 13). The careful selection, placing and compaction of the bed is therefore of primary importance. In view of the obvious difficulty in many soils of obtaining satisfactory uniformity of support with Classes C and D beddings, in general Class B bedding is to be preferred, for both rigid and flexible pipes, with the modifications noted below. Concrete beddings or surrounds are not normally suitable for flexible pipes.

11.1 (h) (i) Class B Granular Bedding (Chart C9)

Where an underbed or protective blanket has already been laid, present limited experience suggests that selected granular bedding material, as described in the notes to Charts C9 and C10 (see also Article 7.4 (*d*) (ii) and Refs. 100, 100*a* and 103) should be evenly spread and carefully compacted up to the underside of the pipe barrel and the surface trued to the required gradient. The depth of the granular bed, exclusive of the underbed or blanket, should not be less than the values given in those notes. Where an underbed or blanket has *not* been laid but the bottom has been undercut and replaced with granular material, the same procedure should be followed.

The essential requirements for the bedding material are that it should be uniform, free-draining and readily compactible, free from large or very fine particles and generally of such particle size, and so disposed, as to prevent penetration of the bedding by wet, fine-grained, foundation soils.[103]

Site experience may suggest modifications in detail to the suggestions made in the notes to Charts C9 and C10, e.g. as regards maximum size of particle see Article 7.4 (*d*) (ii), but the principles governing the design of these beddings are well established and must be observed in order to ensure that the Bedding Factor assumed in the design is obtained in practice, and that the sinking of the pipes into the bedding is minimised.

11.1 (*h*) (ii) *Class A Concrete Beddings or Site Concrete Surrounds* (Chart C9 and Figs. 11.3 and 11.5)

The initial procedure should be as above, but, where necessary, the granular underbed or natural bottom should be finished with a screed of weak concrete 2 in. or more thick, laid to the correct gradient over the full trench width, to act as a blinding course (Chart C9 (*c*) (ii)) and give a good clean, dry surface on which to lay the pipes and place the site concrete. The top of the screed should be at least the specified bedding or surround thickness below the finished underside of the pipe barrel and clear of the lowest point of the pipe sockets.

11.1 (*i*) *Handling, Laying and Jointing Pipes*

11.1 (*i*) (i) *General Notes*

The jointing of rigid pipes with rigid materials, such as cement mortar or other hard and brittle compositions, is a fruitful cause of fracture of the pipes due either to the thermal and moisture movement of the jointing material being greater than that of the pipe material (see Plate 6), or to the pipeline when so jointed being incapable of even slight axial deformation without cracking.

Flexible telescopic joints, on the other hand, permit the pipes to move slightly under the influence of differential settlements, soil subsidence or other soil movements, and of thermal or moisture changes in the pipes themselves, and so prevent the development of some of the secondary loads and forces that might otherwise fracture the pipes (see Chapter 9). However, if the pipes are not uniformly supported throughout the length of their barrels, or if they rest on their sockets or spigots, or bear on local packing or hard spots in the foundation, they will be subjected to local concentrations of reaction which may be high enough to cause extensive fractures in rigid pipes and dangerous flattening of flexible pipes, irrespective of the type of joint used. Such irregularities may also cause leakage in rubber ring type flexible joints.

Plain concrete pipes are sometimes cracked before laying, by exposure to uneven peripheral heating or cooling, uneven wetting or drying, or combinations of these conditions. Concrete and clayware pipes may be similarly cracked by exposure to non-uniform peripheral heating or cooling after laying (e.g. heating by the sun (see Plate 9) or cooling by cold winds or night chill) (see Plate 10). Plain concrete pipes should not be left lying unprotected alongside the trench for long periods before laying, and they should preferably be stacked on end and shaded from the sun, whether in stockyards, on site, or elsewhere.

The methods of handling on site should be such as not to damage the pipes or any protective coatings. Rigid pipes should never be dropped. Coated pipes should not be lifted with chain or rope slings without protective soft wood packing; wide strong canvas slings or pipe hooks are generally to be preferred. If lifting holes are provided in the pipes they should be used, making sure that the beam used inside the pipe is strong enough to carry the weight of the pipe safely. Ovally reinforced pipes must always be kept, and lifted *with slings* the right way up, i.e. with the marked 'TOP' at the top; lifting such large pipes with hooks, or even rolling, may crack them.

Before starting to lay any pipes it is advisable to ensure that the correct strength-class of pipe and size of joint ring, if any, are used, and that cracked pipes or pipes with badly chipped spigots or rough sockets are not accepted into the trench. The joints at each end of the first, or last, pipe connected to a manhole or other structure should always be flexible, and the length of these pipes should not usually exceed about 2 to 3 ft, especially in the smaller sizes Figs. 10.12 and 11.2. Where large differential settlement of the structure relative to the pipeline is possible, small-bore pipes should preferably be provided with a protective casing (see Fig. 11.2 *h–k*).

11.1 (*i*) (ii) *Flexible Joints*

The maker's instructions for the making of flexible joints *and for the storage of rubber rings on site* should be followed carefully,

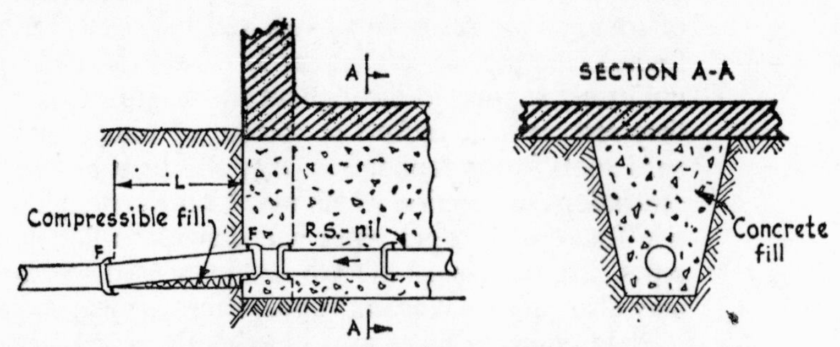

(a) Shallow pipe under monolithic floor - Probable R.S. small.

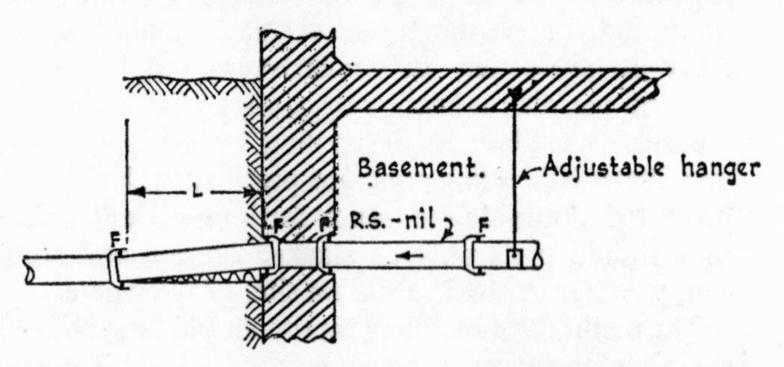

(b) Shallow pipe through basement wall - Probable R.S. small.

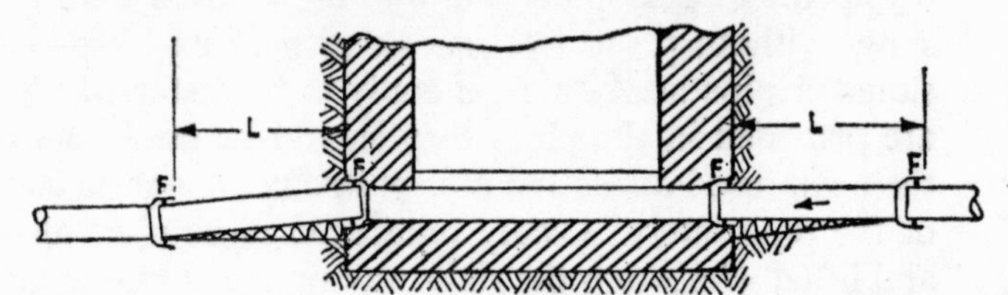

(c) Connection to M.H in fields or gardens - Probable R.S. small.

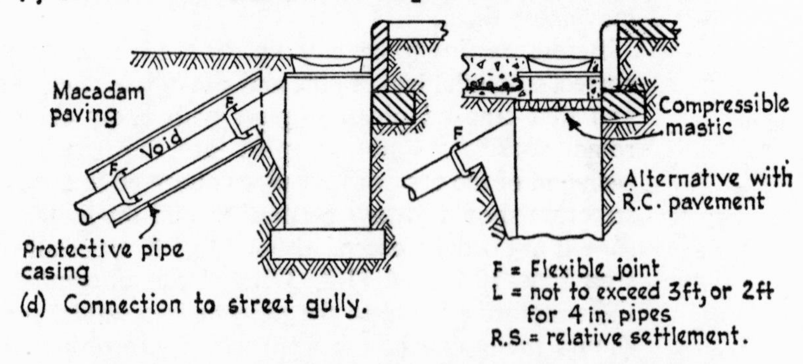

(d) Connection to street gully.

F = Flexible joint
L = not to exceed 3ft, or 2ft for 4 in. pipes
R.S. = relative settlement.

Fig. 11.2 (*a–d*) Methods of alleviating the effects of differential settlement on pipes passing through or under walls or other structures

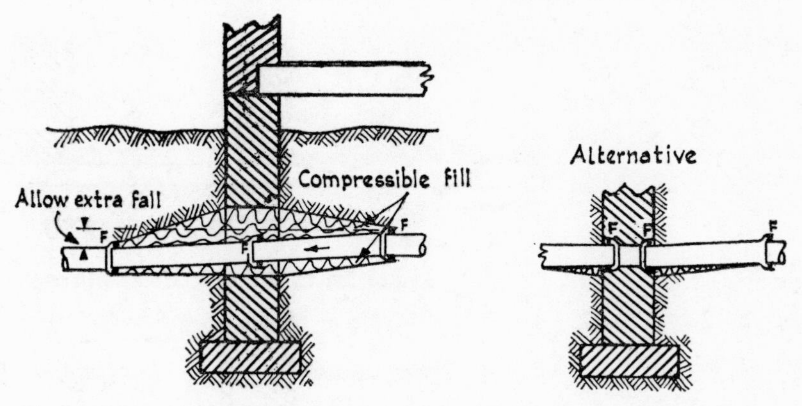

(e) Shallow pipe under suspended or free slab floor - Probable R.S small

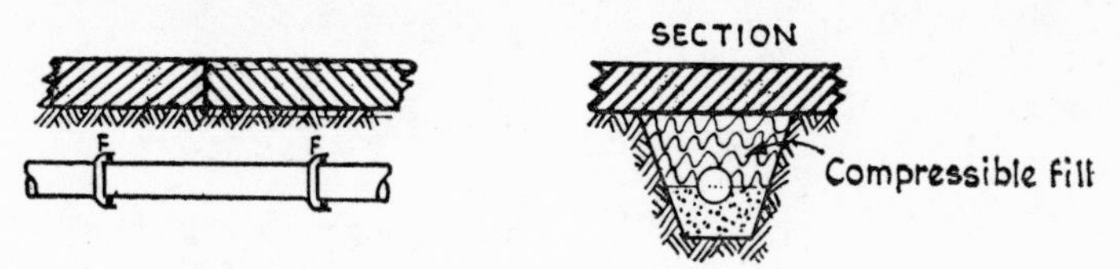

(f) Shallow pipe under floor bays liable to small relative movement.

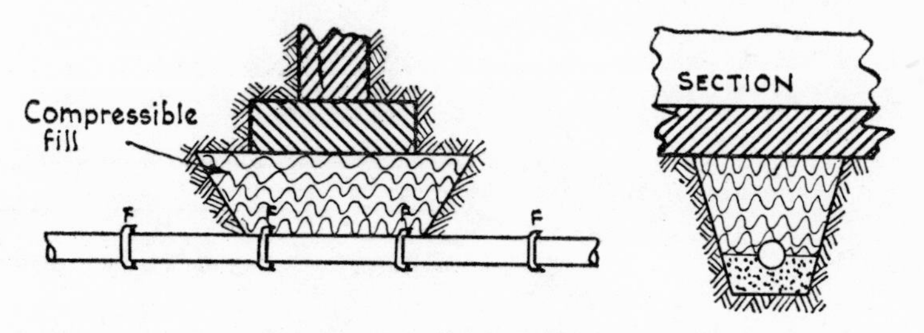

(g) Pipe under wall footing - Probable. R.S. small.

Fig. 11.2 (*e–g*) Methods of alleviating the effects of differential settlement on pipes passing through or under walls or other structures

and precautions should be taken to avoid mixing up natural and synthetic rubber sealing rings or the rings supplied by different pipemakers. All parts of the joint must be clean and, for rolling ring type joints, must also be dry, otherwise the ring may slip out of position when being pulled in. Rings must be correctly placed initially and 'O' rings must not be twisted. Rings of one size must not be stretched for use with a larger size of pipe. Difficult or faulty rings should be rejected. When using a crane, or pipe legs, the pipe should remain suspended until the joint is completed. With most types of joint it is necessary that the pipe

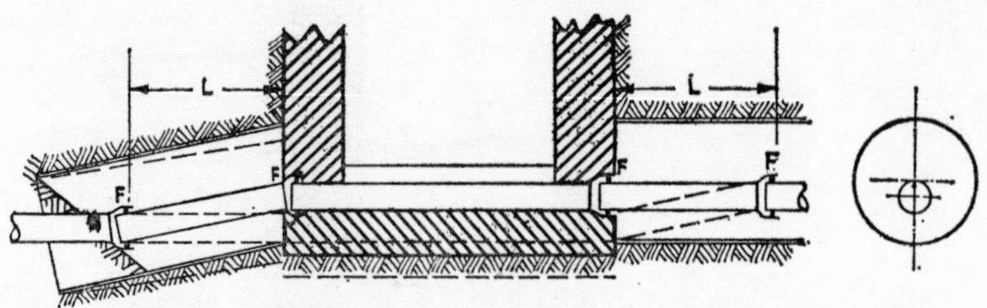

(h) Connection to M.H. in roads or where the probable R.S is large.

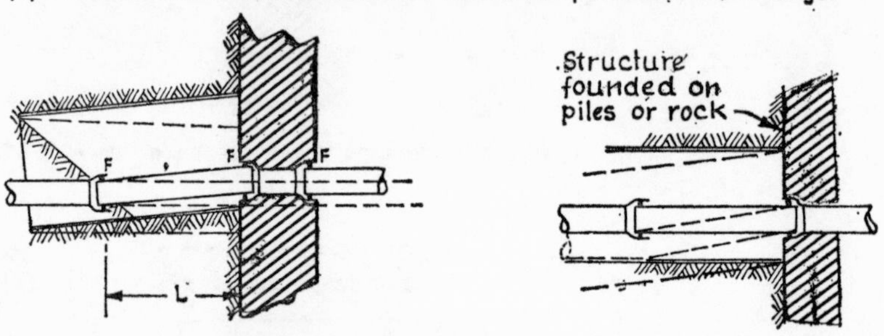

(i) Pipe through basement wall - large probable R.S of structure (left) or of pipeline (right)

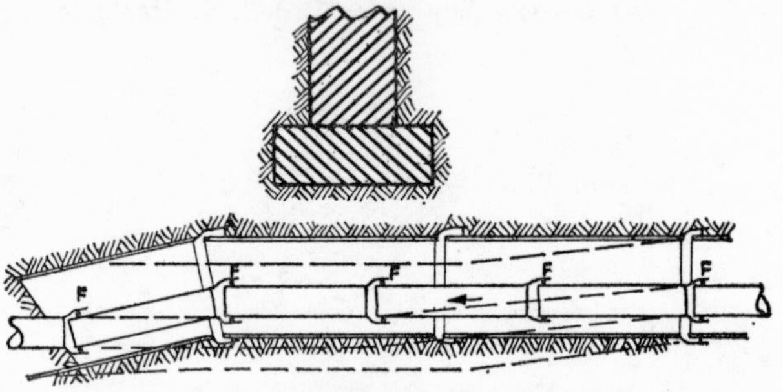

(j) Pipe under wall footing - large probable R.S. of structure

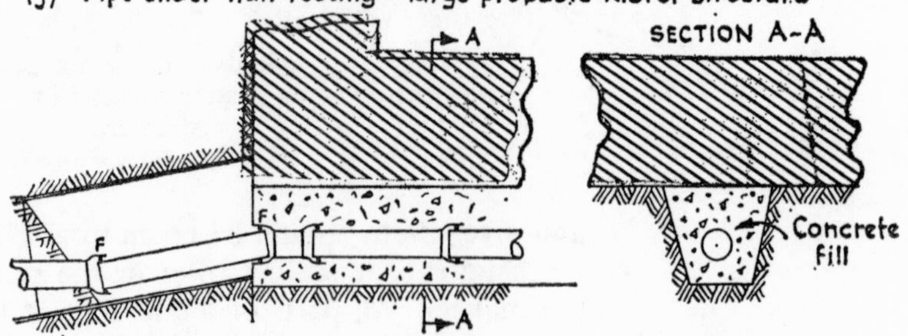

(k) Pipe under monolithic foundation slab - large probable R.S of structure

Fig. 11.2 (*h–k*) Methods of alleviating the effects of differential settlement on pipes passing through or under walls or other structures

being laid should have its axis in line with that of the pipe to which it is being joined during the pulling in of the joint. Similarly, in the smaller sizes of pipe, which are manhandled, some joints require the new pipe to be pushed or pulled straight into the socket or collar and not wriggled sideways or up and down. Other types of joint, in which the sealing ring is positively located and slides, are not so sensitive in this respect but may need lubricating; only the lubricant recommended by the pipemaker should be used and it should be applied to the contact surface over which the ring slides and not to the ring. A gap should be left between the end of the spigot and the back of the adjoining socket (or spigot where collars are used) to enable individual pipes to expand or the line to shorten as well as lengthen (see Fig. 10.12 and 11.3). The width of this gap will depend in part on the temperature at the time of laying. In mining areas it should be about half the total safe draw permitted by the joint when the temperature eventually reaches about 50°F. The correct placing of the joint rings should be checked where necessary by gauging at four or more equally spaced points around the periphery, depending on the type and size of joint, and the width of each joint gap should be checked with a suitable gauge. It is easier and less costly to rectify a displaced ring or a faulty gap at this stage than after testing.

For the smaller sizes of pipe with flexible joints it is possible, when necessary to make small adjustments to both the line and gradient after a section of the line has been jointed, always provided that the bedding is properly re-made where disturbed.

11.1 (*i*) (iii) *Mortar Joints*

Cement mortar jointing produces a rigid pipeline, and is therefore not generally recommended. If it cannot be avoided, and particularly for clayware sockets, the mix should never be stronger than 1 cement and 3 sand. If individual mortar joints are required on an otherwise flexibly jointed pipeline (e.g. at the junction of pipes of different materials, or in repair work), the length of the pipe so jointed should be kept as short as possible, and its other end should be flexibly jointed. Mastic jointing (hot or cold) is to be preferred however, where rubber sealing rings cannot be used.

11.1 (*i*) (iv) *Laying on Class B Granular Bedding*

The bed material should be scraped away at each socket so that the socket does not bear on the bed; care should be taken to ensure that every pipe is in even contact with the bed throughout the length of its barrel, and that the bed is not disturbed whilst positioning and jointing the pipes. If the bed has been properly graded, and has not been disturbed, boning may not be necessary. If for any reason the bed is high or low at any point, it should be carefully re-made to the correct level before placing the pipes, because high or low spots, or casual lowering or raising may result in a broken rigid or excessively deformed flexible pipe after back-filling. Hard packings such as bricks, timber or concrete blocks should *never* be left under the pipes with this (or classes C or D) bedding, as they would produce dangerous hard spots. For the same reason, and to avoid uneven compression of rubber sealing rings and consequent bad joints, spigots should not bear on socket inverts or socket crowns bear on spigots. If they tend to do so the bedding should be so adjusted as to avoid either condition and still retain uniform support of the pipe barrel.

11.1 (i) (v) *Laying on Class A Concrete Bedding or with Site Concrete Surround*

The procedure should be similar to that described above but the pipes should be supported at the correct height above the top of the blinding course, e.g. by means of blocks or short precast concrete cradles placed behind the sockets for small pipes, or on

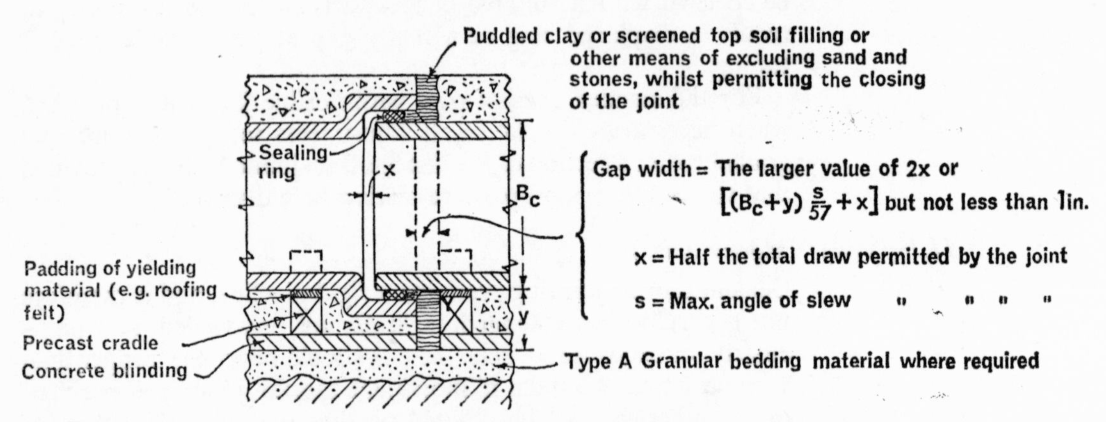

Fig. 11.3 Joint gap in concrete or R.C. bedding or surround for a flexible telescopic joint
Crown Copyright—by permission

each side of the joint for large, long, or heavy pipes. To avoid their becoming hard spots these supports should be provided with soft contact padding which will yield slightly when the pipes are loaded (Fig. 11.3), and they should be set true to line and gradient.

11.1 (j) Temporary Protection and Initial Testing

The possibility, already mentioned, of unreinforced concrete or clayware pipes being cracked by uneven peripheral heating or cooling is increased after jointing, and particularly so if clayware pipes are jointed with cement mortar, because of the radial pressure exerted by the jointing material on the pipe sockets. If, therefore, the next stage of the work cannot proceed immediately after jointing (e.g. if it is awaiting inspection or testing), the exposed length of the pipeline should be protected against hot sunshine or cold air, whether the joints are rigid or flexible, as the laying proceeds, by covering it temporarily with any suitable material (such as hessian, building paper, straw matting, cement bags, loose straw, etc.) or by placing a sun and wind shield over the trench (see Plates 9 and 10).

If the pipeline is to be tested before backfilling, the tests should be carried out as soon as possible after laying the pipes, and in stages, so that the length of line tested is not so large as to delay the next site operation unduly. If the test water is hotter or colder than some parts of the pipes the temperature gradient across the pipe wall will be increased when the line is charged, with consequent increased risk of fracture. Care should be taken therefore not to charge the line when it has been heated on top by the sun or cooled by frost or cold winds or when the test water is appreciably hotter or colder than the pipes. The charging should always be done slowly. To avoid displacement of pipes and the drawing of flexible joints, all plugs, branches and bends should be effectively strutted against the trench ends or sides. The static test head at the lower end of a strongly sloping length of clayware or non pressure concrete pipeline should not be permitted to exceed the maximum recommended by the appropriate Codes of Practice for Sewerage or Drainage (B.S. CECP 5 or B.S. BLCP 301), (1968), viz. 20 feet above the top of the pipe, or the possible submerged head, whichever is greater; this may necessitate testing the line in two or more lengths.

Air is sometimes trapped in flexible joint gaps, so causing false test readings. One method of overcoming this difficulty where rubber joint rings have been used is to press a thin blunt tool under each joint ring at the pipe crown to enable the air to escape, taking care not to damage or displace the ring. Filling with water some hours before the test is made may also be useful in this respect and in avoiding errors due to absorption of water by the walls of the pipes.

The testing of a flexibly jointed pipeline by means of air under a pressure of a few inches of water is not likely to indicate faulty or misplaced sealing rings. For large pipes where it is impracticable to fill the pipeline with water it may be preferable to test each joint hydraulically, as the laying proceeds, by means of a suitable portable inner sealed ring of the type shown in Fig. 11.4.

After testing, leaky pipes and faulty joints should be made good with all possible speed. Any pipes which have been disturbed

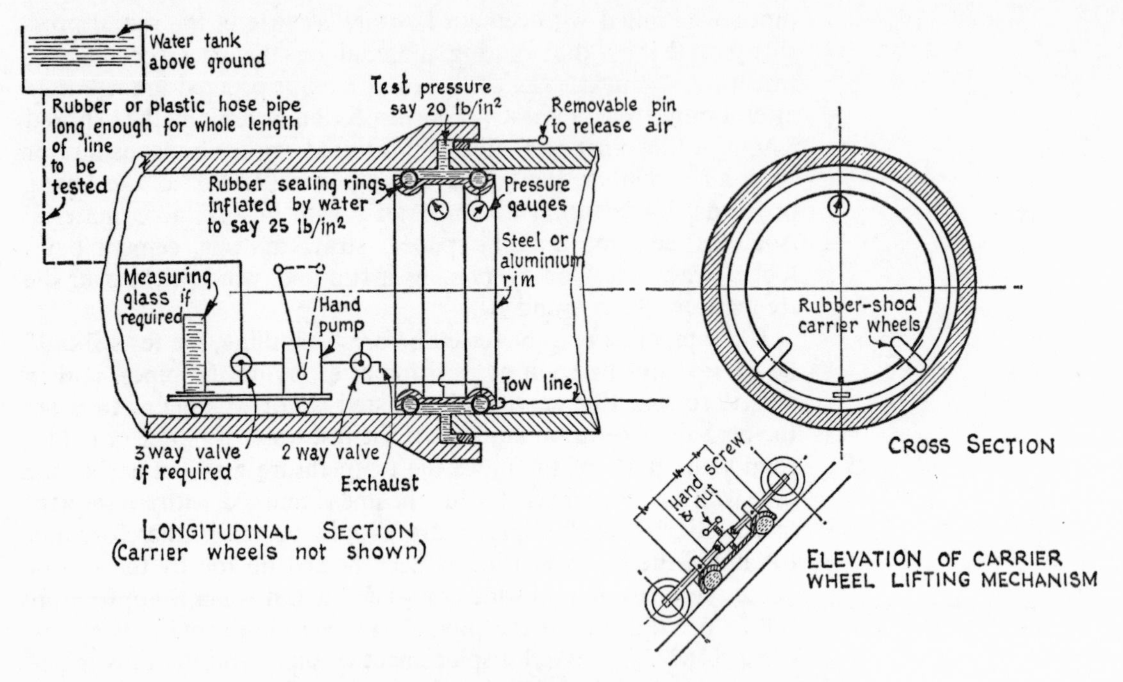

Fig. 11.4 A portable hydraulic test rig for individual joints in large pipes

Crown Copyright—by permission

in this process should be correctly re-seated on their barrels (Class B) or on their cradles (Class A) after retesting. As soon as testing is completed the next stage of the work should proceed without delay, but, if delay is unavoidable, the test water should be left in the pipes, if possible, and the temporary weather protection replaced over them.

11.1 (k) Construction of Bedding—STAGE II

When initial testing is not required, or when the tests and adjustments have been completed, the completion of the bedding, and the placing of the sidefill and of the first foot of selected fill material above the top of the pipe, should proceed without delay. At this stage filling the pipeline with water, or retaining the test water in the line, will help to avoid disturbance of the pipes during subsequent operations, and to prevent flotation of the line in the event of accidental flooding of the trench.

11.1 (k) (i) *Class B Granular Beddings* (Chart C9)

About half the field strength of the pipes with Class B bedding depends on the uniform support of the lower half of the pipe. To maintain flexibility it is essential to avoid jamming of the joints by the intrusion of bedding material, or fill, into the annular gaps outside the sealing rings. These gaps should therefore be sealed by any suitable means (e.g. with puddled clay or damp top soil free of stones), after removing any bedding material which may have entered them at the bottom, and before placing the remainder of the bedding. Some joints have additional rubber rings to serve this purpose.

The appropriate granular bedding material should then be carefully placed in layers of not more than 3 to 6 in. equally on each side of the pipes, and so compacted under and alongside the pipes up to the horizontal diameter (see Chart C9) and laterally up to the undisturbed soil of the trench sides as to eliminate all cavities and ensure as far as possible equality of density with that of the under-bed. Timbering should be withdrawn, whenever possible, as the bedding material is placed so as to avoid disturbance of the bed by its later removal. The remainder of the sidefill and the initial backfill should then be placed without delay (see Article 11.1 (*l*) below).

11.1 (*k*) (ii) *Class A Concrete Beddings, Plain or Reinforced* (Chart C9)

About two-thirds of the field strength of the pipe depends upon the uniformity of its support by a Class A concrete bed; the quality of the concrete and its uniformity are therefore of great importance. Horizontal construction joints in the concrete weaken the bedding and should always be avoided.

The concrete mix should be selected so as to yield a works cube strength of not less than 2000 lb/in.^{2} in the time planned to elapse between placing the concrete and commencing the main backfill, and of not less than 3000 lb/in.^{2} within 28 days. The concrete should be properly and uniformly proportioned and mixed as for other structural work. Table U gives the approximate characteristics of suitable mixes.

Before placing any concrete the annular gaps in flexible joints should be sealed with puddled clay, or otherwise protected from the intrusion of concrete, and the pipes should be washed clean if fouled by mud or clay.

After placing and securing the reinforcement, if any, in position, with the transverse bars as close as possible to the barrels of the pipes, and wetting down the pipes and trench floor, concrete should be evenly placed over the entire width of the trench or bedding, as the case may be (see Chart C9) and to within an inch or so of the bottom of the pipe; then, without stopping, it should be placed gently on one side of the pipeline only and carefully worked under the pipe, ensuring that there are no voids left below the pipe.[103] It should then be brought up equally on each side of the pipe to the required finished height, care being taken not to force the pipes off their supports. The required finished height of the concrete above the bottom of the pipe may be indicated by appropriate paint marks at each socket. Vertical construction joints should always be made at the outer end of a socket or collar. The concrete should be gently and evenly fed into the trench, e.g. through movable chutes or steel pipes, or shovelled in, depending on the width and depth of the trench.

Unless vertical joint gaps have been otherwise formed, the concrete, when set, but before hardening, should be cut out at each pipe joint so as to leave a clear gap between the socket end and the site concrete on the next pipe; this gap should be 1 in. or more, depending on the telescopic movement to be provided for between spigot and socket (see Fig. 11.3). The cut should extend through the blinding. If this is not done the benefits of flexible jointing are lost. The gaps so formed should be sealed by any suitable means (e.g. with fine screened top soil, free from stones and sand, or with puddled clay or rubber or plastic rings), so as to prevent entry of stones or sand into the joint gaps which might jam them and cause fractured sockets. The site concrete and pipes should be kept damp and protected from sun or frost until the concrete is hard enough to take the sidefill safely.

11.1 (*k*) (iii) *Concrete Arch Support, Plain or Reinforced, Placed Above the Pipe* (Chart C9)

The initial procedure up to the required depth of concrete below the top of the pipe should be as for Class B bedding. The annular joint gaps should be protected from the intrusion of concrete or other granular material as in Class A bedding. The concrete, as for Class A, should be placed in one operation and the joint gaps cut out, or formed when placing if preferred. Transverse reinforcement, if any, should be placed as near as possible to the crown of the pipe, and care should be taken to ensure that it is not displaced during concreting. The concrete should be kept damp and protected from the weather until hard enough to take the side fill.

11.1 (*k*) (iv) *Unreinforced Concrete Surrounds*

The initial procedure should follow that outlined for Class A bedding. The distance between vertical construction joints should be so arranged that all the surround concrete can be placed non-stop, i.e. without horizontal construction joints. Care should be taken to keep the concrete the same height at each side of the pipe and to slope it along the trench as necessary to prevent flotation of the pipes. The remaining procedure should follow the recommendations for Class A bedding.

11.1 (k) (v) *Reinforced Concrete Surrounds* (Fig. 11.5)

It may not be necessary to form a contraction gap at every pipe joint; this will depend on soil conditions and the length of the individual pipes. The gap width and the permissible length between gaps will depend upon the draw capacity of a single joint and the anticipated axial shortening or lengthening of the composite length. These gaps, which should always be made at a pipe joint, will then become construction joints, *through which the longitudinal steel should not pass.*

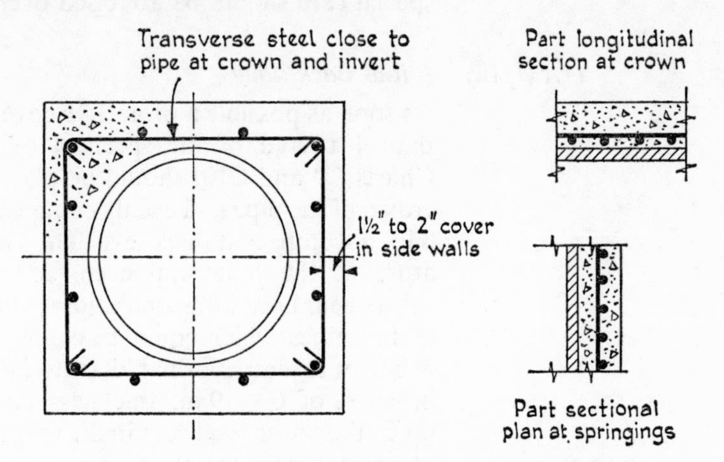

Fig. 11.5 Typical steel arrangement for R.C. surround

Crown Copyright—by permission

Horizontal construction joints are permissible in reinforced surrounds but they should preferably not be placed in the vicinity of the springings and are better avoided. Care should be taken to ensure that the steel is correctly placed (close to the pipe at the crown and invert, and say 1½ to 2 in. from the outside face at the springings of the pipes), and that it is not displaced during concreting. The remaining procedure should follow the recommendations given for Class A bedding and unreinforced surrounds.

If spaces are left between the sides of the trench and the sides of surrounds, side filling as described below will be required.

11.1 (l) Back Filling

11.1 (l) (i) *Side Filling*

As soon as possible with Class B bedding, and as soon as it is safe to do so without damaging site concrete in Class A beddings or surrounds, the specified selected sidefill (see notes to Charts C9 and C10) should be placed and compacted evenly by hand* on each side of the pipe so that the pipes are not displaced either

* Suitable light mechanical appliances are not excluded provided they do not damage or displace the pipes.

laterally or vertically by those operations, timbering or sheeting being withdrawn when possible as the work proceeds so that no voids are left in the filling.

For flexible pipes (e.g. pitch fibre, plastics, steel) in trench, the sidefill should preferably be the same granular material as that used for the Class B type bedding; special care is needed in compacting it thoroughly and evenly at least up to the top of the pipes, since the safe load-carrying capacity of the pipes depends very largely on the lateral support provided by this operation. This special care should be exercised over the full width of the trench.

11.1 (*l*) (ii) *Initial Backfilling*

As soon as possible a protective cover of not less compacted depth than 1 ft and of the specified selected material (see notes to Charts C9 and C10) should be placed gently and evenly over the crown of the pipe and evenly compacted by hand*, in layers of 3 to 6 in. uncompacted thickness. This cover serves to keep the temperature of the pipe wall even, to act as a shock absorber for subsequent backfilling operations, and to avoid the uneven loading of the pipes which could be caused by contact with large lumps of soil or stones. Normal fill material should then be placed evenly in layers of 6 to 9 in. thickness and thoroughly hand tamped* up to the cover level required to protect the pipes from the effects of normal mechanical compaction. (See also Article 11.1 (*l*) (iii) below). The design of thin walled steel pipes may require selected initial fill to be placed and specially compacted alongside and for a height of at least the pipe diameter above the top of the pipe.

11.1 (*l*) (iii) *Main Backfilling of Trenches*

The backfill becomes a loadbearing structure, particularly in trenches under roads. Inattention to its selection, placing and compaction may cause serious damage to the new pipeline, to other existing underground services, or to neighbouring structures. Further, the settlement which occurs in some instances of faulty backfilling under roads adversely affects the riding qualities of the surface and leads to deterioration of adjacent areas of the road structure (Plate 14).

The adequate compaction of backfill implies the use of appropriate compacting equipment and the selection of an appropriate thickness for each compacted layer (see B.S. Code of Practice 2003).[88] In the close proximity of existing pipes or cables where the use of mechanical equipment might cause damage, hand tamping* using thin layers of soil is needed.

* Suitable light mechanical appliances are not excluded provided they do not damage or displace the pipes.

Where spoil from the trench is to be used as backfill under a road, the material should, as far as possible, be replaced in the reverse order (with depth) to that in which it was excavated. It should also be compacted as nearly as possible to the same moisture content and density as those of the undisturbed soil. Care is therefore necessary in stacking the excavated material and in protecting clay soils from rain and excessive evaporation. Dry density and moisture content measurements made on the undisturbed soil in the trench will indicate the degree of compaction necessary in the trench backfill.

In general, the use of a backfill material markedly different in character from the surrounding soil is not recommended where the trench is in a road, since differential volume changes due to moisture migration or to frost action may cause damage to the road bed and unevenness and cracking of the surface.

The fill should be placed gently so as to avoid displacing or cracking the pipes, and spread evenly in layers not exceeding the appropriate thickness. Each layer should be evenly and adequately compacted either by hand or machine to the required density before the next layer is placed.

Timbering or sheeting should be removed as the filling proceeds wherever possible, especially above the lowest permanent water level, to minimise later subsidence of the fill. Below lowest permanent water level, side timbers may be left in place, if need be, but cross timbers above the pipes should be removed, especially if they are near the pipe crown.

The support and bedding of exposed existing pipes should be constructed with the same care as for a new pipe, the object being to restore their original nature as nearly as possible. If the new support is either more or less rigid than the undisturbed soil, rigid pipes may be fractured. A pipe must never be in contact with hard objects such as other pipes either above or below it, or with their concrete support, unless the whole crossing is surrounded with concrete and flexible joints are provided where either new or existing pipes leave the concrete and at the next joint, as for pipes passing through other rigid structures (see Article 9.2 (*g*) and Fig. 11.2). The proper restoration of the side support to existing flexible pipes is essential to their stability. This applies particularly to pitch fibre and plastic pipes. The same precautions as for the new pipe should be taken in backfilling above existing pipes.

In fields the settlement of the fill is not so important as in roads, but it may interfere with existing land drainage or cause objectionable hollows and ponding of water. Existing land drains should be carefully reinstated, or suitably diverted by arrangement with the landowner, and the full depth of top soil finally replaced in fields and gardens.

Sowers has recommended[97] that regardless of the type of conduit, the density of the backfill should be controlled throughout its depth above the final initial fill for various loading conditions as follows:

Class	Type of Loading	B.S. 1377 % Standard Compaction (Proctor Density)
1	Extremely heavy live loads	100+
2	Normal live loads	95–97
3	No live loads	90–93
4	Surface settlement can be tolerated and does no harm	85–90

Note: The compaction ratio is the ratio of the actual soil density in the fill to the British Standard density.

For rigid pipes the bed should be Class 1. The sidefill to within D/6 of the top of the pipe should be Class 1 or 2. The initial backfill to a height of B_c, should be Class 3, and the remainder the Class to suit the live loading condition.

11.1 (m) Road Construction or Reinstatement

Heavy road rollers should preferably not be used over wide trenches with pipes at shallow depth (say less than 4 ft cover). Shallow trenches in an unsurfaced road should be bridged at convenient points for construction traffic, which should not be allowed to pass across or along the filled trench elsewhere. If heavy rollers are to be used in the construction of the road bed, the laying of pipes with cover less than about 3 ft below road formation level should be deferred until the rolling is finished. For very large pipes or shallow cover special measures may be necessary to prevent excessive loading of the pipes.

Where the backfilling has been properly executed the use of temporary surfacing may be neither necessary nor desirable unless the trench is to be shortly reopened. The sub-base, base course and surfacing of the paving should be re-laid to give a structure as closely as possible equivalent to the adjacent road paving. Where the original base course is crushed stone, for example, an equivalent crushed stone (or wet mix) base course should be laid in the reinstatement in preference to lean concrete or some more rigid form of construction. The reverse would apply where the original base material was lean concrete. During excavation, stones used as base course materials may become contaminated with soil and for this reason they should, in general, be replaced rather than re-used in the reinstatement. Great care must be given to the proper compaction of the base and sub-base materials.

The surfacing should also be similar to that used on the adjacent parts of the road, i.e. tar or bitumen macadam should not be used where the prevailing surfacing is rolled asphalt.

If temporary reinstatement is unavoidable, it should be carefully and systematically maintained to ensure a smooth riding surface so minimizing impact effects on underground pipes, vibration in buildings, and nuisance or danger to traffic.

11.2 Construction in Headings

The preparation of the foundation and bedding in headings should follow the procedure that has been outlined for trenches. It will usually be found necessary, however, to pack the heading as the pipe laying proceeds, and this work is simplified by the use of flexible telescopic joints, especially with Class B beddings. The bedding, sidefill and the initial layer of material above the pipe should all be of the same selected material as for trench work, and the final packing above the pipe should also be of selected readily compacted material, free from large lumps or large voids. In surrounded pipes, the provision of joint gaps presents difficulties and for this reason, if for no other, it may be preferable to use stronger pipes with Class B bedding. If surrounds must be provided they should preferably be reinforced both longitudinally and transversely (see Fig. 11.5), in which case horizontal construction joints in the site concrete are permissible. In deep headings the occurrence of a small gap between the top of the packing and the roof of the heading may be of no great consequence. However, in shallow headings, especially where existing pipes pass over the heading, such gaps should be grouted up or filled with well-packed concrete of zero slump; but the creation of hard spots under existing pipes should be avoided (see Article 11.1 (*l*) (iii)). Timber should be withdrawn wherever possible as the packing proceeds, specially above permanent water level. Cills or bottom spreaders which cannot be withdrawn should be sunk well below the bottom of the bedding so as to reduce their effect as hard spots. Small bore pipes may be encased in larger pipes to ease the operation of packing the heading.

11.3 Thrust and Auger Boring [87], [87a], [95]

Where conditions permit, thrust or auger boring methods, which require no timbering other than in the access (manhole) shafts, may be used in preference to construction in heading especially in bad and wet soil. Similar precautions to those applicable to headings should be taken regarding the uniformity of the soil in contact with the pipes. This work is usually undertaken by specialist contractors who are now equipped to place welded steel or reinforced concrete pipes up to 96 in. bore or even more. Concrete pipes for this purpose are socketless but they can, and usually should, be flexibly jointed within the wall thickness. Precast segmental concrete linings can be placed by somewhat similar methods for bores of 48 in. or more. The segments should preferably be so arranged and jointed as to avoid rigidity either longitudinally or circumferentially, and the joints between the segments should be sealed with a suitable pliable material which is watertight and does not harden with age. Jointing material which may become brittle and crack should not be used.

11.4 Construction Under Embankments and Valley Fills

11.4 (a) In Positive Projection Conditions

For pipes in this embankment condition (see Fig. 3.1), the preparation of the foundation and bedding should follow the procedure outlined above for Class A or B bedding in trenches, but with the variations in detail given in Chart C10.

The sidefill and initial overfill should be of the same quality selected material as for trenches, but of sufficient width and depth respectively to ensure uniform loading of the pipes. A minimum width and depth of 12 in. is suggested for rigid pipes, but this should be increased, to say one diameter for pipes over 12 in. I.D. The sidefill should be compacted to not less than 90 per cent Proctor density (B.S. Standard compaction) or any higher value specified by the designer. For flexible pipes the fill surrounding the pipe must be selected and compacted as specified by the designer over the specified width or annular distance surrounding the pipe, and the degree of compaction achieved should be frequently checked by appropriate tests on soil samples in accordance with B.S. 1377. Compacting or earth-moving equipment must not be permitted to impose loads on the pipeline until the designer is satisfied that there is a sufficient depth of fill over the pipes to ensure that their load-carrying capacity (at this depth) is not exceeded (see App. A, Ex. 9, 15, 16). The fill should be brought up equally on each side of the pipe to avoid its displacement by unbalanced lateral pressure. Large variations in the depth of fill along the pipeline should be avoided as far as possible. End tipping should be avoided, especially when manholes are constructed over a sewer or culvert in a valley which is to be filled up, otherwise the manhole shafts may be seriously distorted or so displaced as to require rebuilding.

For pipes crossing embankments with sloping sides founded on compressible soil, allowance should be made in setting out the pipe bedding for the probable variation in settlement of the foundation. The degree of upward camber required may vary from a few inches to several feet, depending on the height of the fill and local soil conditions. As the bank consolidates, there will be a tendency for the fill to spread laterally somewhat, and this movement will tend to stretch a pipeline progressively from the centre

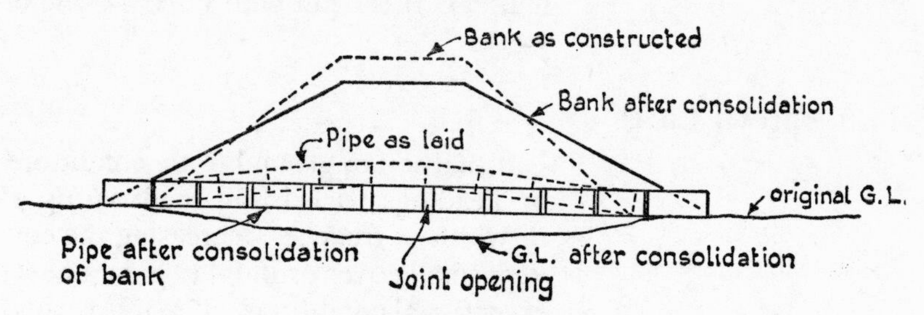

Fig. 11.6 Typical effect of embankment consolidation and lateral spread on a pipeline
After Spangler—by permission

of the bank towards either toe(98) (see Fig. 11.6). Ample draw should therefore be provided in all joints. Spangler reports joint openings of over $1\frac{1}{4}$ per cent of the length between joints in rigid culverts constructed *in situ*.

11.4 (*b*) *In Induced Trench Conditions*(99)

For pipes in the induced trench condition (see Fig. 3.1), the procedure should follow the same lines as just described, up to one or two diameters above the top of the pipe. At this stage a trench of the same overall width B_c as that of the pipe should be excavated in the new fill directly over the pipeline, and refilled with loose uncompacted compressible material such as straw bales. Normal placing of fill should then be resumed, care being taken to avoid over-compacting the loose compressible fill. Compacting or earth moving equipment must not be permitted to impose load on the trench until there is a sufficient depth of fill above its top to form a bridge and ensure that the load-carrying capacity of the pipes (at this depth) is not exceeded. (See App. A, Ex. 9, 15, 16).

For thin walled flexible pipes there should be a thickness of fill of at least B_c between the top of the pipe and the bottom of the sub trench (see Article 6.4 (*c*) (i).

11.4 (*c*) *Negative Projection Conditions*

For this condition (see Fig. 3.1), the procedure should follow that outlined for trenches up to the original ground level, trench type beddings being used (Chart C9). Compacting and earth-moving equipment must not be allowed over the trench until the cover

over the pipes is sufficient to ensure that their load-carrying capacity (at this depth) is not exceeded (see App. A, Ex. 9, 15, 16).

11.4 (d) Construction of the Embankment or Valley Fill

The construction of embankments, and valley fills apart from the special requirements noted above is outside the scope of this manual. It is dealt with in B.S. Code of Practice CP 2003, Earthworks.[88]

11.5 Special Cases

Unusual soil or ground water conditions, the proximity of trenches or headings to existing loadbearing structures, excessive depth or width of trenches, increasing the cover over existing pipes with very small cover or in fill (e.g. in a new bridge approach), or other exceptional conditions, all require expert consideration and are not necessarily covered by the general guidance given in this Chapter. In all such conditions early advice is desirable, since the prevention of trouble is likely to be cheaper and easier than its cure.

11.6 Performance Testing

Failure to provide uniformity in the pipeline foundation, or in the bedding of the pipes, and other faults in construction are not revealed by watertightness tests before backfilling. Moreover, clay and clay-bearing soils shrink or swell with changes in moisture content and display a marked annual cycle in this respect at depths down to about 5 ft, or deeper in the vicinity of trees. The maximum effect on a new pipeline of these movements and of the annual temperature or ground water cycles, therefore cannot be observed in less than about a year after backfilling. There are therefore strong grounds for acceptance tests or performance tests to be carried out after backfilling, and as long after as may be possible without undue inconvenience. As Marston[16] pointed out many years ago, there is frequently a considerable time lag in the development of the load on the pipes.

In general, infiltration is a greater nuisance than exfiltration in the British Isles but it is a lesser menace to public health. Exfiltration is more serious in granular soils and fissured rocks than in clays. Infiltration is most serious in waterlogged coarse-grained soils and in the small pipes of building drainage systems. It cannot be assumed that a pipeline which has proved watertight before backfilling will remain so after backfilling, especially if it is rigidly jointed. Under working conditions a pipeline is always loaded more or less, and where the water table can rise above the pipe crown, the external hydraulic pressure will be greater than the internal pressure in a drain or sewer. Tests for infiltration on pipes laid below the highest water table should be made preferably when the ground water is at its highest annual level. At other times, the test results should be corrected for the reduction in head at the time of test. Conversely, exfiltration tests, if required, should be made when the ground water level is below the pipe invert. More precision is required in performance tests for drain and sewer pipelines than has been usual hitherto. The quality and improved performance of modern joints should make possible the introduction of maximum permissible rates of infiltration, or exfiltration, on the basis of gallons/hour/100 linear ft/in. of diameter of pipe, as is customary for pressure pipes. The maximum, mean and minimum head at the time of test should be recorded, wherever possible, together with the height of the water table if above the pipe invert, and its annual maximum and minimum levels if either is above the pipe invert. This requirement would probably remove some of the grounds for disagreement between inspectors and contractors as to the interpretation of test results, and should enable much closer estimates of maximum leakage to be made by the designer.

The divergence of opinion as to the validity of air tests for drains and sewers can probably also be reduced by the simple provision that failure to pass the air test, for no obvious reason, shall automatically require a water test, and that acceptance or rejection shall be decided by the results of the latter.

Exfiltration tests should take account of the absorption of water by the pipes which varies considerably from one material to another. For instance, concrete needs a much longer period of soaking than, say, cast iron before the test is applied.

By and large, infiltration tests are likely to be of more value as an indication of pipeline performance than exfiltration tests, and they are much easier to carry out. In pipe sizes of 4 in., or more, the sources of infiltration can be located by a closed circuit television survey of the pipelines.

11.7 Safety Precautions

Attention should be given to the safety recommendations contained in B.S CP 2003 Earthworks and to the requirements of Construction (General Provisions) Regulations 1961, Construction (Lifting Operations) Regulations, 1961, and such other Safety, Health and Welfare Regulations as may be promulgated from time to time by the Ministry of Labour, or other responsible Ministries.

A list of the addresses and telephone numbers of contacts with the local authority engineer, public utilities, police, fire brigade, doctors, local hospitals, and museum authority, should be always readily available to site supervisors and foremen.

11.8 The Maintenance of Pipelines

11.8 (a) Cause and Effect of a Fault

The operations involved in maintaining the efficient service of an underground conduit network, however old it may be are subject to the same basic structural principles as the design of a new pipeline. The repairs of a structural fault such as a broken or collapsed pipe requires an investigation of the cause of the fault before a rational method of repair can be devised. Simply to replace the faulty pipe may result in the repetition of the failure and casual 'strengthening' may induce further failure in adjacent sections of the pipeline or of other neighbouring pipelines. To a very useful extent the nature of the fracture of a rigid pipe is an indication of its cause (see Table R).

11.8 (b) General Notes on Maintenance

11.8 (b) (i) Objectives in Repair Work

The behaviour of a pipeline depends very largely on that of the soil in which it is buried. The objective in repair work should therefore always be to ensure an efficient repair without endangering other structures in the vicinity including the road foundations. Note especially Example 3, Article 4.13 (*a*), in this respect. If the fracture is of a type which is eliminated by a flexible pipeline, then flexible joints should be used in the repair work. Replacing rigid joints by rigid joints may result in a repetition of the failure.

11.8 (b) (ii) Detection of Leakage

The major result of inefficiency is leakage, i.e. infiltration of ground water into non-pressure (sewer) pipes or exfiltration from pressure pipes (water, gas, oil).

A broad indication of serious leakage is given by the examination of flow records (meter readings, pumping records, etc.). A nearer approach to the sources of leakage can be obtained by isolating sections of the pipeline and conducting flow tests for infiltration or exfiltration, usually during the night when there is least demand on the service. Gross leaks in pressure pipes are frequently located by wet patches on the surface or by stethoscope. Infiltration and other faults in non-pressure pipes can be actually seen, and located to within a foot or two, by means of a special TV camera towed slowly through conduits of 4 in. or larger diameter.

Isotopes and geiger counters have been used to detect the sources of leakage in water mains. Special gas detectors are used for locating gas leaks.

11.8 (b) (iii) *Locating Existing Pipelines*

The exact location of a pipeline prior to re-excavation is frequently difficult. 'Records' are notoriously unreliable owing to removal of reference points or mistakes in plotting. Magnetic detectors are available, for iron or steel pipes but they do not readily differentiate between different pipes in close proximity. Instruments are also available for detecting any kind of pipe, or cavity, or even, in sensitive hands, of indicating the material and depth of the pipe or the position of a 'lost' manhole. The method resembles water divining and involves a similar element of uncertainty, and some confusion when a number of pipes are close together. Trial holes are usually necessary to confirm the instrumental findings.

11.8 (b) (iv) *General Repairs Without Re-excavation*

Methods are available for the systematic pressure grouting of rigid non-pressure pipes, but the use of a simple cement grout is not entirely satisfactory in soils liable to repeated movements because of its tendency to crack. The presence of sulphate or acids in the groundwater is also a contra indication. In wet conditions a clay grout may be more satisfactory if the individual fractures in the pipes are not too wide.

Mixtures which form a flexible gel after placing have been used with some success but their efficiency over long periods of time is not yet known.

The relining of large old conduits with special precast concrete units has been carried out, mainly on brick sewers, but the problem of the flexibility of joints in such renovations requires consideration in some soils.

Cast iron and steel pipes are successfully relined with cement mortar after thorough descaling and cleaning. More recently the lining of metal and concrete pipes with epoxy resin has been developed with the object of preventing further corrosion and producing a very smooth bore, with less loss of cross sectional area than with mortar. This method is also useful where the effluent is corrosive.

Obviously where draw is likely to occur at existing joints or cracks, continuous brittle linings will crack. For such pipelines a ductile lining is indicated, possibly some epoxy rubber or bitumen-composition. Linings of water pipes must necessarily have no toxic, taste or odour effects.

Small leaky sewers have been restored by placing a smaller flexible pipeline inside them.[49a]

11.8 (b) (v) *Local Repairs after Excavation*

In using site concrete for local repairs, care must be exercised to avoid creating hard spots in the pipeline or in any neighbouring or crossing pipelines. Rigid contact between pipelines should be carefully avoided and some form of resilient packing introduced both above and below, where necessary, to prevent load transmission from one pipeline to another. The soft packing should not be of such a large extent, however, as to cause a dangerous soft spot for any of the pipes involved, cf. Article 11.1 (*l*) (iii). In aggressive soil or groundwater a special cement may be needed as for new work.

11.8 (*b*) (vi) *Cleaning*

The removal of silt and grit from non-metallic sewers is best done hydraulically wherever possible, by means of a hydraulic ferret or an inflated rubber or plastic ball. Metallic scrapers are liable to damage the surface of the pipes especially at the joints. The fairly frequent passage of the ball is also cheaper than the infrequent use of buckets and scrapers, and it maintains a higher hydraulic efficiency in the pipeline. It is not suitable however for small pipes or pipes where the necessary waterhead to push the ball along is not available. Under these conditions the ferret is preferable, after rodding to ensure that there is no hard obstruction (scrubbing brushes, childrens' toys, tools, brickends) in the pipes. The careless or over-generous use of rods especially with metallic head attachments, may seriously damage or puncture pitch-fibre or plastics pipes. Fortunately these pipes are smooth and do not retain considerable deposits if laid at normal gradients.

11.8 (*b*) (vii) *Eliminating Backfalls*

Serious backfalls or 'swamps' in concrete or asbestos-cement sewer pipes should be eliminated, if possible, since, by checking the velocity of flow of the effluent, they may, under certain conditions, give rise to septicity in the sewage and consequent acid corrosion of the pipes above the water line, downstream of the backfall. Similar effects can occur at the outlet of a rising main if the velocity in the riser is too slack or if a self-cleansing velocity occurs too infrequently.

11.8 (*b*) (viii) *Strength of New Pipes*

Any new rigid pipes required in the course of repairs must be of a strength class which is adequate for the loads and pressures to be carried and the bedding to be used (see Chapter 5).

For thin flexible pipes the same attention to the adequate compaction of the sidefill and initial backfill as for newly installed pipes (see Articles 6.1, 11.1 (*l*) (i) and (ii) and 11.4) is imperative. This also applies to any existing flexible pipes which have been exposed in the course of the repair work.

For either type of pipe the new bedding must be of the appropriate class and workmanship and of the same uniformity as for new pipelines.

11.8 (*b*) (ix) *Backfilling*

Especially in roads, the backfilling and reinstatement of the opening requires the same care and consideration as for a new pipeline (see Article 11.1 (*l*) and (*m*)). This rule also applies to the reinstatement of the bedding of existing pipes exposed in the excavation (see Article 11.1 (*h*) and (*k*)).

Chapter 12

Failures—Inspection—Economics—Future Trends

12.1 Failures

Apart from calamities of nature, a pipe breaks or a joint leaks because of some fault in either:

(i) Design by the engineer;
(ii) Construction by the contractor;
(iii) The pipe or its joint components by the pipemaker;
(iv) Because of changes in its bedding or other environment produced by later underground works in its vicinity, without proper regard for its safety;
(v) Because of an unforeseen increase in the load conditions;
(vi) Because of faulty operation of pressure pipes.

Thus the occurrence of a failure is not just an expensive misfortune. It provides an opportunity to learn why the failure occurred, where the responsibility lies, and how to avoid the repetition of a similar fracture or fault in repairs or future work. Plates 2, 3, 6, 8, 9, 10, 13 are photographs of typical actual fractures.

The adoption of rational principles of design and construction, and the making of pipes to a recognised strength standard under proper statistical quality control, enables responsibility for a failure to be allocated with reasonable justice to all parties and, it is to be hoped, will greatly reduce the incidence of failures and the controversies arising from them.

It is not reasonable to hold the contractor responsible for faulty design, or the pipemaker for faulty construction. The designer's specifications should not become stereotyped but should be progressively amended in the light of field experience. Likewise the pipemaker should not be content merely to supply more pipes 'free of charge' but should be prepared, either to seek out and rectify the fault in the pipes, or to repudiate the claim when it is clear that the fault lay with the designer or the contractor. If the contractor is at fault he should not blame the pipemaker as a matter of course, but should ascertain the cause of the fault and make sure that it is not repeated in any future work.

For all parties, the keeping of a record of all failures, giving the evidence and the findings as to cause and effect, and possibly the cost of rectification, would be both enlightening and profitable.

12.2 The Relation of the Quality of Inspection and Workmanship to the Factor of Safety

The efficiency of inspectors, and the quality of inspection, vary widely, as do the experience, skill and reliability of contractors.

The finished quality of the workmanship is therefore likely to vary considerably from job to job, or even on the same job. Logically this variation should be reflected in the factor of safety adopted by the engineer for any particular job. Assuming that he is satisfied with his estimate of the total effective load to be carried by the conduit, the risk lies with the field strength of the pipes and the higher this risk, either from inadequate inspection or the inexperience of the contractor, the higher the factor of safety should be, with consequently stronger and more expensive pipes or bedding.

There is a dilemma here since at the time of writing the specification the efficiency of the contractor is usually unknown, and to leave the selection or ordering of the pipes until the contract is let may cause unavoidable and unacceptable delay in delivery, particularly in the larger sizes. It may be possible to overcome this difficulty, however, by providing for alternative beddings in the specification, and calling for tenders on the alternatives, whilst ordering a suitable class of pipe for the lower bedding class and a normal factor of safety before the contract is advertised. The decision as to which bedding to adopt may then be left until the tenders are received, when the weighting of doubtful low bids by the more expensive bedding may produce a more equitable comparison of the tenders and a safer pipeline.

Looking to the future, amalgamations, or consortia, of reputable pipemakers and pipelaying contractors, whereby a group could offer to supply and lay the pipes for the larger schemes, would have the advantage of undivided responsibility, experienced supervision, properly trained operatives, early knowledge of requirements, and avoidance of delay in supplies. Direct labour, with properly trained gangs and supervision, would have much the same advantages for smaller schemes and repairs.

12.3 Staff Training

In view of the interdependence of site work and design, it would evidently be of advantage if all members of supervisory and executive staffs were made aware of the basic principles involved in the rational design and construction of buried conduits, in order to ensure that the essential contribution of workmanship to the field strength of a pipeline is understood and achieved. A colour film illustrating these principles has been prepared and copies are available for purchase or on loan.*

12.4 Economics

Cost comparisons cannot safely be made on the basis of tenders alone, since these vary with the contractor and with his experience

* *Modern Methods of Underground Pipelaying*. Produced by the Building Research Station. Available on application to the Station or on loan from the Film Officer, Ministry of Public Building and Works, Lambeth Bridge House, London, S.E.1.

of the method of construction. Tenders are not costs; neither do they include the future cost of maintenance which depend very much on the quality of the original work. Still less can fair comparisons be made on designs and specifications involving very different degrees of risk of failure.

Especially with the small bore pipes used in building drainage, the cost, ex works, of pipes with factory made flexible joints is likely to be higher than that of the pipes alone. Similarly the use of shorter lengths of pipe in mining areas generally, and in the smaller water and gas mains anywhere, may increase initial cost to some extent. This extra cost and that of the more careful site work and inspection required by modern methods of pipeline design and construction may be offset to some extent, however, by the lower site labour costs of jointing and the increased speed and lower overheads with which the work as a whole can be carried out, having in mind the elimination of handwork and the fact that work need not stop in frosty weather where site concrete has been eliminated.

Another important consideration is the probable saving which, depending on local conditions, is likely to be achieved in operation and maintenance costs over the system as a whole by the adoption of modern methods and materials. Thus in sewerage systems, the probable additional costs of pumping and disposal of ground-water should be considered, whilst in water and gas distribution systems, the probable losses from leakage should be evaluated over the expected lifetime of the pipeline. In any system, the incidence and the probable cost of replacing or repairing fractured pipes, recaulking joints, and compensation for consequential damage to other property owners, all enter into the real costs of the system.

Comparative data are not yet readily available and engineering judgement is therefore needed in estimating the probabilities in any particular scheme. It should not, however, be taken for granted that the lowest initial cost is the most economical solution of a given problem. *The economic objective should evidently be the lowest TOTAL cost over the lifetime of the pipeline.*

12.5 Future Trends in the Design and Construction of Buried Pipelines

Rational methods of pipeline design and construction are spreading rapidly throughout the civilised world and promise to become as universal as structural theory and practice. The application of soil mechanics analysis will probably refine the load theory but the variable nature of the soil will remain. The possible excessively conservative factors in the Marston theory may be brought to light and result in a closer approximation to the actual loads, but a more rational theory taking into account all the practical hazards, is not yet in sight.

'late 2. Overload fracture of concrete pipes – *'ailure of an 18 in. concrete pipe in very soft ground aused by overload due to the rigidity of the piled foundaion, the absence of bedding between the pipes and the upporting beams, and excessive effective width of trench.*

3 (*a*) Cavity caused beneath the roadway.

3 (*b*) Exposed pipeline showing overload cracking.

3 (*c*) Condition after removing pipe fragments.

'late 3. Overload fracture of clayware pipes – *Caused by lack of suitable bedding for the pipe strength. 12 in. pipe vith 10.5 ft cover under a road. All pipes in 100 ft of pipeline similarly fractured with the results shown in Plates* (*a*), (*b*), *nd* (*c*). *Pipeline completely blocked by the infiltration of eroded soil from the cavity above.*

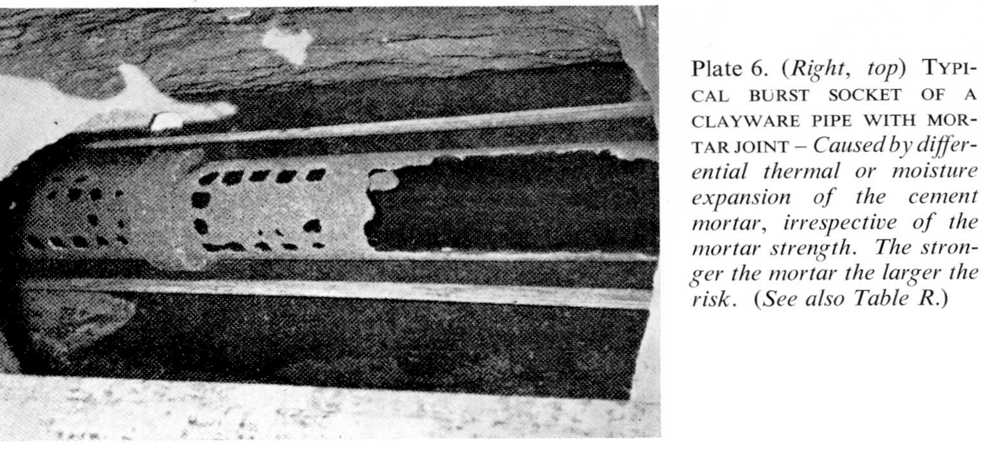

Plate 4. CORROSION OF A 14 YEAR OLD CONCRETE PIPE BY H_2SO_4. (after Stutterheim)[115]. *This is the effect of the generation of H_2S in septic sewage sludge or slime which can occur under favourable conditions of temperature and turbulence. The H_2S is oxidised to H_2SO_4 by bacterial activity in the presence of air above the water line in the pipes. Warm conditions in air or effluent encourage the bacterial activity*

Plate 6. (*Right, top*) TYPICAL BURST SOCKET OF A CLAYWARE PIPE WITH MORTAR JOINT – *Caused by differential thermal or moisture expansion of the cement mortar, irrespective of the mortar strength. The stronger the mortar the larger the risk. (See also Table R.)*

Plate 7. (*Right, below*) AXIAL BUCKLING OF A STEEL PIPE BY GROUND MOVEMENT – *Gas main laid with shallow cover subjected to compression by mining subsidence.*

Plate 5. (*Left*) EFFECT OF VACUUM PRESSURE ON THIN STEEL PRESSURE PIPE – *Buckling has occurred at a crest in the pipeline due to vacuum induced by faulty operation*

Plate 8. (*Below*) THRUST FRACTURE OF A CLAYWARE PIPE JUNCTION – *Caused by moisture expansion of unweathered pipes in conjunction with rigid mortar joints.* (*See also Table R.*)

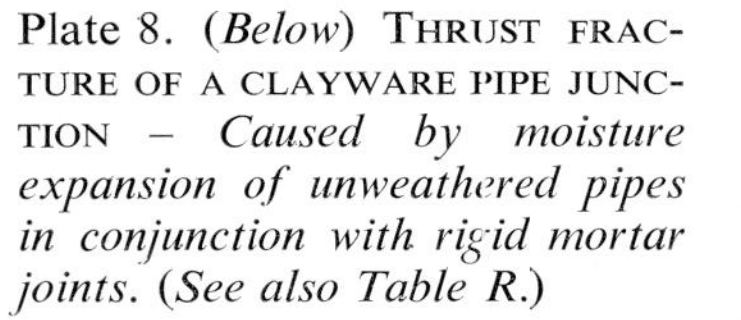

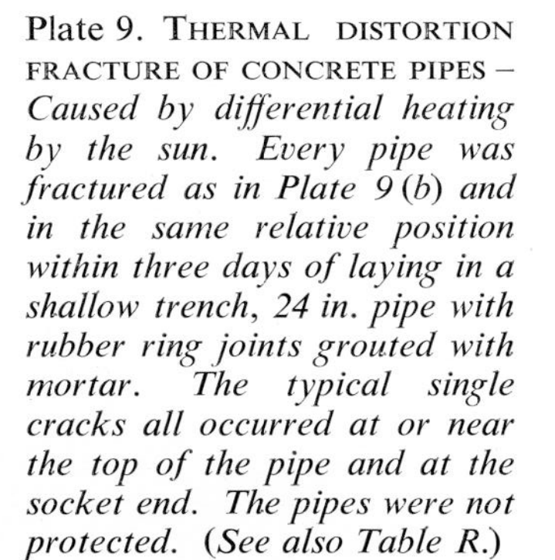

9 (*b*) CLOSE-UP VIEW OF THE CRACKS IN THE PIPES OF PLATE 9 (*a*) AFTER REMOVAL OF MORTAR.

Plate 9. THERMAL DISTORTION FRACTURE OF CONCRETE PIPES – *Caused by differential heating by the sun. Every pipe was fractured as in Plate 9* (*b*) *and in the same relative position within three days of laying in a shallow trench, 24 in. pipe with rubber ring joints grouted with mortar. The typical single cracks all occurred at or near the top of the pipe and at the socket end. The pipes were not protected.* (*See also Table R.*)

9 (*a*) GENERAL VIEW OF THE AFFECTED LENGTH OF PIPELINE.

Plate 10. THERMAL DISTORTION FRACTURE OF CONCRETE PIPES – *Caused by differential cooling by cold winds. 42 in. pipe with flexible joints (bitumen) laid above ground. All pipes similarly cracked on the windward side. Note the similarity to Plate 9 (b) and again the typical single crack at the socket end.*

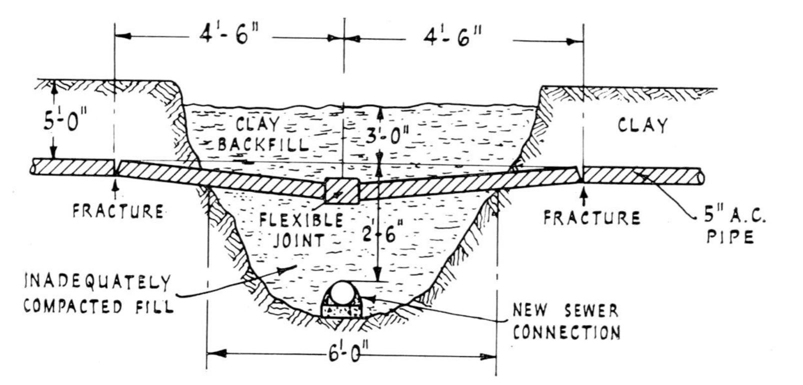

Plate 11. FRACTURE OF AN EXISTING A.C. PIPELINE BY CARELESS INSTALLATION OF A NEW PIPELINE BELOW AND ACROSS IT – *Note the excessive trench width, absence of timbering and slings, and the inadequate compac-*

Plate 12. BAD PRACTICE IN PIPE BEDDING – *Note the soft wet trench bottom, with bricks or timber supporting each pipe. The probable effect*

Plates 13. CLAYWARE PIPELINE SHATTERED BY HARD SPOTS IN THE BEDDING – *9 in. pipes laid on a soil bed with a brick behind each socket. Every pipe in the line was fractured.*

13 (*a*) (*Above*) THE PIPELINE UNCOVERED SHOWING TYPICAL HARD SPOT SHATTER FRACTURES FROM ABOVE

13 (*b*) (*Left*) A RECONSTRUCTED PIPE SHOWING THE FRACTURE FROM BELOW. *Note the typical star fracture originating at the brick.*

Plate 14. BAD PRACTICE IN BACKFILLING A TRENCH – *The large lumps of broken concrete are very likely to produce uneven loading of the pipeline and uneven settlement of the fill. In a roadway they could also cause irregularities in the road bed and serious obstruction to future excavations across the line of the trench.*

Delays in site work, from whatever cause, are expensive and especially so in urban areas where they are a fruitful cause of traffic congestion. Measures to speed up the work will therefore always be welcome and economical, provided that proper standards of workmanship are ensured. Thus the laudable trend already evident towards the use of factory-made joints is likely to accelerate. This will shortly make the standardisation of the external diameter of pipes essential. It is already desirable and has been achieved for cast iron, spun iron and asbestos-cement, which are interchangeable, so far as their external diameters are concerned. It is not suggested, however, that thick wall pipes, such as concrete, can be made interchangeable with the relatively thin wall metal pipes. They could, however, be made interchangeable amongst themselves.

Mechanisation may be expected to continue and to replace hand labour wherever possible and to improve the quality of backfilling compaction.

Speed will also be improved by the elimination of site concrete wherever possible. The trend towards factory strengthened pipes makes this possible and is likely to accelerate when its advantages are more widely appreciated.

A more rational attitude towards the testing of pipelines for watertightness and a much higher standard of efficiency in this respect, are to be expected with beneficial effects on operating costs.[100]

The attitude towards accuracy of gradients is likely to be rationalised as the effects of ground movements subsequent to backfilling and the effect of the dimensional tolerances on pipes are more clearly understood. It has already been shown[116,119] that the irregularities at joints have a far greater influence on the hydraulic capacity of a pipeline than the nature of the surface of the pipes.

In town streets where underground services are congested, some form of tunnelling for new main sewers would appear to have a number of advantages in suitable soils. This would also greatly affect the layout of trunk sewers which need no longer slavishly follow the street pattern. For the larger sewers there will probably be a greater use of non-metallic segmental tunnel linings.

Under embankments flexible pipes have many advantages over rigid pipes, and their use is likely to increase as the knowledge of their design becomes more widespread. They are already being used sparingly for culverts and underpasses on motorways and as protective sleeves to other conduits under railways.

Prestressed concrete pipes seem likely, because of their relative lightness, to displace high strength reinforced concrete pipes of large diameter. They can be made longer for equal weight so saving labour and jointing costs.

Glass fibre reinforced plastics are likely to be more widely used for the smaller pipes as their physical properties and durability become established. The use of plastic linings and coatings to pipes, as anti-corrosive or anti-permeability measures, may be expected to increase if and as they become cheaper.

Ductile cast iron has properties similar to mild steel and is likely to replace normal cast or spun iron for pipes of small and medium diameter, where cast iron is preferred to steel, if its cost differential decreases with increasing output.

Greater use of cathodic protection for all metal pipes is to be expected.

Clayware sewer and drain pipes are likely to dispense with glazing and with sockets and possibly to have a higher strength-weight ratio. With more precise dimensional control, segmental construction may become possible for large pipes either as a prestressed lining or as the complete conduit.

A performance standard for flexible joints is greatly needed and may not be delayed much longer.

The sooner the targets for these various changes can be fixed, the sooner can they become common practice to the advantage of the whole community.

References and Short Bibliography

References and Short Bibliography

General

1. A.S.C.E. and Water Poll. Control Fed., Design and construction of sanitary and storm sewers, *Manual 37* (1961) New York.
2. American Concrete Pipe Association, Concrete pipe handbook, Chicago (1965 or later).
3. B.S. CP.301, Building drainage, 2nd Edition. Drafting in Progress (1968).
4. B.S. CP.2005, Sewerage, 2nd Edition. Drafting in Progress (1968).
5. Clarke, N. W. B., Some problems in underground pipe design and in pipe laying, *P.W. Congress* (*1956*), London.
6. Spangler, M. G., Analysis of loads and supporting strengths and principles of loads and design of highway culverts, *Proc. U.S. Highway Res. Bd.* (1946).
7. Spangler, M. G., Underground conduits—an appraisal of modern research, *Trans. A.S.C.E.*, **113** (1948).
8. Spangler, M. G., Soil engineering (1960). Int. Textbook Co., Scranton, Penn., U.S.A.

Primary Loads

9. Binnie and Thackrah, Water hammer in a pumping main and its prevention, *Proc. I. Mech. E., Hyd. Gp.* **165** (1951).
10. Clarke, N. W. B., Simplified tables of loads on underground pipelines, *Nat. Bldg. Studies Special Report No. 32* (1962). H.M.S.O.

10*a*. Second report of the working party on the design and construction of underground pipe sewers. H.M.S.O., Oct. 1967.

11. Clarke, N. W. B. and Young, O. C., Loads on underground pipes caused by vehicle wheels, *Proc. I.C.E.* (Jan. 1962).
12. Clarke, N. W. B., The wide trench condition, *Proc. I.C.E.* (Sept. 1963).
13. Clarke, N. W. B., Loading charts for the design of buried rigid pipelines, *Nat. Bldg. Studies, Special Report No. 37* (1966). H.M.S.O.
14. Clarke, N. W. B., The loads imposed on conduits laid under embankments or valley fills, *Proc. I.C.E.* (Jan. 1967). London.
15. Costes, N. C., Factors affecting vertical loads on underground ducts due to arching, *U.S. Highway Res. Bd. Bull 125.* (1955)
16. Marston, A. and Anderson, A. O., The theory of loads on pipes in ditches, *Bulletin 31* (1913). Iowa Eng. Exp. Stn. Ames, Iowa. (Historical).
17. Marston, A., The theory of external loads on closed conduits in the light of the latest experiments, *Bulletin 96* (1930). Iowa Eng. Exp. Stn.

18. Newmark, N. M., Simplified computation of vertical pressures in elastic foundations, *Circ. 24* (1935). Eng. Exp. Stn. Univ. Illinois.

19. Schlick, W. J., Loads on pipes in wide ditches, *Bulletin 108* (1932). Iowa Eng. Exp. Stn.

20. Schlick, W. J., Loads on negative projecting conduits, *Proc. U.S. Highway Res. Bd.* (1952). **31,** 308.

21. Spangler, Mason and Winfrey, Experimental determination of static and impact loads transmitted to culverts, *Bulletin 79* (1926). Iowa Eng. Exp. Stn.

22. Spangler, M. G. and Hennessy, A method of computing live loads transmitted to underground conduits, *Proc. U.S. Highway Res. Bd.* (1946).

23. Spangler, M. G., Field measurements of settlement ratios of various Highway culverts, *Bulletin 170* (1950). Iowa Eng. Exp. Stn.

24. Spangler, M. G., A theory of loads on negative projecting conduits, *Proc. U.S. Highway Res. Bd.*, **30** (1950).

25. Terzaghi, K., Theoretical soil mechanics (Appendix pp. 485–7). John Wiley & Sons Inc. (New York), 1944 Edition.

26. Timmers, J. N., Load study of flexible culverts under high fills, *Bulletin 125* (1955). U.S. Highway Res. Bd.

27. Wetzorke, M., The calculation of the resistance of sewers to fracture, *Report of Public Works Congress* (1964). London.

27*a* Page, J., Impact tests on pipes buried under roads, *Report No. 35* (1966). Road Research Lab., Crowthorne.

27*b* Hainsworth, I. H., *et al.*, Pilot experiments to determine the loads causing failure of sewer pipes under roads, *I.C.E. and I.Mun.E.* (1967). London.

Rigid Pipes

28. Abernethy, L. L., Effect of trench condition and arch encasement on load bearing capacity of vitrified clay pipe, *Bulletin 158* (Nov. 1955). Ohio Eng. Exp. Stn.

29. American Standards Association, Manual for the computation of strength and thickness of cast iron pipe (1957). Am. Waterworks Assoc., New York, ASA/A21.1—1957.

30. British Standards Inst., Various standard specifications for pipes (see Table H).

31. Clarke, N. W. B. and Young, O. C., Some structural aspects of the design of concrete pipelines, *Proc. I.C.E.* (Sept. 1959).

31*a* Clarke, N. W. B., Concrete pipes—a review of recent developments in Great Britain and some of the problems encountered, *Proc. Fifth Int. Conf. P.C. Ind.* (May 1966). London.

32. Heger, F. J., Structural behaviour of circular reinforced concrete pipe—development of theory, *J. Am. Conc. Inst.* (Nov. 1963).

32*a* Heger, F. J., A theory for the structural behaviour of reinforced concrete pipe, *D. Sc. Thesis, Mass. Inst. Tech.* (Jan. 1962).

33. Roark, R. J., Formulas for stress and strain (1943). McGraw-Hill, New York, 1943.

34. Rowe, R. R., Rigid culverts under high overfills, *Proc. A.S.C.E.*, **81** (1955).

35. Schlick, W. J. and Johnson, Concrete cradles for large pipe conduits, *Bulletin 80* (1926). Iowa Eng. Exp. Stn.

36. Schlick, W. J., Supporting strength of concrete encased clay pipe, *Bulletin 93* (1929). Iowa Eng. Exp. Stn.

37. Schlick, W. J., Supporting strengths of cast iron pipe for water and gas service, *Bulletin 146* (1940). Iowa Eng. Exp. Stn.

38. Spangler, M. G., The supporting strength of rigid pipe culverts, *Bulletin 112* (1933). Iowa Eng. Exp. Stn.

39. Young, O. C., High-strength beddings for unreinforced concrete and clayware pipes, *Nat. Bldg. Studies Special Report No. 38* (1967). H.M.S.O.

Flexible Pipes

40. ARMCO Int. Corp., Handbook of drainage and construction products, (1958 or later).

41. A.W.W.A., Steel pipe design and installation, *A.W.W.A. Manual M.11* (1964).

42. Costes and Proudly, Performance study of multi-plate corrugated pipe culvert under embankment—N. Carolina, *U.S. Highway Res. Bd. Bull. 125* (1955).

43. Meyerhof, G. G. and Blaikie, L. D., Strength of steel culvert sheets bearing against compacted sand backfill, *Proc. U.S. Highway Res. Bd. 42nd Ann. Mtg.* (1963).

44. Meyerhof, G. G. and Fisher, C. L., Composite design of underground steel structures, *Eng. Inst. Canada Eng. Jnl.* (Sept. 1963).

45. Sears, E. L., Ductile pipe design, *J.A.W.W.A.* (Jan. 1964).

45*a* Boden, W. and Morgan, E., Ductile iron pipes, *Pub. 671, I. Gas E.* (Nov. 1964). London.

46. Spangler, M. G., The structural design of flexible pipe culverts, *Bulletin 153* (1941). Iowa Eng. Exp. Stn.

46*a* Spangler, M. G., Stresses in pressure pipelines and protective casing pipes (steel pipes), *Proc. Struct. Div. A.S.C.E.* (Sept. 1956).

47. Spangler, M. G. and Donovan, Application of the Modulus of passive resistance of soil in the design of flexible pipe culverts, *Proc. U.S. Highway Res. Bd.* (1957).

48. White, H. L. and Layer, J. P., The corrugated metal conduit as a compression ring, *Proc. U.S. Highway Res. Bd.* (1960).

49. Whitman, R. V. and Luscher, U., Basic experiment into soil-structure interaction, *J. Soil Mech. and Found. Div. A.S.C.E.* (Dec. 1962).

49*a* Anon, Unusual repair to a 'live' sewer (6 in. P.F. pipe inserted in 9 in. G.V.C. pipeline), *Plumber and J. of Heating* (May 1960).

Prestressed Pipes

50. Abeles, P. W., Prestressed concrete pipes and circular tanks (Parts I and II), *Civ. Eng. and Public Works Review*, *1955*, 50: 1235–8, 1375–8. *1956*, 51: 83–4, 203–5, 319–22, 522–54, 671–2, 895–6, 1008–10.

51. Bold, R. E., Standardisation of design procedure for prestressed concrete pipes (A.W.W.A. C.301 cylinder type), *J. Am. Waterworks Assn.* (Nov. 1960).

52. Cowan, H. J., The design of prestressed concrete tanks and pipes, *R.C. Review* (1955), 3: 571–85.

53. Evans, R. H., Applications of prestressed concrete to water supply and drainage, *Proc. Inst. C.E., Part III* (Dec. 1955), 4 (3): 725–53. London.

54. Evans, R. H., Circumferential stresses in prestressed pipes, *Proc. Inst. C.E., Part III* (Dec. 1955), 4 (3): 776–83. London.

55. Foster, H. A., R.C. water pipe, steel cylinder type, prestressed. Tentative standard specification No. A.W.W.A. C.301–55T, *Am. Waterworks Assn.* (Sept. 1955). New York.

56. Guerrin, A., Prestressed concrete pipes for fluids under pressure, *Inst. Tech. du Bat. et des Trav. Pub.* (1943). Paris. Circulair Sc. F. No. 12, 24 pp. (in French).

*57. Gandenberger, W., The behaviour of air in pipelines, *Die Wasserverlerlung* (12–13 Feb. 1953), 40–6 (in German).

58. Hasker, J., The manufacture of prestressed concrete pipes, *Construction Review* (Nov. 1953), 26 (7): 16–22. Sydney, Australia.

59. Hendrickson, J. G., Prestressed concrete pipe, *Proc. First U.S. Conf., Prestressed concrete* (1952), 21–4.

60. Hendrickson, J. G., Prestressed concrete pipe, *Am. Conc. Pressure Pipe Assn.* (1954). Chicago, U.S.A.

61. Hubbard, S. R., Non cylinder prestressed concrete pipes, *Proc. A.S.C.E. Paper No. 2240* (Oct. 1959), (see also *J. Pipeline Div. A.S.C.E.* (Sept. 1961)).

62. Kennison, H. F., Design of prestressed concrete cylinder pipe, *J. Am. Waterworks Assn.* (Nov. 1950), 42: 1049–64. New York.

63. Kennison, H. F., Prestressed concrete pipe (discussion), *Am. Concrete Pressure Pipe Assn.* (1954). Chicago, U.S.A.

64. Kennison, H. F., Water hammer allowances, *J. Am. Waterworks Assn.* (March 1958). New York.

65. Kennison, H. F., Tests on prestressed concrete embedded cylinder pipe, *Proc. A.S.C.E.* (1960), 86 (HY9) Part 1, 7798 (combined internal and external load tests on 60 in. pipe).

*Refs. 57 and 71 are applicable to pressure pipes generally.

66. Leggett, R. F., Failure of P.C. pipe at Regina, Saskatchewan, *Proc. Inst. C.E.* (May 1962), 11–20.

67. Monfore, G. E. and Verbeck, G. J., Corrosion of prestressed wire in concrete (Regina), *J. Am. Conc. Inst.* (Nov. 1960), 491–515.

68. Nicol, A., Cas de rupture par corrosion d'une frette en acier dans des Tuyaux en beton, *Revue des materiaux de Construction* (Oct. 1952) (in French). Paris.

69. Ooykaas, G. A. P., Prestressed concrete pipes, *J. Inst. Water Engineers* (March 1952) 6: 85–103. London.

70. Ooykaas, G. A. P., Discontinuity in concrete pipes, *Mag. Concrete Res*, No. 9 (1952) 3: 131–8. London.

*71. Richards, R. T., Air binding (air lock) in pipelines, *J.A.W.W.A.* (June 1962).

72. Seaman, F. E., Tests of prestressed concrete cylinder pipe, *J. Am. Waterworks Assn.* (Nov. 1950) 42: 1065–82. New York.

73. Shaw, J. A., Concrete pressure pipe for major projects, *Civil Engineering* (*A.S.C.E.*) (1961), 31 (2): 46–9. Describes five notable pipelines.

74. Turazza, G., Characteristics of prestressed concrete pipes, *Giornale del Genio Civile* (1955), 93: 606–8 (in Italian).

75. Woinowsky, Kreiger and Potvin, Strains in prestressed concrete pipes, *Eng. Journal of Canada* (March 1953), 36: 230–5.

Pipe Joints

76. Anon, Infiltration into sewers, *Report by Tech. Comm. N.Z. Eng. Inst. N.Z. Engineer* (July 1959).

77. A.S.T.M., Vitrified clay pipe joints using materials having resilient properties, C.425–60T.

78. A.S.T.M., Joints for circular concrete sewer and culvert pipe using flexible watertight rubber type gaskets, C.443–60T.

79. Barracos, Hurst and Leggett, Effects of physical environment on cast iron pipe, *J.A.W.W.A.* (May 1955).

80. Lindley, P. B., Engineering design with natural rubber, *N.R.P.R.A. Tech. Bulletin 8* (1964). London.

80*a* Lindley, P. B., Compression characteristics of laterally unrestrained rubber 'O' rings, *Jnl. I.R.I.* (July/Aug. 1967).

80*b* Lindley, P. B., Rubber 'O' rings as rolling seals—Design Considerations. *J. Inst. P.H. Eng.* 1968 (Publication pending).

80*c* Staff Report, Degradation of Rubber Gaskets, *Jnl. A.W.W.A.* (Dec. 1964). New York.

81. Clarke, N. W. B., The causes and prevention of fractures in salt-glazed ware or other ceramic pipelines, *P.W. Congress 1958.*

81*a* Clarke, N. W. B., Modern flexible joints, *J. Inst. P.H. Eng.* (April 1964).

81*b* Clarke, N. W. B., Rubber sealing rings in flexible joints for underground pipelines, *Use of Rubber in Engineering*, Ed. Allen *et al.*, MacLaren and Sons (1967). London.

82. M.W.B., Report of the sub committee on burst water mains, *Metropol. Water Bd.* (July 1930). London.

83. Orchard, R. J., The effect of mining subsidence upon public health engineering works, *J. Inst. P.H. Eng.* (July 1957).

83*a* I.C.E., Report on mining subsidence, *Inst. C.E.* (1959). London.

84. Santry, I. W., Infiltration in sanitary sewers, *J.W.P.C.F.* (Oct. 1964).

85. Various authors, Experiences with main breaks in 4 large cities, *J.A.W.W.A.* (Aug. 1960).

86. Velsey and Sprague, Infiltration specifications and tests, *Sewage and Industrial Wastes* (March 1955). U.S.A. See also Refs. 100, 31*a*.

Site Work and Testing

87. Anon, Thrust boring an 11 ft steel tunnel at Bristol, *C.E. and P.W. Rev.* (Jan. 1963).

87*a* Thompson, J. C., Horizontal earth boring, *Proc. I.C.E.* (April 1967). London.

88. British Civil Eng.: Codes of Practice.
Site investigations, *Civ. Eng. Code of Practice No. 1* (1950 or later edition). I.C.E. London 1950, or B.S.I.
Earth retaining structures, *Civ. Eng. Code of Practice No. 2* (1951 or later edition). Inst. Struct. E. London, 1951.
Earthworks, *B.S. Code of Practice C.P.2003*. B.S.I. London, 1959.
Foundations, *Civ. Eng. Code of Practice No. 4* (1954). Inst. C.E., London, 1954.
See also Refs. 3 and 4.

89. A.S.T.M., Recommended practice for installing clay sewer pipe, C.12–54, (1954).

90. A.W.W.A., Installation of cast iron water mains, C.600–54T, (1954).

91. A.W.W.A., Trenching, pipelaying and backfilling steel pipes, *J.A.W.W.A.* (Sept. 1962).

92. Becker, E. W., Eventual effects of faulty pipelaying in Wisconsin, *J.A.W.W.A.* (Dec. 1955).

93. Clarke, N. W. B., Pipe laying principles, *Nat. Bldg. Studies Special Report No. 35* (1964). H.M.S.O.

94. Fairbank, J. R., The testing of newly laid water mains, *Water and Water Eng.* (July 1959).

95. Price, H. A., Multiple jacking of concrete pipe at National City, 120 in. pipe jacked 1300 ft, *J.A.W.W.A.* (Feb. 1961).

96. Roske, K., The imperviousness of concrete in R.C. pipes and pipelines, *Int. Cong. Precast Concrete Ind.* (June 1957). Weisbaden.

97. Sowers, G. F., Trench excavation and brickfilling, *J.A.W.W.A.* (July 1956).

98. Spangler, M. G., Influence of compression and shearing strains in soil foundations on structures under earth embankments, *Bulletin 126, U.S. Highway Res. Bd.* (1955).

99. Spangler, M. G., A practical application of the imperfect ditch method of construction, *Proc. U.S. Highway Res. Bd.* (1958).

100. Storey, J. B., Flexibly jointed strengthened concrete pipes laid on granular beds, *Chart. Mun. Eng.* (Aug. 1964).

100*a* Storey, J. B., Some factors affecting the construction of extra strength pipe sewers, *J. I. Mun. E.* (Jan. 1966).

101. Ward, W. H., Effect of vegetation on the settlement of structures, *Proc. Conf. Biology and Civil Eng. Inst. C.E.* (1948).

102. Ward, W. H., The use of simple relief wells in reducing water pressure beneath a trench excavation, *Geotechnique V.* (Sept. 1957), viii.

103. Young, O. C., Pipeline design—the relationship between structural theory and laying practice, *Report of Public Works Congress (1964)*.

Corrosion and Protection

104. Anon, Concrete in sulphate bearing clays and ground waters, *B.R.S. Digest No. 31* (1951). H.M.S.O.

105. Anon, Internal corrosion of concrete sewer pipes, *B.R.S. Digest No. 79* (1955). H.M.S.O.

106. Anon, Principles of corrosion, protective coatings (steel pipe), etc., *J.A.W.W.A.* (April/July 1962).

107. Beaton, J. L. and Stratfield, R. F., Corrosion of corrugated metal culverts in California, *Bull. U.S. Highway Res. Bd.* (1959), 223.

108. Butlin and Vernon, Underground corrosion in metals, causes and prevention, *J. I. Water E.* (1949).

109. Douglas, A. M., Corrosion investigations at Burton-on-Trent, *Chart. Mun. Eng.* (April 1964).

110. Jones and Latham, A survey of the behaviour in use of asbestos-cement pressure pipe, *Nat. Bldg. Studies, Sp. Report No. 15* (1952). H.M.S.O.

111. Lea, F. M. and Davey, N., The deterioration of concrete in structures, *J. Inst. C.E.* (May 1949).

112. Nat. Bur. Stan., Study of causes and effects of underground corrosion, *J.A.W.W.A.* (Dec. 1958).

113. Parker, C. D., Mechanics of corrosion of concrete sewers by H_2S, *Sewage and Indust. Wastes* (1951), V. 23, No. 12.

114. Santry, J. W., Hydrogen sulphide in sewers, *J. Water Poll. Control Fed.* (Dec. 1963). U.S.A.

115. Stutterheim, N. *et al.*, Corrosion of concrete sewers, *S. Africa C.S.I.R. Report Series DR 12* (1958).

Hydraulics

116. Ackers, P., Hydraulic resistance of drainage conduits, *Proc. I.C.E.* (July 1961).

117. Ackers, P., Charts for the design of channels and pipes, *H.R.L. Paper No. 1*, 2nd Edition (1963). H.M.S.O. London.

118. Ackers, P., Tables for the hydraulic design of storm drains, sewers and pipelines, *H.R.L. Paper No. 4*, (1963). H.M.S.O. London.

119. Ackers, P., *et al.*, Effects of use on the hydraulic resistance of drainage conduits (slime effects), *Proc. I.C.E.* (July 1964).

120. Ackers, P., Hydraulic Research of interest to public health engineers, *J.I.P.H.E.* (Autumn 1965).

See also Refs. 9, 57, 64, 71.

Tables

TABLE A

THEORETICAL VALUES OF FILL LOAD COEFFICIENTS

Installation condition	Coefficient and its theoretical value	Derivation	Remarks
'Narrow' trench	$C_d = \dfrac{1 - e^{-2K\mu' H/B_d}}{2K\mu'}$	See Art. 4.1(*a*)	$K\mu'$ varies from 0·11 to 0·19 C_d reduces to $1/(2K\mu')$ when H/B_d is very large (say > 20)
Positive projection in embankment or 'wide' trench	Complete projection or complete trench condition ($H < H_e$) $C_c = \dfrac{e^{\pm 2K\mu H/B_c} - 1}{\pm 2K\mu}$	See Art. 4.4(*b*) (i)	$K\mu = 0{\cdot}19$ (or less) when r_{sd} is positive (complete projection) using upper signs $K\mu = 0{\cdot}13$ (or more) when r_{sd} is negative (complete trench) using lower signs
	Incomplete projection or trench condition ($H > H_e$) $C_c = \dfrac{e^{\pm 2K\mu H_e/B_c} - 1}{\pm 2K\mu} + \left(\dfrac{H}{B_c} - \dfrac{H_e}{B_c}\right) e^{\pm 2K\mu H_e/B_c}$	See Art. 4.4(*b*) (ii)	$K\mu = 0{\cdot}13$ (or more) when r_{sd} is negative (incomplete proj.) using lower signs $K\mu = 0{\cdot}13$ (or more) when r_{sd} is negative (incomplete trench) using lower signs
Negative projection or induced trench in embankment or 'wide' trench with 'narrow' sub-trench	Complete projection ($H < H_e$) $C_n = \dfrac{1 - e^{-2K\mu H/B_d}}{2K\mu}$	See Art. 4.8(*b*) (i)	$K\mu = 0{\cdot}13$ (or more)
	Incomplete projection ($H > H_e$) $C_n = \dfrac{1 - e^{-2K\mu H_e/B_d}}{2K\mu} + \left(\dfrac{H}{B_d} - \dfrac{H_e}{B_d}\right) e^{-2K\mu H_e/B_d}$	See Art. 4.8(*b*) (ii)	$K\mu = 0{\cdot}13$ (or more) r_{sd} is always negative

TABLE B

SETTLEMENT RATIOS (r_{sd})

For *positive projecting* conduits

$$r_{sd} = \frac{(S_m + S_g) - (S_f + d_c)}{S_m}$$

$$= \frac{\text{Settlement of critical plane} - \text{settlement of pipe crown}}{\text{Settlement of soil in the height of } pB_c}$$

For *negative projecting* conduits

$$r_{sd} = \frac{S_g - (S_d + S_f + d_c)}{S_d}$$

$$= \frac{\text{Settlement of natural ground} - \text{settlement of critical plane}}{\text{Settlement of soil in the height } p'B_d \text{ (or } p'B_c \text{ for induced trench)}}$$

Where S_d = settlement of soil in the height $p'B_d$ (negative projection) or $p'B_c$ (induced trench)

S_f = settlement of the pipe invert

S_g = settlement of the natural ground surface

S_m = settlement of soil in the height pB_c (positive projection)

d_c = vertical deflection of pipe

VALUES OF r_{sd} RECOMENDED FOR USE IN DESIGN (after Spangler)[1,23,24]

Conduit type	Construction	Projection condition	Foundation soil condition	Recommended value
Rigid	Wide trench	Positive	Any	+1·0
	Embankment	Positive	Rock or hard unyielding soil	+1·0
			Ordinary soils	+0·5 to 0·8
			Yielding soils	+0·0 to 0·5
	Any	Positive	Pipeline supported on piles in soft soil	Uncertain Use complete projection value of C_c
	Wide trench	Negative	Any protem (value not well established)	−0·3
	Embankment	Negative	Any protem (value not well established)	−0·3
	Induced trench	Negative	Any protem (value not well established)	0·0 to −0·3
Flexible	Wide trench Embankment	Positive Positive	Any with poorly compacted side fill	0·0 to −0·4
			Any with well compacted side fill	0·0
	Wide trench Embankment Induced trench	Negative Negative Negative	Any with well compacted side fill	0·0

TABLE C

NEWMARK COMPUTATIONS

Influence Values I (Newmark)* for the *Total Load* PI on a Rectangular Horizontal Area of Width B and Length L and at a Depth H below the Surface caused by a Point Load P on the Surface when Vertically above one Corner of the Rectangle (see Figs. 4.26 and 4.27).

or

For the Load per Unit Area at a Depth H below the Surface at a Point Vertically below one Corner of a Uniformly Loaded Rectangle on the Surface (see Figs. 4.29 and 4.30).

$m = \frac{B}{H}$	$n = \frac{L}{H}$											
	0·1	0·2	0·3	0·4	0·5	0·6	0·7	0·8	0·9	1·0	1·2	1·4
0·1	0·00470	0·00917	0·01323	0·01678	0·01978	0·02223	0·02420	0·02576	0·02698	0·02794	0·02926	0·03007
0·2	0·00917	0·01790	0·02585	0·03280	0·03866	0·04348	0·04735	0·05042	0·05283	0·05471	0·05733	0·05894
0·3	0·01323	0·02585	0·03735	0·04742	0·05593	0·06294	0·06858	0·07308	0·07661	0·07938	0·08323	0·08561
0·4	0·01678	0·03280	0·04742	0·06024	0·07111	0·08009	0·08734	0·09314	0·09770	0·10129	0·10631	0·10941
0·5	0·01978	0·03866	0·05593	0·07111	0·08403	0·09473	0·10340	0·11035	0·11584	0·12018	0·12626	0·13003
0·6	0·02223	0·04348	0·06294	0·08009	0·09473	0·10688	0·11679	0·12474	0·13105	0·13605	0·14309	0·14749
0·7	0·02420	0·04735	0·06858	0·08734	0·10340	0·11679	0·12772	0·13653	0·14356	0·14914	0·15703	0·16199
0·8	0·02576	0·05042	0·07308	0·09314	0·11035	0·12474	0·13653	0·14607	0·15371	0·15978	0·16843	0·17389
0·9	0·02698	0·05283	0·07661	0·09770	0·11584	0·13105	0·14356	0·15371	0·16185	0·16835	0·17766	0·18357
1·0	0·02794	0·95471	0·07938	0·10129	0·12018	0·13605	0·14914	0·15978	0·16835	0·17522	0·18508	0·19139
1·2	0·02926	0·05733	0·08323	0·10631	0·12626	0·14309	0·15703	0·16843	0·17766	0·18508	0·19584	0·20278
1·4	0·03007	0·05894	0·08561	0·10941	0·13003	0·14749	0·16199	0·17389	0·18357	0·19139	0·20278	0·21020
1·6	0·03058	0·05994	0·08709	0·11135	0·13241	0·15028	0·16515	0·17739	0·18737	0·19546	0·20731	0·21510
1·8	0·03090	0·06058	0·08804	0·11260	0·13395	0·15207	0·16720	0·17967	0·18986	0·19814	0·21032	0·21836
2·0	0·03111	0·06100	0·08867	0·11342	0·13496	0·15326	0·16856	0·18119	0·19152	0·19994	0·21235	0·22058
2·5	0·03138	0·06155	0·08948	0·11450	0·13628	0·15483	0·17036	0·18321	0·19375	0·20236	0·21512	0·22364
3·0	0·03150	0·06178	0·08982	0·11495	0·13684	0·15550	0·17113	0·18407	0·19470	0·20341	0·21633	0·22499
4·0	0·03158	0·06194	0·09007	0·11527	0·13724	0·15598	0·17168	0·18469	0·19540	0·20417	0·21722	0·22600
5·0	0·03160	0·06160	0·09014	0·11537	0·13737	0·15612	0·17185	0·18488	0·19561	0·20440	0·21749	0·22632
6·0	0·03161	0·06201	0·09017	0·11541	0·13741	0·15617	0·17191	0·18496	0·19569	0·20449	0·21760	0·22644
8·0	0·03162	0·06202	0·09018	0·11543	0·13744	0·15621	0·17195	0·18500	0·19574	0·20455	0·21767	0·22652
10·0	0·03162	0·06202	0·09019	0·11544	0·13745	0·15622	0·17196	0·18502	0·19576	0·20457	0·21769	0·22654
∞	0·03162	0·06202	0·09019	0·11544	0·13745	0·15623	0·17197	0·18502	0·19577	0·20458	0·21770	0·22656

$m = \frac{B}{H}$	$n = \frac{L}{H}$										
	1·6	1·8	2·0	2·5	3·0	4·0	5·0	6·0	8·0	10·0	∞
0·1	0·03058	0·03090	0·03111	0·03138	0·03150	0·03158	0·03160	0·03161	0·03162	0·03162	0·03162
0·2	0·05994	0·06058	0·06100	0·06155	0·06178	0·06194	0·06199	0·06201	0·06202	0·06202	0·06202
0·3	0·08709	0·08804	0·08867	0·08948	0·08982	0·09007	0·09014	0·09017	0·09018	0·09019	0·09019
0·4	0·11135	0·11260	0·11342	0·11450	0·11495	0·11527	0·11537	0·11541	0·11543	0·11544	0·11544
0·5	0·13241	0·13395	0·13496	0·13628	0·13684	0·13724	0·13737	0·13741	0·13744	0·13745	0·13745
0·6	0·15028	0·15207	0·15326	0·15483	0·15550	0·15598	0·15612	0·15617	0·15621	0·15622	0·15623
0·7	0·16515	0·16720	0·16856	0·17036	0·17113	0·17185	0·17185	0·17191	0·17195	0·17196	0·17197
0·8	0·17739	0·17967	0·18119	0·18321	0·18407	0·18469	0·18488	0·18496	0·18500	0·18502	0·18502
0·9	0·18737	0·18986	0·19152	0·19375	0·19470	0·19540	0·19561	0·19569	0·19574	0·19576	0·19577
1·0	0·19546	0·19814	0·19994	0·20236	0·20341	0·20417	0·20440	0·20449	0·20455	0·20457	0·20458
1·2	0·20731	0·21032	0·21235	0·21512	0·21633	0·21722	0·21749	0·21760	0·21767	0·21769	0·21770
1·4	0·21510	0·21836	0·22058	0·22364	0·22499	0·22600	0·22632	0·22644	0·22652	0·22654	0·22656
1·6	0·22025	0·22372	0·22610	0·22940	0·23088	0·23200	0·23236	0·23249	0·23258	0·23261	0·23263
1·8	0·22372	0·22736	0·22986	0·23334	0·23495	0·23617	0·23656	0·23671	0·23681	0·23684	0·23686
2·0	0·22610	0·22986	0·23247	0·23614	0·23782	0·23912	0·23954	0·23970	0·23981	0·23985	0·23987
2·5	0·22940	0·23334	0·23614	0·24010	0·24196	0·24344	0·24392	0·24412	0·24425	0·24429	0·24432
3·0	0·23088	0·23495	0·23782	0·24196	0·24394	0·24554	0·24608	0·24630	0·24646	0·24650	0·24654
4·0	0·23200	0·23617	0·23912	0·24344	0·24554	0·24729	0·24791	0·24817	0·24836	0·24842	0·24846
5·0	0·23236	0·23656	0·23954	0·24392	0·24608	0·24791	0·24857	0·24885	0·24907	0·24914	0·24919
6·0	0·23249	0·23671	0·23970	0·24412	0·24630	0·24817	0·24885	0·24916	0·24939	0·24946	0·24952
8·0	0·23258	0·23681	0·23981	0·24425	0·24646	0·24836	0·24907	0·24939	0·24964	0·24973	0·24980
10·0	0·23261	0·23684	0·23985	0·24429	0·24650	0·24842	0·24914	0·24946	0·24973	0·24981	0·24989
∞	0·23263	0·23686	0·23987	0·24432	0·24654	0·24846	0·24919	0·24952	0·24980	0·24989	0·25000

* After N.M. Newmark 'Simplified Computation of Vertical Pressures in Elastic Foundations', Circ. 24, *Eng. Exp. Stn.*, University of Illinois, 1935.

TABLE D

MAGNITUDE OF SURCHARGES APPLICABLE IN ENGLAND AND WALES

As recommended in the Second Report of the MOHLG Working Party issued in October 1967 (see Ref. 10*a*, para. 235 (*a*) to (*d*)).

Position	Minimum Concentrated Surcharge*
Under main traffic routes or roads liable to temporary diversions of heavy traffic.	B.S. 153, Type HB loading as shown in Fig. 4.24 with an impact factor of 1·3, i.e. the equivalent of 8 static wheel loads, each of 26,000 lb. acting simultaneously.
Under other roads, except access roads used only for very light traffic.	Two 16,000 lb. wheel loads spaced 3 ft. apart acting simultaneously with an impact factor of 1·5, i.e. the equivalent of 2 static wheel loads, each of 24,000 lb.†
Under lightly trafficed access roads and private carriageways, fields and gardens.	Two 7,000 lb. loads spaced 3 ft. apart acting simultaneously, with an impact factor of 2·0, i.e. the equivalent of 2 static wheel loads each of 14,000 lb.†
Under railways.	The British Transport Commission considers (1961) that the equivalent railway loading shown in Fig. 4.25, Case III represents the maximum envisaged for main lines in this country.‡
	Uniformly Distributed Surcharge
Any.	No special provision need be made for loads which are not likely to exceed 500 lb/ft^2 and 15 ft × 15 ft, in extent in plan. Otherwise the maximum load to be provided for.

Notes * All impact factors are independent of the depth of cover.

† The 3 ft spacing of the loads is regarded as the minimum distance between the adjacent wheels of passing vehicles. The axle spacing of the wheels is assumed to be not less than 6 ft.

‡ The loads imposed by this surcharge on pipes crossing the rail tracks have been computed and prove to be somewhat less severe than the B.S. 153 HB loading with an impact factor of 2·0 for cover depths of 3 ft or more. This loading was not specified in the above Report by the Working Party.

TABLE E

MAXIMUM EFFECTIVE DESIGN FLUID PRESSURES

Pipe type	Service Condition	Effective fluid pressures				Effective design fluid pressure
		External		Internal		
		$p_{e_{max}}$	$p_{e_{min}}$	$p_{i_{max}}$	$p_{i_{min}}$	
Rigid	Non-pressure	$p_{g_{max}}$	0	0	0	$p_{g_{max}}$, which may be ignored unless large with respect to W_e
	Pressure	$p_{g_{max}} + p_{vac}$	0	$-p_t$	0	$p_{g_{max}} + p_{vac}$ (which may be ignored unless large with respect to W_e) when empty; $-p_t$ when under pressure
Flexible	Non-pressure	$p_{g_{max}}$	0	0	0	$p_{g_{max}}$
	Pressure	$p_{g_{max}} + p_{vac}$	0	$-p_t$	0	$p_{g_{max}} + p_{vac}$ when empty; $-p_t$ when under pressure and uncovered

TABLE F

PIPE CLASSIFICATION

Rigid				Flexible			
Non-pressure	B.S. No.	Pressure	B.S. No.	Non-pressure	B.S. No.	Pressure	B.S. No.
Concrete	556	—	—	Steel (smooth wall)	534, 3601 and 1387	Steel (smooth wall)	534, 3601 and 1387
Concrete OGEE (Surface water)	4101	Prestressed concrete	—	Corrugated steel	—	—	—
Reinforced concrete	556	Reinforced concrete	556	Ductile iron	—	Ductile iron	—
Porous concrete (Field drains)	1194	—	—	Pitch fibre	2760	—	—
Clayware	65 and 540	—	—	uPVC	3505 and 3506	Unplasticised PVC	3505 and 3506
Clayware field drains	1196	—	—				
Cast iron	78	Cast iron	78				
Spun iron	1211	Spun iron	1211				
Asbestos-cement	3656	Asbestos-cement	486				

Note. Oval or egg shaped or flat bottomed circular tubes, horseshoe, semicircular, 'D' and rectangular sections are classed as 'non circular' conduits, for which there are no British Standards.

TABLE G

MINIMUM FACTORS OF SAFETY (F_s) FOR RIGID NON-PRESSURE PIPES HAVING A B.S. (OR OTHERWISE GUARANTEED) MINIMUM PROOF AND/OR ULTIMATE CRUSHING TEST LOAD

Pipe Type	F_s min for pipes in trench applicable to specified minimum crushing test load				Anticipated quality of site work and supervision
	Ultimate load		0·01 crack or proof load		
	Cover < 10 ft	Cover ⩾ 10 ft	Cover < 10 ft	Cover ⩾ 10 ft	
Any rigid pipe for which both minimum ultimate and minimum proof crushing test loads are specified, and for which the specified 0·01 in. crack, or proof, load is not more than 80 per cent of the specified ultimate load	1·25	1·25 to 1·5	1·0	1·0 to 1·20	First class
Any for which only the minimum safe or ultimate crushing test strength is specified	1·25	1·25 to 1·5	—	—	First class

Notes 1. Applicable to circular pipes only of plain or reinforced concrete, asbestos-cement, clayware, cast or spun iron laid in trenches, heading or thrust bore, or in positive or negative or induced trench conditions under embankments or valley fills. See also Article 5.2 (*d*).

2. Not applicable to rigid pressure pipes, for which see below.
   Not applicable to flexible pipes, for which see Article 6.2.
3. For any rigid circular pipes for which no B.S. (or otherwise guaranteed) minimum or ultimate crushing test strength is specified, but for which a few individual test values or average test values are known, and/or where the quality of site work and supervision is doubtful, the above values with respect to ultimate strength should be 1·5 to 2·25 at all cover depths.
4. Where a range of values is suggested above, the actual value to be used within the range, will depend on the degree of confidence in the particular local conditions.
5. For pipes laid under embankments or high fills, $F_{s_{min}} = 1{\cdot}5$ (Article 5.2 (*d*).

MINIMUM FACTORS OF SAFETY (F_{se}, F_{si}) FOR RIGID PRESSURE PIPES HAVING A B.S. MINIMUM ULTIMATE CRUSHING TEST LOAD AND BURSTING PRESSURE

(or minimum specified characteristics from which they can be derived)

Material	B.S.	Minimum Factor of Safety for	
		Crushing F_{se}	Bursting F_{si}
Asbestos-Cement	486/1956	2·5 on ultimate load	4·0 on bursting pressure
Cast Iron Spun Iron	78/1961 1211/1958	2·0 to 2·5 on ultimate load	2·0 to 2·5 on bursting pressure
Steel Reinforced Concrete	556/1966	1·6 on $\frac{1}{100}$ in. crack load	2·0 on bursting pressure
Prestressed concrete, cylinder type	—	1·0 on $0{\cdot}90 \times \frac{1}{1000}$ in. crack load	1·0 on zero compression pressure
Prestressed concrete, non-cylinder type	—	Not yet determined	

All factors assume first class site work.

Standard Minimum Crushing Test Loads (W_T) for Rigid Pressure and non Pressure Pipes

Pipe Size I.D. in.	Clayware Pipes B.S. 65/1966 and 540/1966		Concrete Pipes B.S. 556/1966										Asbestos-cement Sewer Pipes B.S. 3656/1963		
			Standard		Extra Strength										
	Stan. Str. lb/ft	Extra Str. lb/ft	Class 1 lb/ft		Class 2 lb/ft		Class 3 lb/ft		Class 4 lb/ft		Class 5 lb/ft		Class 1 lb/ft	Class 2 lb/ft	Class 3 lb/ft
min. ult.	min. ult.	min. ult.	min. ult.	min. proof	min. proof	min. ult.	min. proof	min. ult.	min. proof	min. ult.	min. proof	min. ult.	min. ult.	min. ult.	min. u .
3	—	—	—	—	—	—	—	—	—	—	—	—	—	—	—
4	1350	1500	1350	1600	—	—	—	—	—	—	1600	2000	—	—	2600
5	—	—	—	—	—	—	—	—	—	—	—	—	—	—	2600
6	1350	1500	1350	1600	—	—	—	—	—	—	1750	2190	—	—	2600
7	—	—	—	—	—	—	—	—	—	—	—	—	—	—	2700
8	—	—	—	—	—	—	—	—	—	—	—	—	—	—	2700
9	1350	1850	1350	1600	—	—	—	—	—	—	2100	2630	—	2000	2700
10	—	—	—	—	—	—	—	—	—	—	—	—	—	2200	3000
12	1450	2250	1350	1600	—	—	1850	2310	—	—	2800	3500	2000	2500	3600
15	—	—	1350	1600	1630	2030	2100	2630	2880	3600	3750	4700	2600	3800	5200
18	—	—	1350	1600	1950	2440	2600	3250	3300	4120	4350	5450	2800	4200	5800
21	—	—	1350	1600	2280	2840	3000	3750	3850	4810	4900	6130	3100	4600	6500
24	—	—	1350	1600	2600	3250	3400	4250	4400	5500	5600	7000	3600	5400	7600
27	—	—	1350	1600	2930	3660	3800	4750	4950	6190	6300	7880	—	—	—
30	—	—	1350	1600	3250	4060	4150	5190	5250	6560	6750	8440	—	—	—
33	—	—	1350	1600	3580	4470	4540	5690	5780	7220	7430	9280	—	—	—
36	—	—	1350	1600	3900	4880	4960	6200	6300	7880	8100	10130	—	—	—
39	—	—	1350	1600	4230	5280	5370	6710	6830	8530	8780	10970	—	—	—
42	—	—	1350	1600	4550	5690	5780	7250	7350	9190	9450	11830	—	—	—
45	—	—	1350	1600	4880	6090	6190	7740	7880	9840	10130	12660	—	—	—
48	—	—	1350	1600	5200	6500	6600	8250	8400	10500	10800	13500	—	—	—
51	—	—	1350	1600	5530	6900	7010	8760	8930	11160	11480	14340	—	—	—
54	—	—	1350	1600	5850	7310	7430	9290	9450	11810	12150	15190	—	—	—
57	—	—	1350	1600	6180	7720	7840	9800	9980	12470	12830	16030	—	—	—
60	—	—	1350	1600	6500	8130	8250	10300	10500	13130	13500	16880	—	—	—
63	—	—	1350	1600	6830	8530	8670	10840	11030	13780	14180	17720	—	—	—
66	—	—	1350	1600	7150	8940	9080	11370	11550	14440	14850	18560	—	—	—
69	—	—	1350	1600	7480	9340	9500	11870	12080	15090	15530	19410	—	—	—
72	—	—	1350	1600	7800	9750	9900	12570	12600	15750	16200	20250	—	—	—

Asbestos-cement Pressure Pipes B.S. 486/1956

Classes: A B C D

Compute W_T and bursting pressure p_{ult} from the formulae:

$W_T = 47000\, t^2/R$ lb/lin. ft

$= 3200\, 2t/D$ lb/in.²

Where t = wall thickness (in.
R = mean radius (in.)
D = Internal diameter (in.)

both given in B.S. 486.

See B.S. 486 also for maximum test and working hydraulic pressures.

CAST AND SPUN IRON PRESSURE PIPES

B.S. 78 and B.S. 1211 do not specify the data necessary for computation of W_T and p_{ult}.
Refer to makers.
See B.S. 78 or B.S. 1211 also for wall thickness, and maximum test and working hydraulic pressures.

NOTE: Unreinforced concrete pipes are not normally available in sizes and strength classes to the right of and below the heavy dotted line on the table.

FLEXIBLE PIPES

MINIMUM YIELD STRESS (f_y) AND ELASTIC MODULUS (E) FOR FLEXIBLE PRESSURE AND NON PRESSURE PIPES

1. For the wall thickness of smooth walled steel pipes see B.S. 534.
   For the wall thickness of galvanised steel tubes see B.S. 1387.
   Neither standard specifies the minimum yield stress or the elastic modulus of the steel, for which refer to makers.
2. For the wall thickness of corrugated steel pipes and segments, made in England to U.S. specifications see Table K. For these pipes the specified minimum yield stress of the metal is 40,000 lb/in.², and the elastic modulus is 30×10^6 lb/in.².
3. For the wall thickness of P.V.C. pipes see B.S. 3505 and B.S. 3526. Neither standard specifies the long term yield stress or elastic modulus of the material.
4. There is no B.S. yet for DUCTILE IRON, **refer to the makers for data.**

TABLE H

TABLE J

COMPARATIVE STANDARDS

N.B. These Standards deal with similar articles but may vary widely in detail.

Material	Subject	British B.S. No.	American A.S.T.M. No.	International I.S.O. No.
Aggregates	Aggregates from natural sources concrete	882, 1201	C 33	—
	Building sands from natural sources	1198–1200	C 33	
Bricks	Bricks and blocks of fired brick earth, clay and shale	3921	C 32	—
Jointing materials	Rubber joint rings for gas, water and sewage	2494	C 425, C 443, D 1869	—
	Oil resisting joint rings	3514	?	—
Loads	Girder bridge loads and stresses (1954)	963 153 Part 3A	?	—
Pipes	Clay drain and sewer pipes	65, 540, 1196	C 4, C 13, C 200, C 211, C 301	—
	Cast iron spigot and socket pipes	78	A 377, A 142, ASA 121·1	
	Asbestos-cement pressure pipes	486	C 296, C 500	R.160
	Asbestos-cement pipes and fittings for sewerage and drainage	3656	C 428	—
	Steel pipes for water, gas and sewage	534, 3601	ASA.B36.10	—
	Concrete cylindrical pipes and fittings	556	C 14, C 76, C 361, C 506, C 507	—
	Concrete porous pipes for under drainage	1194	C 444	—
	Concrete tubes and fittings with OGEE joints	4101	C 118	—
	Spun iron pressure pipes for water, gas and sewage	1211	?	R.13
	Pitch impregnated fibre drain and sewer pipes	2760	D 1861	—
	Unplasticised PVC pipes for water and industrial uses	3505–6	D 2241	—
Soil	Standard soil tests	1377	?	—
Steel reinforcement	Hot rolled bars and hard drawn wire	785	A 82, A 431	—
	Cold twisted bars	1144	?	—
	Steel woven fabric	1221	A 185	—
Steel tubes	Screwed ends	1387	ASA.B36.10	R.65
	Plain ends	3601–2	?	R.64

TABLE K

PHYSICAL PROPERTIES OF CORRUGATED STEEL PLATES U.S. PATTERN[(40)]

Dimensions of corrugations	U.S. Gauge No.	Thickness, t in.	Moment of inertia, I in.^{4}/in.	Area of section A in.^{2}/lin. in.
6 in. pitch × 2 in. depth	12	0·1046	0·0604	0·1297
	10	0·1345	0·0781	0·1669
	8	0·1644	0·0961	0·2041
	7	0·1838	0·1080	0·2283
	5	0·2145	0·1270	0·2666
	3	0·2451	0·1463	0·3048
	1	0·2758	0·1659	0·3432
2·66 in. pitch × 1/2 in. depth	16	0·0598	0·0020	0·0649
	14	0·0747	0·0025	0·0810
	12	0·1046	0·0035	0·1134
	10	0·1345	0·0045	0·1459
	8	0·1644	0·0055	0·1783

Min. Yield Stress, 40,000 lb/in.^{2}, $E = 30 \times 10^6$ lb/in.^{2}

TABLE L

COEFFICIENTS OF DEFLECTION AND BENDING MOMENTS FOR FLEXIBLE PIPES WITH NO SIDE SUPPORT (after Spangler)[(46a)]

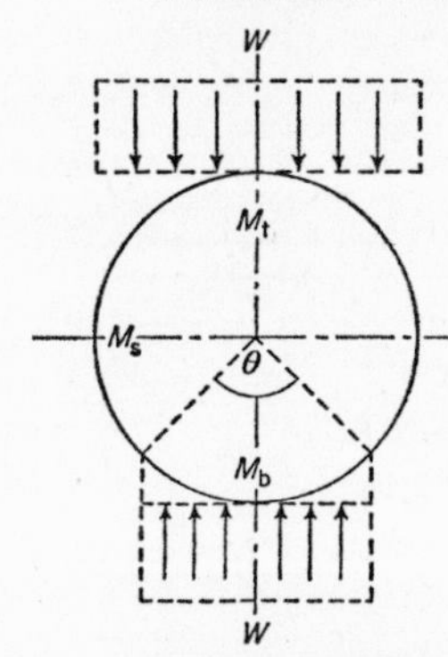

As in Equation (6.3),

$$\Delta_x{}^* = \frac{K_\theta W R^3}{12EI} \text{ in.} \qquad \text{(i)}$$

$$M_b = \frac{K_b W R}{12} \text{ in. lb/in.} \qquad \text{(ii)}$$

Substituting the value of W from Eqn. (i) in Eqn. (ii),

$$M_b = \frac{K_b}{K_\theta}\,\frac{\Delta_x EI}{R^2} \text{ in. lb/in.} \qquad \text{(iii)}$$

where W is the total effective external load in lb. per linear foot and the other quantitites are in lb. and in. units.

θ Degrees	K_b	K_t	K_s	K_θ Deflection
	Bending moments			
0	0·294	0·150	0·153	0·110
30	0·235	0·148	0·152	0·108
60	0·189	0·143	0·147	0·103
90	0·157	0·137	0·140	0·096
120	0·138	0·131	0·133	0·089
150	0·128	0·126	0·127	0·085
180	0·125	0·125	0·125	0·083

* N.B. For the deflection of thin walled pipes supported by the surrounding soil see Article 6.4 (*c*) and Equations (6.10) and (6.11).

TABLE M

SUMMARY OF LOAD FORMULAE FOR RIGID AND FLEXIBLE

Construction condition	Bedding factor	Source of load	Load formula	Chart
(i) TRENCH LOADING—RIGID CONDUITS				
'*Narrow*' parallel-sided trench or Parallel-sided, or 'V', trench with 'narrow' sub-trench	F_m	Trench fill	$W_c = C_d \gamma_s B_d^2$	C1 for C_d use H/B_d
		Uniform surcharge of large extent	$W_{us} = C_{us} B_d U_s$	C2 for C_{us}
'*Wide*' parallel-sided, or V, trench without sub-trench; or with 'wide' sub-trench; or in '*very wide*' trench *without* sub-trench	F_m	Trench fill	$W'_c = C_c \gamma_s B_c^2$	C3 for C_c and C'_c
	Use F_p only for 'very wide' trench	Uniform surcharge of large extent	$W'_{us} = (C'_c - C_c) \gamma_s B_c^2$	
(ii) HEADING OR THRUST BORE LOADING*—RIGID CONDUITS				
Heading or thrust bore	F_m	Fill, i.e. overburden	$W_c = C_d \gamma_s B_c^2$	C1 for C_d
			$W_c = C_d \gamma_s B_t^2$	
		Uniform surcharge of large extent	$W_{us} = C_{us} B_c U_s$	C2 for C_{us}
			$W_{us} = C_{us} B_t U_s$	
* N.B. These formulae, and the Charts ignore soil cohesion in the overburden soil.				
(iii) WATER LOAD—RIGID OR FLEXIBLE CONDUITS				
All conditions	As for fill load	Weight of water in full conduit	$W_w = 0{\cdot}75 \left(\frac{\pi}{4}\right) D^2 \gamma_w$	—
			$W_w = 0{\cdot}5 \left(\frac{\pi}{4}\right) D^2 \gamma_w$	
			$W_w = 0$	—
(iv) EMBANKMENT LOADING—RIGID CONDUITS				
Positive projecting conduit without sub trench or in 'wide' sub trench	F_p	Fill	$W'_c = C_c \gamma_s B_c^2$	C3 for C_c and C'_c
		Uniform surcharge of large extent	$W'_{us} = (C'_c - C_c) \gamma_s B_c^2$	
Negative projecting conduit in 'narrow' sub trench	F_m	Fill	$W_c = C_n \gamma_s B_d^2$	C4 for C_n and C'_n if Ku = 0·13 or more
		Uniform surcharge of large extent	$W_{us} = (C'_n - C_n) \gamma_s B_d^2$ $W_{us} = C_{us} B_d U_s$	C2 for C_{us}

TABLE M

PIPES LAID UNDER VARIOUS CONSTRUCTION CONDITIONS

Remarks
Use when $W_c < W'_c$ (see below under wide trench) i.e. where B_d is less than the transition width.
Use only when W_c prevails for the fill load.
Use when $W'_c < W_c$, i.e. B_d is $\geqslant$ transition width. Use $r_{sd}p = 1{\cdot}0$ for C_c or C'_c and concrete bedding. Use $r_{sd}p = 0{\cdot}5$ for C_c or C'_c for Class B type beddings extending over full width of trench or sub-trench. Use F_p when $B_d \geqslant B_c + 2(H + mB_c)/\mu$.
Use only when W'_c prevails for the fill load. Use $H' = H + U_s/\gamma_s$ to obtain W'_c with same chart as for the fill load W'_c.
Use B_c in place of B_d in Chart C1 for solid packing.
Use B_t in place of B_d in Chart C1 for compressible packing. Use $K\mu$ for undisturbed soil for W_c and W_{us}.
Use B_c in place of B_d in Chart C2 for solid packing.
Use B_t in place of B_d in Chart C2 for compressible packing.
Use for rigid pipes when not permanently submerged. Use for flexible pipes when not permanently submerged.
When any pipes are permanently submerged.
Select r_{sd} by reference to prevailing foundation soil conditions, Table B. Select p by reference to Fig. 4.7 and Bedding Class, Chart C10.
Use $H' = (H + U_s/\gamma_s)$ for C'_c.
Determine the minimum value of p' from Fig. 4.7 and site contour and use that value in obtaining C_n and C'_n. Use $r_{sd} = -0{\cdot}3$.
Use for incomplete projection condition. Use for complete projection condition.

TABLE M—*continued*

Construction condition	Bedding factor	Source of load	Load formula	Chart
(iv) EMBANKMENT LOADING—RIGID CONDUITS (*Continued*)				
Induced trench with trench width = B_c	F_p	Fill	$W_c = C_n \gamma_s B_c^2$	C4 for C_n and C'_n
		Uniform surcharge of large extent	$W_{us} = (C'_n - C_n)\gamma_s B_c^2$ $W_{us} = C_{us} B_c U_s$	C2 for C_{us}
(v) TRENCH LOADING—FLEXIBLE CONDUITS				
'*Narrow*' parallel-sided trench or Parallel-sided, or 'V' trench with 'narrow' sub-trench	$K\theta$	Trench fill	$W_c = C_d \gamma_s B_c B_d$ (or $C_d \gamma_s B_d^2$) (see remarks)	C1 for C_d use H/B_d
		Uniform surcharge or large extent	$W_{us} = C_{us} B_c U_s$ (or $C_{us} B_d U_s$)	C2 for C_{us}
'*Wide*' parallel-sided, or 'V', trench without sub-trench, or with 'wide' sub-trench	$K\theta$	Trench fill	$W'_c = H \gamma_s B_c$	—
		Uniform surcharge or large extent	$W'_{us} = B_c U_s$	—
(vi) HEADING OR THRUST BORE LOADING†—FLEXIBLE CONDUITS				
Heading or thrust bore	$K\theta$	Fill	$W_c = C_d \gamma_s B_c^2$	C1 for C_d
			$W_c = C_d \gamma_s B_c B_t$	C1 for C_d
		Uniform surcharge or large extent	$W_{us} = C_{us} B_c U_s$	C2 for C_{us}
† These formulae and the charts ignore soil cohesion in the overburden soil.				
(vii) EMBANKMENT LOADING—FLEXIBLE PIPES				
Positive projecting pipe in '*wide*' sub-trench or in '*very wide*' trench without sub-trench	$K\theta$	Fill	$W'_c = H \gamma_s B_c$	—
		Uniform surcharge of large extent	$W'_{us} = U_s B_c$	—
True positive projection with no sub-trench	$K\theta$	Fill	$W'_c = C_c \gamma_s B_c^2$	C3 for C_c and C'_c
		Uniform surcharge of large extent	$W'_{us} = (C'_c - C_c)\gamma_s B_c^2$	

TABLE M

Remarks
Use $r_{sd} = -0{\cdot}3$ to $0{\cdot}0$ for C_n and C'_n.
Use for 'incomplete' condition. Use for 'complete' condition.
Use H/B_d for C_d. Assumes well compacted side and initial fill. Use $C_d\gamma_s B_c B_d$ and $C_{us} B_c U_s$ for pipes up to 48 in. diameter or very flexible pipes of larger sizes.[41] Use $C_d\gamma_s B_d^2$ and $C_{us} B_d U_s$ for larger sizes generally.
Assumes well compacted side and initial fill, and $r_{sd} = 0$. Compute W'_c.
Compute W'_{us}.
Assumes well compacted fill in all conditions. Use for solid packing with B_c in place of B_d in Chart C1.
Use for compressible packing with B_t in place of B_d in chart C1.
Use B_c in place of B_d in chart C2. Use the value of $K\mu$ for undisturbed soil for both W_c and W_{us}.
Assumes well compacted side and initial fill and $r_{sd} = 0{\cdot}0$. Compute W'_c and W'_{us}.
Assumes poorly compacted side and initial fill and $r_{sd} = -0{\cdot}4$ to $0{\cdot}0$.

TABLE M—*continued*

Construction condition	Bedding factor	Source of load	Load formula	Chart
(vii) EMBANKMENT LOADING—FLEXIBLE PIPES (*Continued*)				
Negative projection with '*narrow*' sub trench	K_θ	Fill	$W_c = C_n \gamma_s B_d^2$	C4 for C_n and C'_n
		Uniform surcharge of large extent	$W_{us} = (C'_n - C_n)\gamma_s B_d^2$ $W_{us} = C_{us} B_d U_s$	C2 for C_{us}
		Fill	$W_c = C_n \gamma_s B_c B_d$	C4 for C_n and C'_n
		Uniform surcharge or large extent	$W_{us} = C_{us} B_c U_s$ $W_{us} = (C'_n - C_n)\gamma_s B_c B_d$	C2 for C_{us}
Induced trench with trench width $= B_c$	K_θ	Fill	$W_c = C_n \gamma_s B_c^2$	C4 for C_n and C'_n
		Uniform surcharge of large extent	$W_{us} = C_{us} B_c U_s$ $W_{us} = (C'_n - C_n)\gamma_s B_c^2$	C2 for C_{us}
(viii) CONCENTRATED SURCHARGE LOADING—RIGID OR FLEXIBLE CONDUITS				
All conditions	As for W_c or W'_c for rigid pipes. K_θ for flexible pipes	Concentrated transient surcharge (wheel loads)	$W_{csu} = F_i C_t F_L P/3$	C5 for I_σ C6 for F_L
			$W_{csu} = F_i W_L$ $W_{csu} = F_i W_L P/_{7000}$	C7 for W_L
			$W_{csu} = F_i W_0$	C8 for W_0
(ix) UNIFORMLY DISTRIBUTED SURCHARGE OF KNOWN POSITION AND LIMITED EXTENT				
All conditions	As for W_c or W'_c for rigid pipes. K_θ for flexible pipes	Uniform surcharge of limited extent, static	$W_{sus} = \Sigma I_\sigma U_{sus} B_c$ $= C_{sus} U_{sus} B_c$	C5 for I_σ
		Transient (tracked vehicles)	$W_{sust} = \Sigma I_\sigma U_{sust} B_c F_i$ $= C_{sus} U_{sust} B_c F_i$	

Remarks
Use for large pipes generally. Assumes well compacted side and initial fill and $r_{sd} = 0{\cdot}0$.
Use for 'incomplete' condition. Use for 'complete' condition.
Use for pipes up to 48 in. and thin smooth walled pipes of larger sizes. Assumes well compacted side and initial fill and $r_{sd} = 0{\cdot}0$.
Use H/B_d in Chart C2, for C_{us} and 'complete' condition. Use for 'incomplete' condition. Use H/B_d for C_n and C'_n.
Assumes well compacted side and initial fill and $r_{sd} = 0{\cdot}0$. Use H/B_c in place of H/B_d in Chart C4.
Use H/B_c in place of H/B_d in Chart C2 for C_{us} and 'complete' condition. Use H/B_c in place of H/B_d in Chart C4 for C_n and C'_n and 'incomplete' condition.
Use for any single or multiple surcharges other than those below (see Article 4.17 (*b*) (iii) for method).
Use for field or other conditions with 2 loads of 7000 lb. each, 3 ft apart and wheel track width 6 ft or more. Use for road conditions with 2 loads of Plb each, 3 ft apart and back wheel width 6 ft or more (see Ref. 10*a*). Use Chart C7 (*b*) for concrete arch support.
Use for main traffic routes (or railway loading) in England and Wales subject to BS 153 Type HB Loading (See Ref. 10*a*). Use Chart C8 (*b*) for concrete arch support.
(L × B)—RIGID OR FLEXIBLE CONDUITS
Use only for known or anticipated local static surcharges. Construct diagram as Fig. 4.30. Compute $C_{sus} = \Sigma I_\sigma$ from chart C5. Compute W_{sus}.
Use for tracked vehicles used during construction or later. Compute W_{sust}.

TABLE N

MAXIMUM ASSUMED EXTERNAL DIAMETERS OF RIGID PIPES AND EFFECTIVE MAXIMUM TRENCH WIDTHS (FOR PURPOSES OF LOAD ESTIMATION WHEN THE ACTUAL VALUES ARE NOT KNOWN)

Nominal internal diameter, D in.	Assumed external diameter, $B_{s_{max}}$ in.	Assumed maximum effective trench width, $B_{d_{max}}$ in.	Nominal internal diameter, D in.	Assumed external diameter, $B_{c_{max}}$ in.	Assumed maximum effective trench width, $B_{d_{max}}$ in.
4	5	21	36	45¾	76
6	8	24	39	49	79
9	11¼	27	42	52½	82
12	16	30	45	56	86
15	19½	41	48	59½	89
18	23	45	54	66¼	96
21	26½	48	60	73½	103
24	32	54	66	80½	110
27	35¼	57	72	87½	117
30	38½	60	78	94½	124
33	42	64	84	101½	131

$B_d = B_c + 16$ in. (+6 in. for sheeting) for pipes of 15 to 33 in. I.D.
$B_c + 24$ in. (+6 in. for sheeting) for pipes of 36 in. I.D. or more.

Use lower values of B_d wherever soil and working space permit, in order to reduce the fill and uniform surcharge loads.

B_c – Use actual values wherever possible in order to reduce the concentrated surcharge load in 'narrow' trench and negative projection conditions and all the loads in wide trench and positive projection conditions.

TABLE O

SOIL CHARACTERISTICS

Soil type	Nature of soil	Saturated density γ_s lb/ft³	$K\mu'_{max}$
Non-cohesive (granular)	Crushed rock		
	granites and shales	120–140	0·192
	basalts and dolerites	130–160	
	limestones and sandstones	100–140	
	Broken brick	100–120	
	Ashes	80–90	
	Coarse gravels	120–140	0·165
	Sand, or sand and gravel mixtures	120–140	
	Chalk	80–100	variable
Cohesive	Saturated top soil or mixed soils with some clay	120–140	0·150
	Ordinary clay soils	110–140	0·130
	Soft clays	100–120	0·110

N.B. If the drained density of a coarse grained granular soil (e.g. a gravel) is γ_d and its void ratio (i.e. volume of voids/total volume) is n, $\gamma_s = \gamma_d + n\gamma_w$.

TABLE P

SELECTION AND SUMMATION OF LOADS TO OBTAIN THE MAXIMUM TOTAL EFFECTIVE EXTERNAL DESIGN LOAD W_c

Load	Values of component loads											
	H_{min}						H_{max}					
	If $W'_c < W_c$ (wide trench or sub-trench)			If $W_c < W'_c$ (narrow trench or sub-trench)			If $W'_c < W_c$ (wide trench or sub-trench)			If $W_c < W'_c$ (narrow trench or sub-trench)		
	$W'_{us} > W_{csu} > W_{sust}$	$W_{csu} > W'_{us} > W_{sust}$	$W_{sust} > W'_{us} > W_{csu}$	$W_{us} > W_{csu} > W_{sust}$	$W_{csu} > W_{us} > W_{sust}$	$W_{sust} > W_{us} > W_{csu}$	$W'_{us} > W_{csu} > W_{sust}$	$W_{csu} > W'_{us} > W_{sust}$	$W_{sust} > W'_{us} > W_{csu}$	$W_{us} > W_{csu} > W_{sust}$	$W_{csu} > W_{us} > W_{sust}$	$W_{sust} > W_{us} > W_{csu}$
Fill	W'_c	W'_c	W'_c	W_c	W_c	W_c	W'_c	W'_c	W'_c	W_c	W_c	W_c
Water	W_w	W_w	W_w	W_w	W_w	W_w	W_w	W_w	W_w	W_w	W_w	W_w
Static limited surcharge (if any)	W_{sus}	W_{sus}	W_{sus}	W_{sus}	W_{sus}	W_{sus}	W_{sus}	W_{sus}	W_{sus}	W_{sus}	W_{sus}	W_{sus}
Other surcharge	W'_{us}	W_{csu}	W_{sust}	W_{us}	W_{csu}	W_{sust}	W'_{us}	W_{cus}	W_{sust}	W_{us}	W_{csu}	W_{sust}
TOTALS												

W_e = maximum total load =

TABLE Q—RESISTANCE OF PIPE JOINTING AND BEDDING MATERIALS TO CHEMICAL ATTACK

Group	B.S. No.	Material	Normal domestic sewage	Trade effluent containing: At normal temperature		Organic solvents	Oils and fats		At sustained high temperature		Soil environment containing		Remarks
				Acids	Alkalis		vegetable	Mineral	Acids	Alkalies	Sulphates	Acids	
Ceramic (Clayware)	65 540	Clay drain and sewer pipes and fittings	O	S	S	S	S	S	O	O	S	S	Mainly available in England and Wales Mainly available in Scotland
	3921	Bricks and blocks of fired brickwork, clay, and shale	O	O	O	O	O	O	O	O	O	O	Resistance of brickwork depends on jointing material (see below)
Cement† and Concrete†	556	Concrete: Portland cement	O	X	O	O	X	O	X	O	X	X	
		Concrete: Sulphate resisting Portland cem.	O	X	O	O	X	O	X	O	O	X	
		Concrete: High alumina cement	O	X	X	O	X	O	X	X	O	X	
		Concrete: Super sulphated cement	O	X	O	O	X	O	X	O	O	X	Resists weak acids e.g. peat
	486	Asbestos cement pressure	O	X	O	O	X	O	X	O	O	X	
	3656	Asbestos cement drain and sewer	O	X	O	O	X	O	X	O	O	X	
METAL† Cast Iron Steel	 78 1211 437 534 1387, 3601 —	Ductile Iron Cast Iron vertically cast Cast iron centrifugally cast Cast iron S. and S. drain S. and S. welded (not galvanised) Steel tubes, galvanised or not galvanised Corrugated	O	X	O	O	P	O	X	X	P	P	Assumes pipes coated ex factory
Plastics	3505 3506	U.P.V.C. for cold-water supply U.P.V.C. for general purposes	O	S	S	X	O	O	X	X	S	S	
Pitch fibre	2760	Pitch fibre drain and sewer	O	O	O	X	X	X	X	X	O	O	

TABLE Q—*continued*

Jointing materials		Portland cement mortar†	O	X	O	O	X	O	X	O	X	X	Other cements as for concrete pipes
		*Bituminous compositions	O	O	O	X	X	X	X	X	O	O	
		Iron or steel components†	O	X	O	O	P	O	X	X	P	P	As for iron and steel pipes
	2494	Rubber compositions: natural—Sealing rings	O	O	O	X	X	X	X	O	O	O	
		Rubber compositions: Butyl—Sealing rings	O	S	S	O	O	O	O	O	S	S	
	Part II	Rubber compositions: Neoprene—Sealing rings	O	S	O	X	O	O	O	O	S	S	
	3514	Rubber compositions: nitrile (oil resistant) —Sealing rings	O	O	O	X	O	O	O	O	O	O	
		*Plastisols—Fairings	O	S	S	X	X	O	X	X	S	S	
		*Polyester Resin—Fairings	O	S	O	X	O	O	X	X	S	S	
		*Polypropylene—Fairings	O	S	S	O	S	O	O	O	S	S	
		*Polyeurethane—Fairings	O	S	S	O	S	O	O	O	S	S	
		*Epoxy resin. Fairings or adhesives	O	O	S	X	X	O	X	S	S	O	
Bedding materials		Concrete (Portland Cement)†	INAPPLICABLE								X	X	Other cements as for concrete pipes
		Gravels and Sand, Broken Stone, Broken Brick									O	O	Excluding Limestones
		Limestones (all)									X	X	

O = Normally suitable. X = May not be suitable. Chemist or maker should be consulted. P = May require special protection.
S = Specially suitable for active industrial wastes.
† = Liable to sulphuric acid attack in the presence of septic sewage.
* There is no suitable B.S. for these jointing materials. This classification assumes that the formulation and methods of curing are appropriately chosen for pipe jointing.

18

TABLE R

TYPES AND CAUSES OF RIGID PIPE FRACTURES

Type of failure	Appearance	Cause	Preventive measures
(*a*) Overload fracture		Excessive vertical load or inadequate bedding	Higher bedding class, or stronger pipe, or concrete surround
(*b*) Burst socket	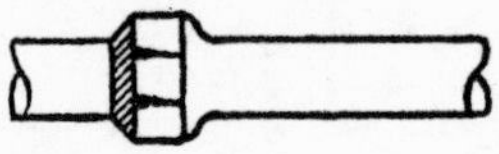	Differential thermal or moisture expansion of jointing mortar	Resilient jointing material which does not cause excessive radial pressure on the socket
(*c*) Distortion fracture	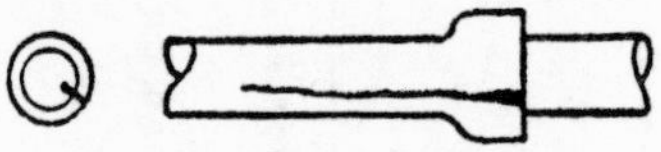	Differential heating or cooling or moisture content	Protection of uncovered pipes against sun or cold night (or drying wind, with concrete pipes)
(*d*) Beam fractures	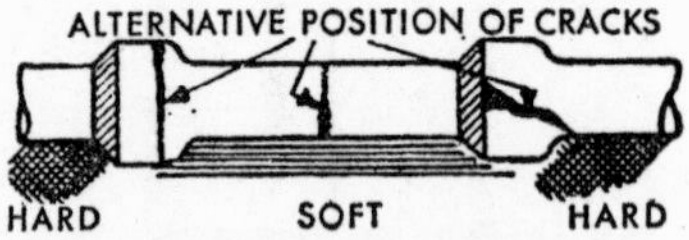 ALTERNATIVE POSITION OF CRACKS HARD SOFT HARD	Uneven resistance of foundation, or soil movement, or differential settlement	Flexible joints and uniform hardness of foundation
(*e*) Pull fractures	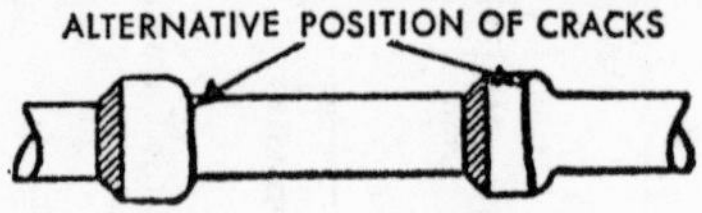 ALTERNATIVE POSITION OF CRACKS	Thermal or drying shrinkage of pipe or site concrete, drying shrinkage of clay soil	Flexible-telescopic joints and gaps in site concrete at pipe joints
(*f*) Shear fractures	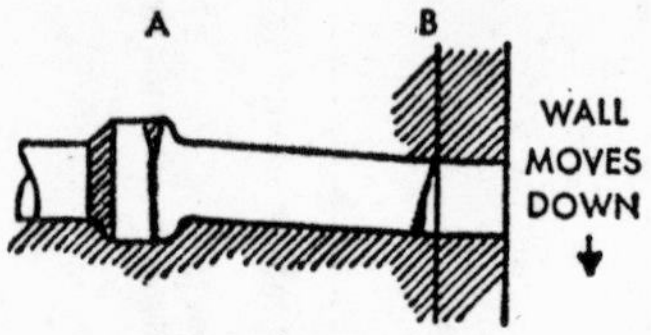 A B WALL MOVES DOWN	Differential settlement of wall relative to pipe or/vice versa	Flexible joints at least at A and B and making AB not more than 3 ft.
(*g*) Bearing fracture	 HARD OBJECT	Hard spots in pipe bed	Elimination of hard spots
(*h*) Thrust fracture	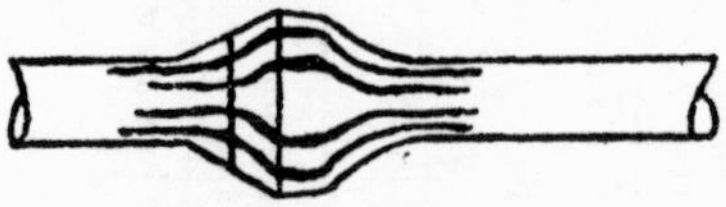	Restrained thermal or moisture expansion of pipe or compression due to subsidence	Flexible-telescopic joints which do not cause excessive radial pressure on the sockets. Spigot end not hard up in socket
(*k*) Leverage fracture	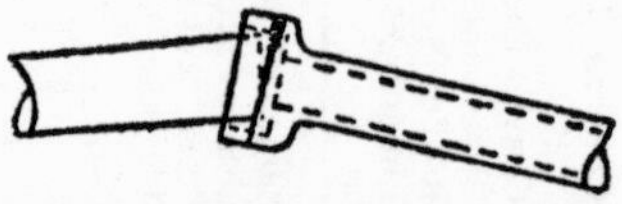	Excessive angular displacement	Avoidance of excessive slew when laying

Types of failure *b*, *d*, *e* and *f* may occur with rigid (e.g. cement mortar) joints.
Type *k* may occur with flexible (e.g. rubber ring) joints.
The other failures are uninfluenced by type of joint.

APPROXIMATE COEFFICIENTS OF THERMAL EXPANSION OF VARIOUS MATERIALS AT SOIL TEMPERATURE OF ≂ 50°F = 10°C

Material	Approx. Coefficient of expansion in./in./°C × 10^{-6}
Clayware	5
Concrete and cement mortar	10
Asbestos-cement	8·5
Cast and spun iron	8·5
Ductile iron	11·8
Steel	11·9
Pitch fibre	40
uPVC	50

TABLE Ta

TYPES OF MODERN FLEXIBLE-TELESCOPIC JOINTS FOR VARIOUS KINDS OF PIPES

Type of Pipe	Pipe material B.S. No. and Type of ends*	Trade name of Joint	Joint Type	Pipe Size Range * Min. in.	Max. in.	* Max. Test Pressure p.s.i.	Fig. No.
Rigid, Non-Pressure (Sewer and Drain Pipes)	Clayware B.S. 65 and 540 (S and S)	Hepseal P.S.R. Ellflex Vitriflex and others	Push-in sliding rubber 'O' ring retained by die cast hard plastic fairings on spigot and socket	4 4 4 4	12 12 15 12	20 20 20 20	10.1 (*a*)
		Drawflex	Push-in sliding ringless in die cast resilient plastic fairings on spigot and socket	4	9	20	10.1 (*b*)
		Oanco	Hot poured bituminous compound retained by precast tar-sulphur rings	4	18	20	10.1 (*c*)
	Concrete B.S. 556 (S and S)	*Cornelius Type* Stanton Trocoll Redland Spun concrete S. Coast etc.	Push-in rolling rubber 'O' ring. Spigots and sockets variously shaped by different makers	6	78	35 20 20 20 20	10.2 (*a*) 10.2 (*b*) 10.2 (*b*) (*c*) (*d*) as (*d*) as (*c*)
		Rocla		12	54	35	10.2 (*h*)
		Tylox A	Push-in sliding multi-finned rubber ring on spigot	6	18	20	10.2 (*e*)
		Tylox B	Do. on socket	21	72	20	10.2 (*f*)
		Deckon	Push-in rolling rubber 'O' ring in hard glass fibre-reinforced plastic socket.	6	66	20	10.2 (*g*)
	Asbestos-cement B.S. 3656 (D.S.)	Turnall	Rolling rubber 'O' rings in A.C. Collar (4 rings per joint)	4	36	25	10.3 (*a*)
Rigid, Pressure	Cast Iron B.S. 78 Spun Iron B.S. 1211 (S and S)	Tyton	Push-in sliding composite rubber ring fixed in socket groove	3	27	347	10.4 (*a*)
		Bolted Gland	Triangular rubber ring axially compressed by bolted C.I. gland	3	48	347	10.4 (*b*)
	Do. (D.S.)	Viking-Johnson	Triangular rubber rings axially compressed in steel collar by bolted steel glands	3 3	72 48 (spun)	347 347	10.5
	Cast Iron B.S. 78 (S and S)	Screwed Gland	Triangular rubber ring axially compressed by C.I. screwed-in gland	3	6	347	10.4 (*c*)
	Prestressed Concrete, No B.S. yet (S and S)	Lock-Joint	Push-in sliding rubber 'O' ring retained in steel fairing rings	27	48	325	10.7

* S and S = Spigot and Socket: D.S. = Double Spigot.
** The size range may be extended from time to time by individual makers.

TABLE Ta—continued

Type of Pipe	Pipe material, B.S. No. and Type of ends*	Trade name of Joint	Joint Type	Pipe Size Range* Min. in.	Max. in.	* Max. Test Pressure p.s.i.	Fig. No.
Rigid, Pressure —*contd.*	Asbestos-Cement B.S. 486 (D.S.)	Widnes Everite	Push-in sliding composite rubber rings fixed in grooves in A.C. Collar	2	24	347	10.3 (*b*)
		Eternit N.A.C.	Push-in sliding rubber rings fixed in grooves in A.C. Collar. Collar has stop rings	2	24	347	10.3 (*e*)
		Eternit Comet	Push-in sliding rubber rings with single grooved fin, fixed in grooves in A.C. collar	2	15	347	10.3 (*f*)
		Everite and Eternit Detachable	Sliding rubber 'O' rings axially compressed in C.I. Collar by bolted glands	2	24	347	10.3 (*c*)
		Everite Screwed	Sliding rubber 'O' rings axially compressed in A.C. Collar by screwed A.C. glands	2	9	347	10.3 (*d*)
		Viking-Johnson	As for Cast Iron	2	24	347	10.5
Flexible Non Pressure	Pitch-Fibre B.S. 2760 (D.S.)		Push-in sliding neoprene 'O' ring in special plastics or P.F. Collar	4	6	10 ft	10.8 (*a*) and (*b*)
Flexible Pressure	Steel B.S. 534 (D.S.)	Viking-Johnson	As for Cast Iron	2	72	270–1000	10.5
	Do. (S and S)	Fastite	Push-in sliding rubber 'O' ring fixed in socket groove	3	30	320–520	10.10
	Ductile Iron	Tyton	As for Cast Iron	4	27	900	10.4 (*a*)
	No B.S. yet (S and S)	Bolted Gland	Do.	4	48	900	10.4 (*a*)
	Do. (D.S.)	Viking-Johnson	As for Cast Iron	4	48	900	10.5
	Unplasticised PVC B.S. 3505, 3506 (D.S.)	Wavintite	Push-in sliding rubber 'O' rings fixed in grooves in moulded PVC sleeve	2	6	347	10.9 (*b*)
		Viking-Ehrimuffe	Push-in sliding rubber composite ring fixed in groove in moulded PVC socket. Socket solvent-welded to pipe	2	5	347	10.9 (*c*)
		Viking-Johnson	As for Cast Iron	2	6	347	10.5
	Do. (S and S)	Chemidor Yorcoring Wavin Extrudex Formica	Push-in sliding rubber 'O' ring retained in socket-groove produced by heating and shrinking socket end on site.	2	6	347	10.9 (*a*)

* S and S = Spigot and Socket: D.S. = Double Spigot.

TABLE Tb

TYPES OF MODERN FLEXIBLE POSITIVE (NON-TELESCOPIC) JOINTS FOR VARIOUS KINDS OF PIPES

Type of Pipe	Pipe Material, B.S. No. and Type of ends	Trade name of Joint	Joint Type	Pipe Size Range Min. in.	Pipe Size Range Max. in.	Max. Test Pressure p.s.i.	Fig. No.
Rigid Pressure	Cast Iron B.S. 78 (D.S. Special)	Victaulic	Rubber U ring restrained by C.I. split bolted collar	2	36	347	10.6
	Spun Iron B.S. 1211 (D.S. Special)	Do.	Do.	2	36	347	10.6
Flexible Pressure	Steel B.S. 534 (D.S. Special)	Do.	Do.	2	36	270–1000	10.6
	Unplasticised PVC B.S. 3505 B.S. 3506 (D.S. Special)	Do.	Do.	2	6	347	10.6

TABLE U

STANDARD CONCRETE MIXES FOR CLASS A *in-situ* CONCRETE BEDDINGS

Concrete Grade	Minimum 28 day works cube strength	Weight of dry, Zone 2*, sand per 112 lb of cement	Weight of dry coarse aggregate per 112 lb cement						Time required to reach a strength of 2000 lb/in.^{2}†			
			$\frac{3}{4}$ in. max. size, Slump			$1\frac{1}{2}$ in. max. size, Slump			At normal temp. above 50°F		At low temp. 38°F–50°F	
			$\frac{1}{2}$–1 in.	1–2 in.	2–5 in.	1–2 in.	2–4 in.	4–7 in.	Ordinary Portland cement,	Rapid hardening Portland cement,	Ordinary Portland cement,	Rapid hardening Portland cement,
	lb./in.^{2}	lb.	lb.	lb.	lb.	lb.	lb.	lb.	days	days	days	days
A	3000	200	425	350	300	500	425	375	7	4	11	6
B	3750	175	375	300	250	450	375	325	4	2	6	4
C	4500	150	325	250	200	375	300	250	3	$1\frac{1}{2}$	$4\frac{1}{2}$	3

NOTES:
1. If crushed sand is used reduce the coarse aggregate by at least 25 lb. per 112 lb. of cement.
2. If Zone 3 sand is used increase the coarse aggregate by at least 25 lb per 112 lb of cement and reduce the sand by the same weight.
3. If Zone 1 sand is used reduce the coarse aggregate by at least 25 lb per 112 lb of cement and increase the sand by the same weight.

* See B.S. 882.

† These times are for concrete which just complies with the minimum 28 day strength requirements for its grade. For the same grade but strengths greater than the required minimum they may be appropriately reduced, if and preferably the strength is determined by test cubes made and cured under the same conditions as the site concrete.

Design Charts

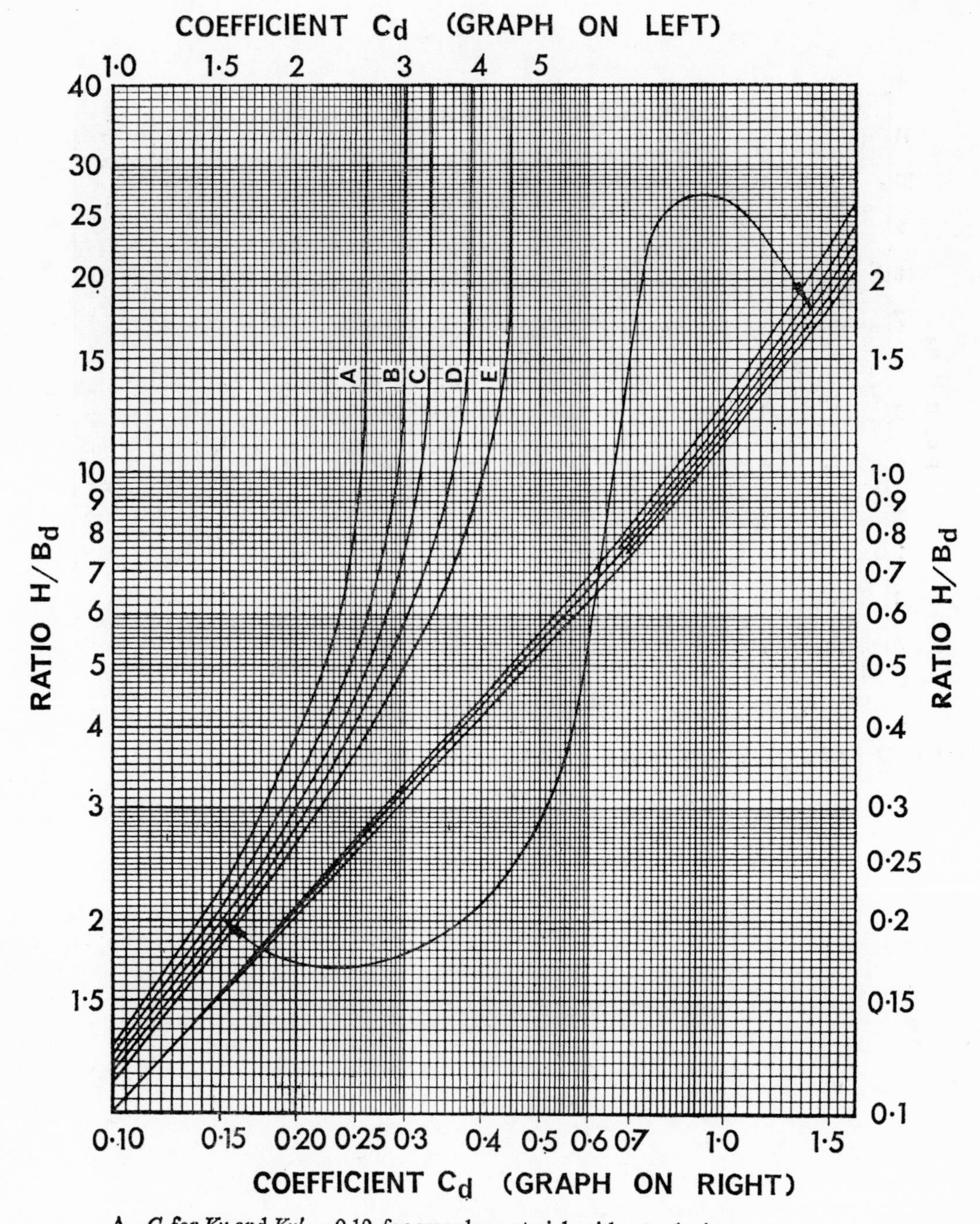

A—C_d for $K\mu$ and $K\mu' = 0.19$, for granular materials without cohesion
B—C_d for $K\mu$ and $K\mu' = 0.165$ max. for sand and gravel
C—C_d for $K\mu$ and $K\mu' = 0.150$ max. for saturated top soil
D—C_d for $K\mu$ and $K\mu' = 0.130$ ordinary max. for clay
E—C_d for $K\mu$ and $K\mu' = 0.110$ max. for saturated clay

CHART C1 *Values of load coefficient C_d (trench fill)*

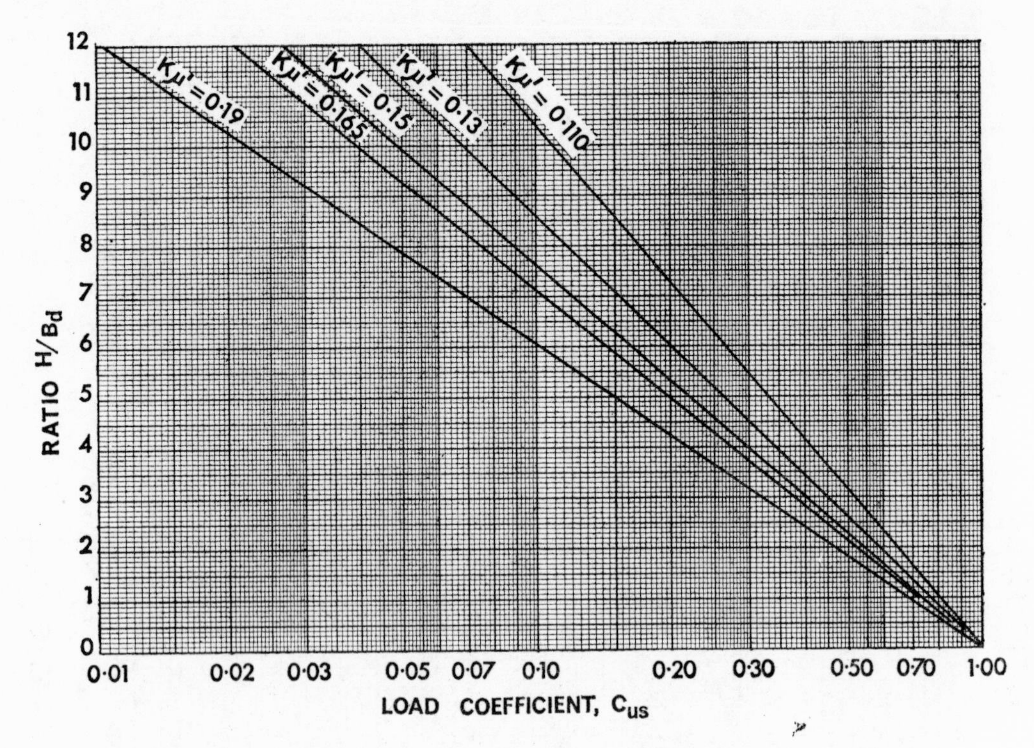

CHART C2 *Values of load coefficient C_{us} (trench uniform surcharge)*

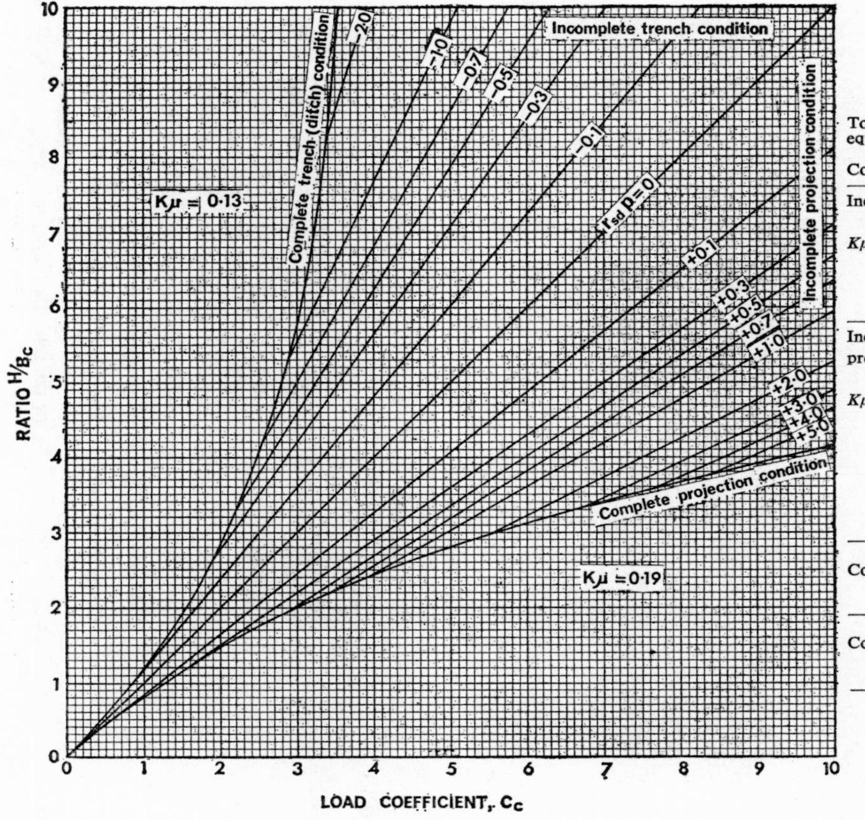

CHART C3 *Values of load coefficient C_c (positive projection fill)*

To extrapolate values of C_c for higher values of H/B_c use the following equations

Condition	$r_{sd}p$	Equation of curve
Incomplete trench	− 0·1	$C_c = 0{\cdot}82(H/B_c) + 0{\cdot}05$
	− 0·3	$C_c = 0{\cdot}69(H/B_c) + 0{\cdot}11$
$K\mu = 0{\cdot}13$	− 0·5	$C_c = 0{\cdot}61(H/B_c) + 0{\cdot}20$
	− 0·7	$C_c = 0{\cdot}55(H/B_c) + 0{\cdot}25$
	− 1·0	$C_c = 0{\cdot}47(H/B_c) + 0{\cdot}40$
	− 2·0	$C_c = 0{\cdot}30(H/B_c) + 0{\cdot}91$
Incomplete	0	$C_c = H/B_c$
projection	+ 0·1	$C_c = 1{\cdot}23(H/B_c) - 0{\cdot}02$
	+ 0·3	$C_c = 1{\cdot}39(H/B_c) - 0{\cdot}05$
$K\mu = 0{\cdot}19$	+ 0·5	$C_c = 1{\cdot}50(H/B_c) - 0{\cdot}07$
	+ 0·7	$C_c = 1{\cdot}59(H/B_c) - 0{\cdot}09$
	+ 1·0	$C_c = 1{\cdot}69(H/B_c) - 0{\cdot}12$
	+ 2·0	$C_c = 1{\cdot}93(H/B_c) - 0{\cdot}17$
	+ 3·0	$C_c = 2{\cdot}08(H/B_c) - 0{\cdot}20$
	+ 4·0	$C_c = 2{\cdot}19(H/B_c) - 0{\cdot}21$
	+ 5·0	$C_c = 2{\cdot}28(H/B_c) - 0{\cdot}22$
Complete trench	Use the formula $C_c = C_d = \frac{1 - e^{-2K\mu}}{2K\mu}$ c and $K\mu = 0{\cdot}13$	
Complete projection	Use the formula $C_c = \frac{e^{2K\mu H/B_c} - 1}{2K\mu}$ and $K\mu = 0{\cdot}19$	

CHART C4 *(a)*

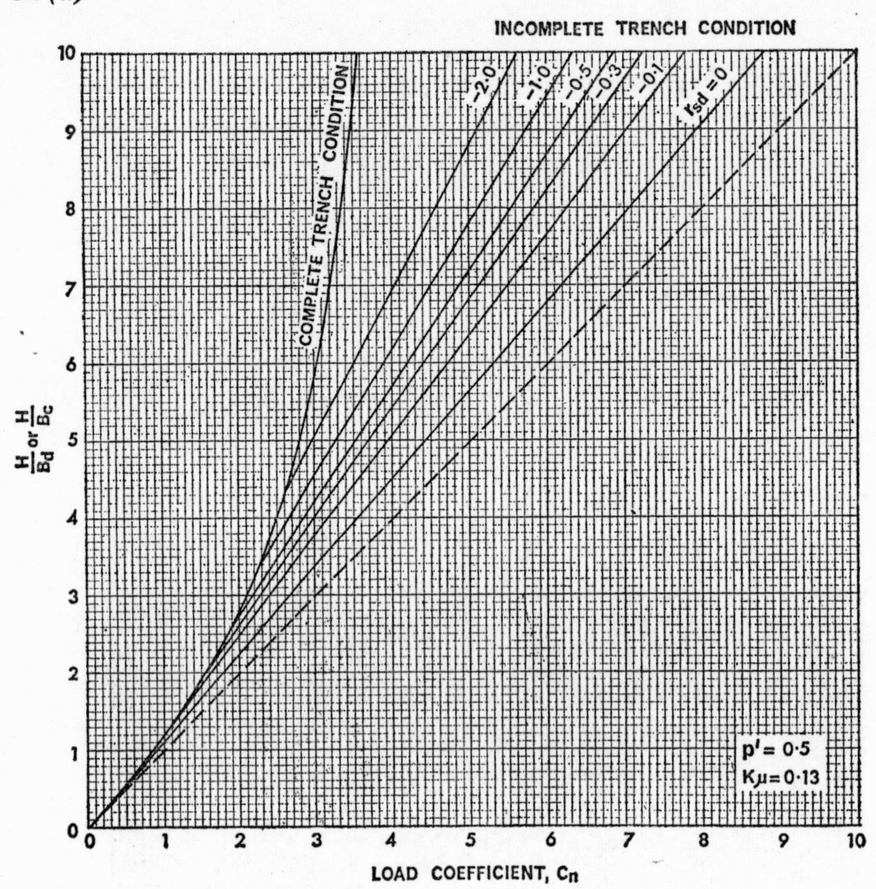

Chart C4 ($p' = 0{\cdot}5$) *Negative Projection—Chart for C_n, $p' = 0{\cdot}5$*

To extrapolate values of C_n for values of $H/B_d > 10{\cdot}0$ use the following equations:

Condition	r_{sd}	Equation of curve	p'
Incomplete	0	$C_n = 0{\cdot}88\ (H/B_d) + 0{\cdot}03$	0·5
Negative	−0·1	$C_n = 0{\cdot}77\ (H/B_d) + 0{\cdot}09$	
Projection	−0·3	$C_n = 0{\cdot}71\ (H/B_d) + 0{\cdot}14$	
	−0·5	$C_n = 0{\cdot}67\ (H/B_d) + 0{\cdot}17$	
	−1·0	$C_n = 0{\cdot}61\ (H/B_d) + 0{\cdot}23$	
	−2·0	$C_n = 0{\cdot}53\ (H/B_d) + 0{\cdot}33$	

Complete trench — Use Chart C.1 for C_d when $K\mu = 0{\cdot}13$ or the formula $C_n = \frac{1 - e^{-2K\mu H/B_d}}{2\,K\mu}$ and $K\mu = 0{\cdot}13$

Charts 4 (a), (b), (c) and (d) are reproduced by courtesy of the Council of the Institution of Civil Engineers

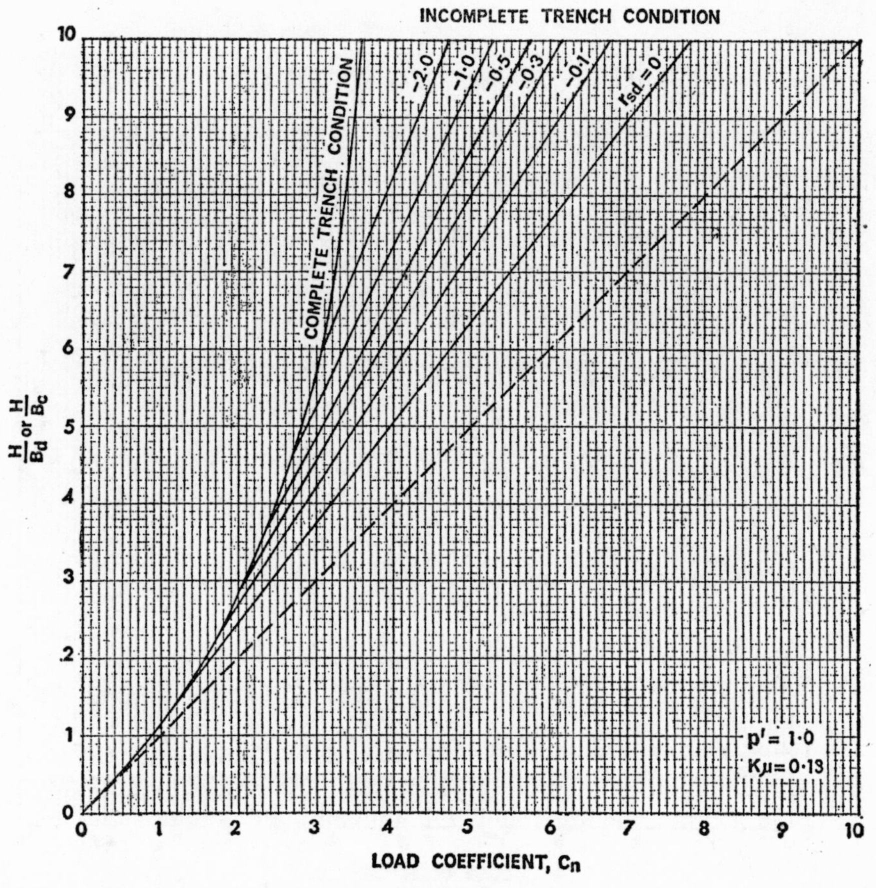

Chart C4 ($p' = 1{\cdot}0$) *Negative Projection—Chart for C_n, $p' = 1{\cdot}0$*

To extrapolate values of C_n for values of $H/B_d > 10{\cdot}0$ use the following equations:

Condition	r_{sd}	Equation of curve	p'
Incomplete	0	$C_n = 0{\cdot}77\,(H/B_d) + 0{\cdot}11$	1·0
Negative	−0·1	$C_n = 0{\cdot}65\,(H/B_d) + 0{\cdot}25$	
Projection	−0·3	$C_n = 0{\cdot}58\,(H/B_d) + 0{\cdot}34$	
	−0·5	$C_n = 0{\cdot}53\,(H/B_d) + 0{\cdot}41$	
	−1·0	$C_n = 0{\cdot}47\,(H/B_d) + 0{\cdot}52$	
	−2·0	$C_n = 0{\cdot}40\,(H/B_d) + 0{\cdot}69$	
Complete trench	Use Chart C1 for C_d when $K\mu = 0{\cdot}13$ or the formula $C_n = \dfrac{1 - e^{-2K\mu H/B_d}}{2\,K\mu}$ and $K\mu = 0{\cdot}13$		

CHART C4 *(c)*

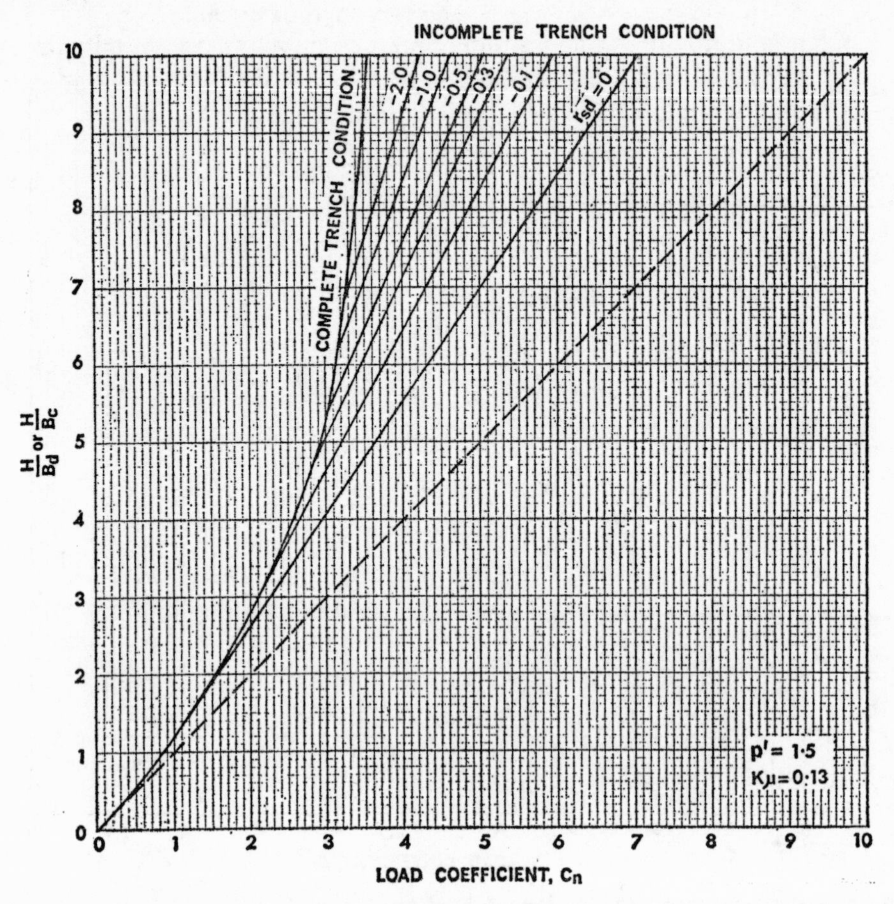

Chart C4 ($p' = 1{\cdot}5$) *Negative Projection—Chart for C_n, $p' = 1{\cdot}5$*

To extrapolate values of C_n for values of $H/B_d > 10{\cdot}0$ use the following equations:

Condition	r_{sd}	Equation of Curve	p'
Incomplete	0	$C_n = 0{\cdot}68\,(H/B_d) + 0{\cdot}23$	1·5
Negative	−0·1	$C_n = 0{\cdot}55\,(H/B_d) + 0{\cdot}44$	
Projection	−0·3	$C_n = 0{\cdot}48\,(H/B_d) + 0{\cdot}58$	
	−0·5	$C_n = 0{\cdot}44\,(H/B_d) + 0{\cdot}66$	
	−1·0	$C_n = 0{\cdot}38\,(H/B_d) + 0{\cdot}81$	
	−2·0	$C_n = 0{\cdot}31\,(H/B_d) + 1{\cdot}15$	
Complete trench	Use Chart C1 for C_d when $K\mu = 0{\cdot}13$ or the formula $C_n = \dfrac{1 - e^{-2K\mu H/B_d}}{2\,K\mu}$ and $K\mu = 0{\cdot}13$		

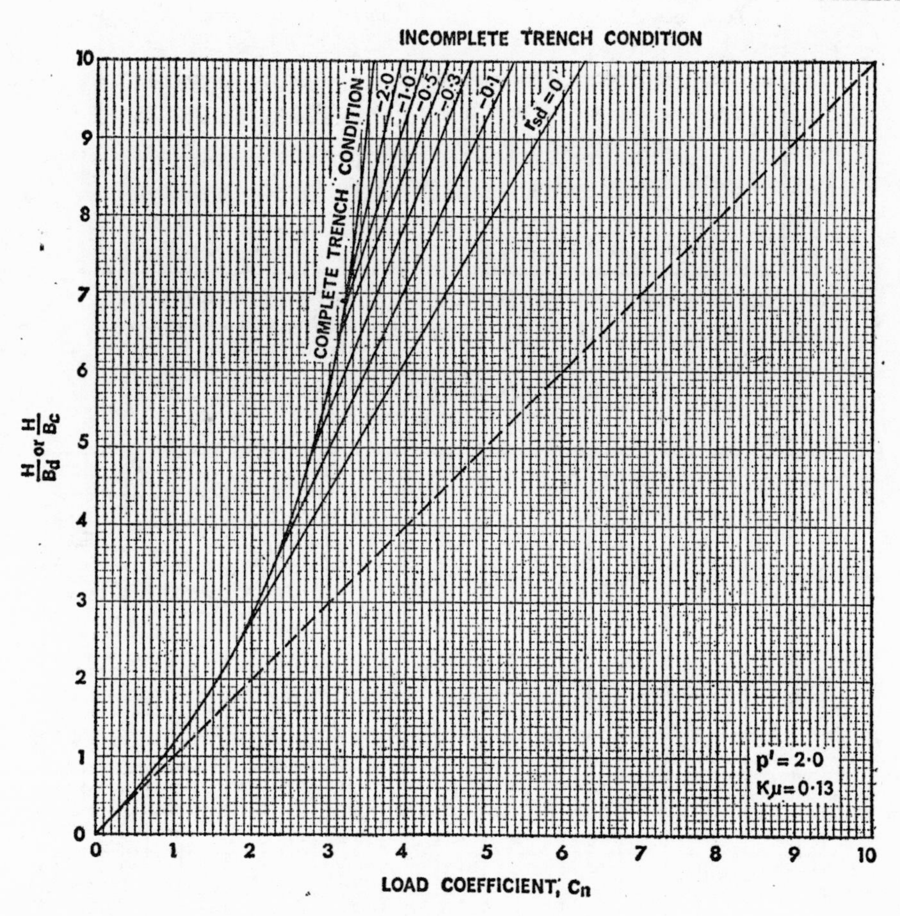

Chart C4 ($p' = 2{\cdot}0$) *Negative Projection—Chart for C_n, $p' = 2{\cdot}0$*

To extrapolate values of C_n for values of $H/B_d > 10{\cdot}0$ use the following equations:

Condition	r_{sd}	Equation of Curve	p'
Incomplete	0	$C_n = 0{\cdot}59\,(H/B_d) + 0{\cdot}37$	2·0
Negative	−0·1	$C_n = 0{\cdot}47\,(H/B_d) + 0{\cdot}65$	
Projection	−0·3	$C_n = 0{\cdot}40\,(H/B_d) + 0{\cdot}82$	
	−0·5	$C_n = 0{\cdot}36\,(H/B_d) + 0{\cdot}92$	
	−1·0	$C_n = 0{\cdot}31\,(H/B_d) + 1{\cdot}11$	
	−2·0	$C_n = 0{\cdot}24\,(H/B_d) + 1{\cdot}52$	
Complete trench	Use Chart C1 for C_d when $K\mu = 0{\cdot}13$ or the formula $C_n = \dfrac{1 - e^{-2K\mu H/B_d}}{2\,K\mu}$ and $K\mu = 0{\cdot}13$		

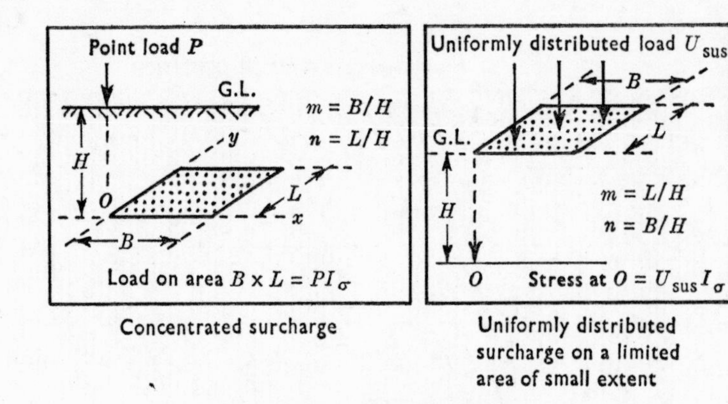

Concentrated surcharge

Uniformly distributed surcharge on a limited area of small extent

N.B. The area $B \times L$ is rectangular in true plan in both cases

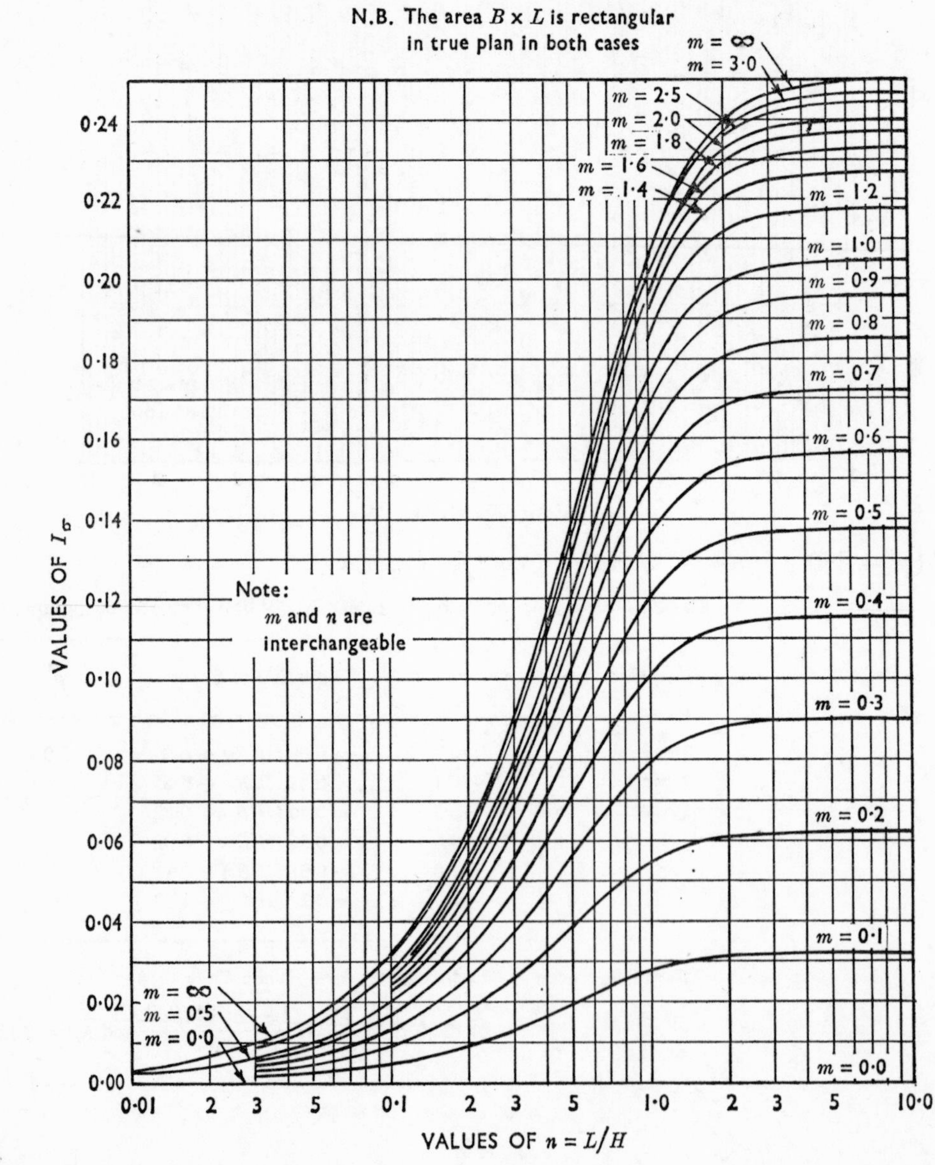

CHART C5 *Values of Influence value I_σ for concentrated surcharge or uniform surcharge of limited extent*

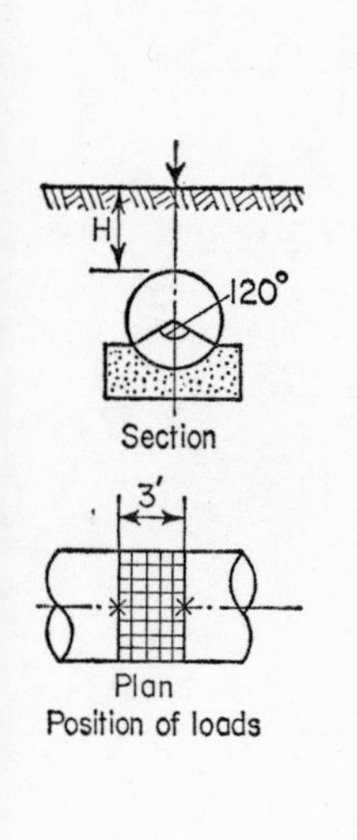

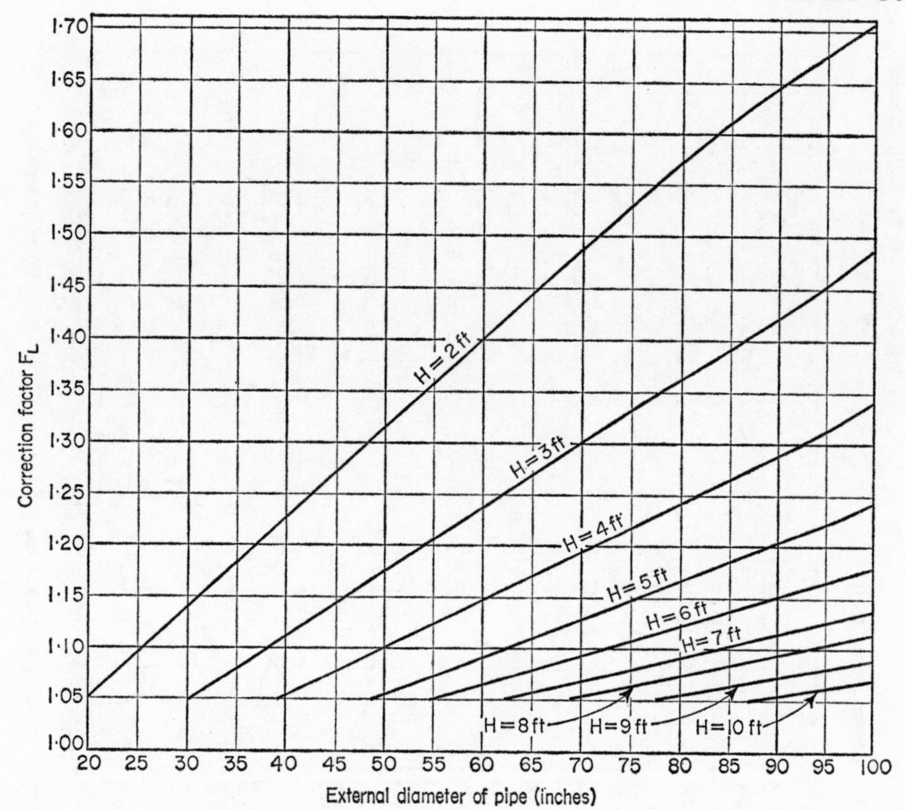

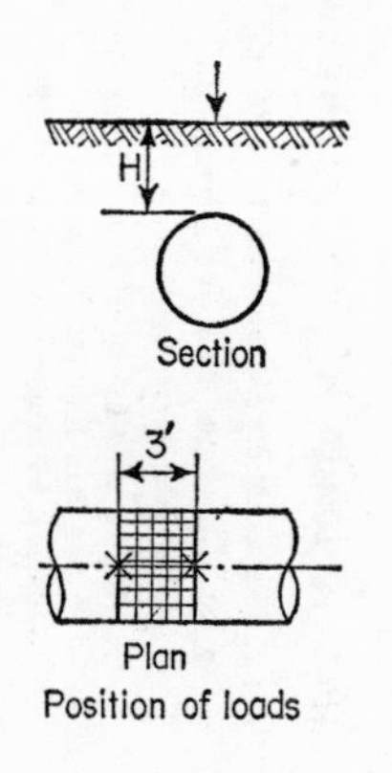

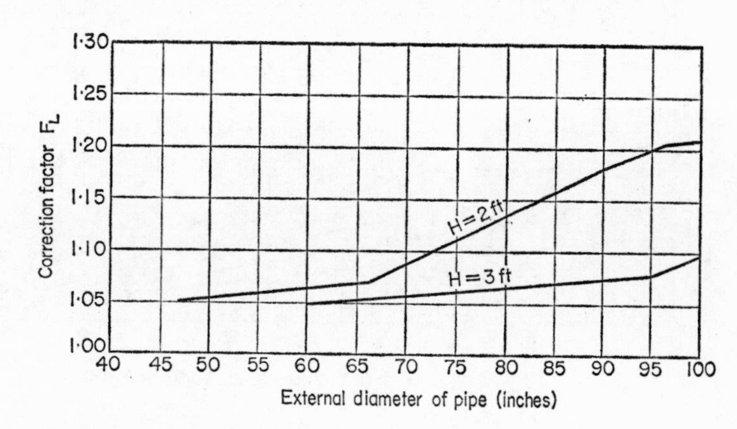

CHART C6 ***(a)*** *Correction factor F_L (R.C. Bedding, 2 wheels)*

CHART C6 ***(b)*** *Correction factor F_L (Class B Bedding, 2 wheels)*

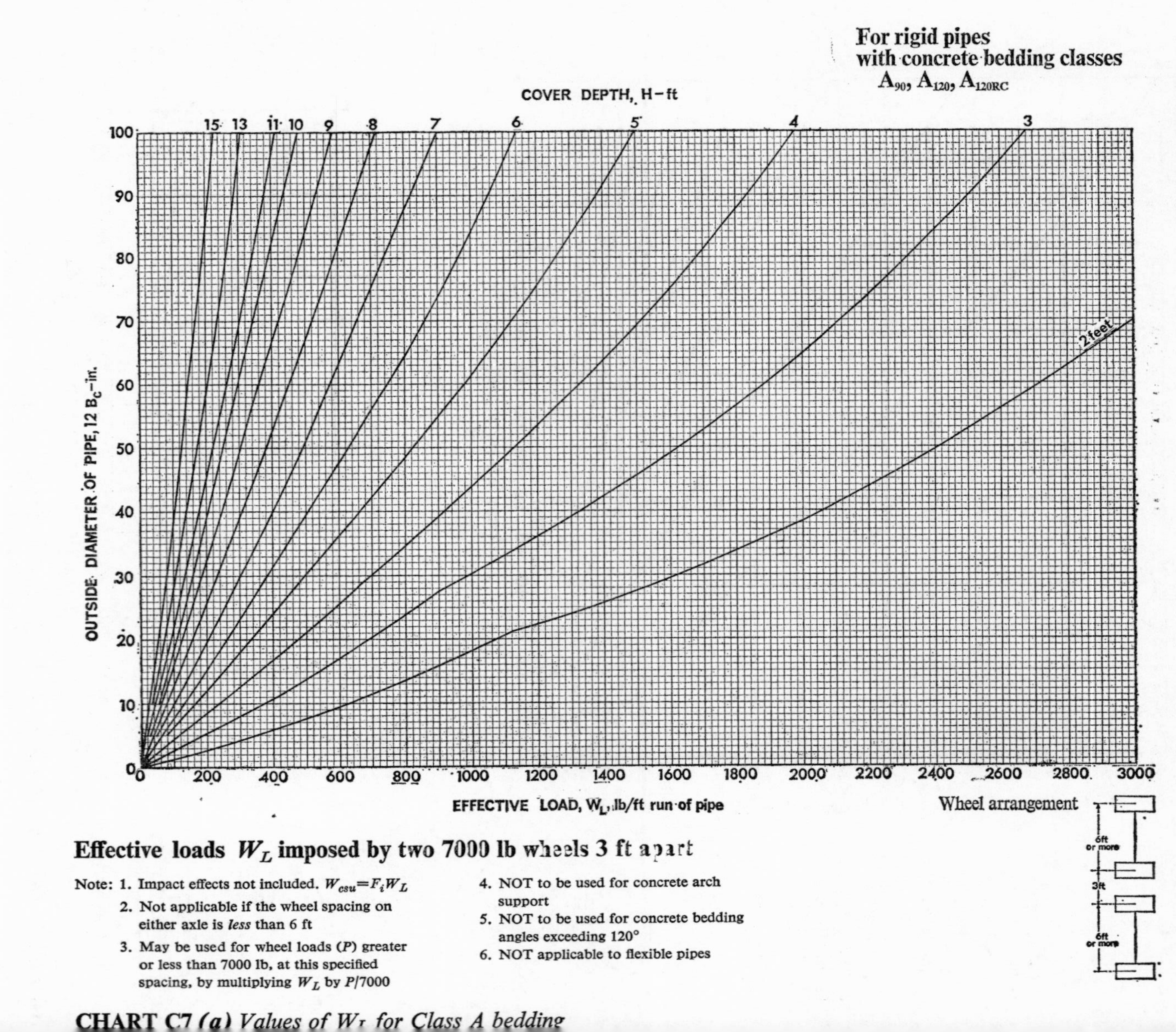

Effective loads W_L imposed by two 7000 lb wheels 3 ft apart

Note:
1. Impact effects not included. $W_{csu}=F_i W_L$
2. Not applicable if the wheel spacing on either axle is *less* than 6 ft
3. May be used for wheel loads (*P*) greater or less than 7000 lb, at this specified spacing, by multiplying W_L by $P/7000$
4. NOT to be used for concrete arch support
5. NOT to be used for concrete bedding angles exceeding 120°
6. NOT applicable to flexible pipes

CHART C7 *(a)* *Values of W_L for Class A bedding*

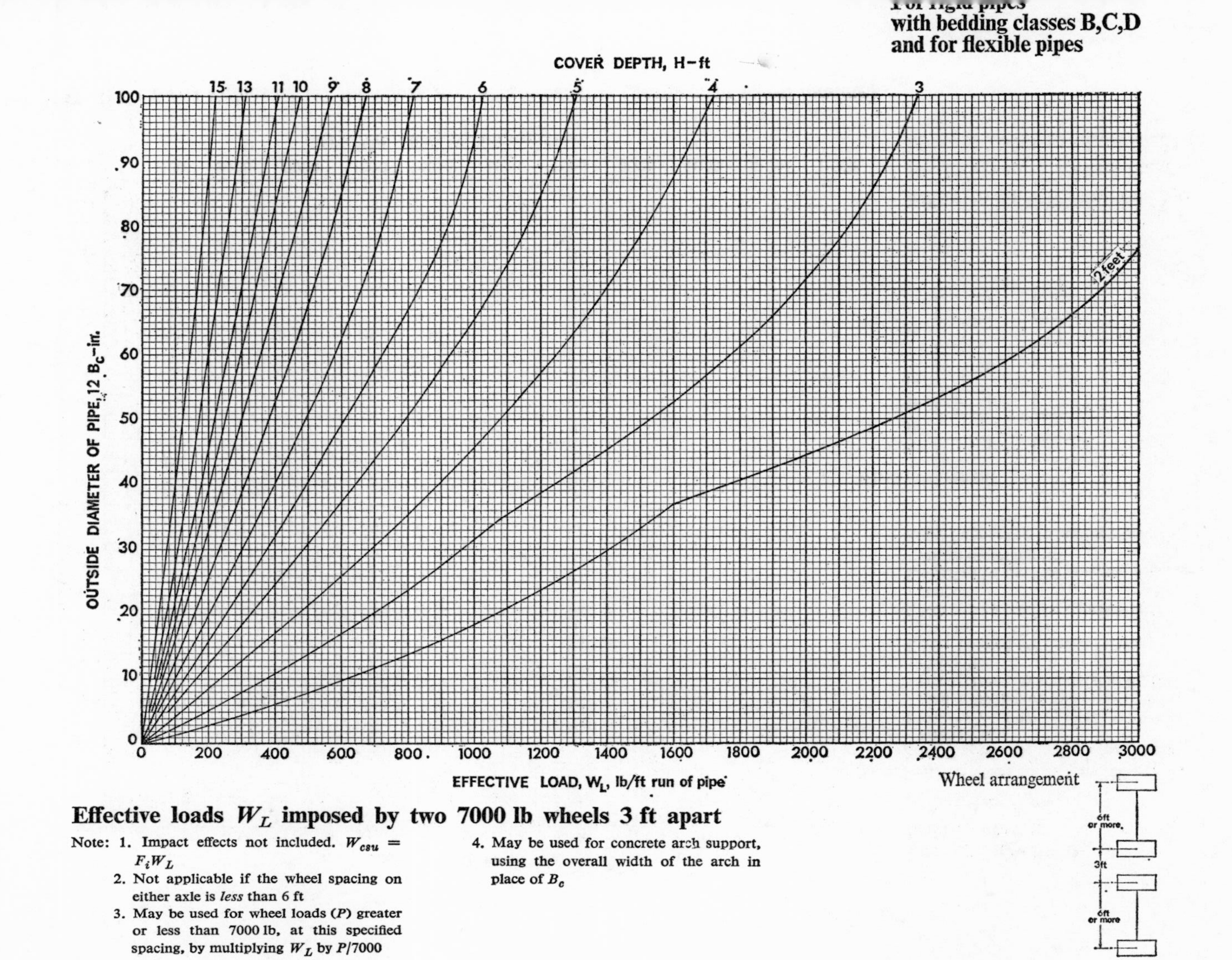

Effective loads W_L imposed by two 7000 lb wheels 3 ft apart

Note:
1. Impact effects not included. $W_{csu} = F_i W_L$
2. Not applicable if the wheel spacing on either axle is *less* than 6 ft
3. May be used for wheel loads (P) greater or less than 7000 lb, at this specified spacing, by multiplying W_L by $P/7000$
4. May be used for concrete arch support, using the overall width of the arch in place of B_c

CHART C7 *(b)* *Values of W_L for Class B, C, D beddings*

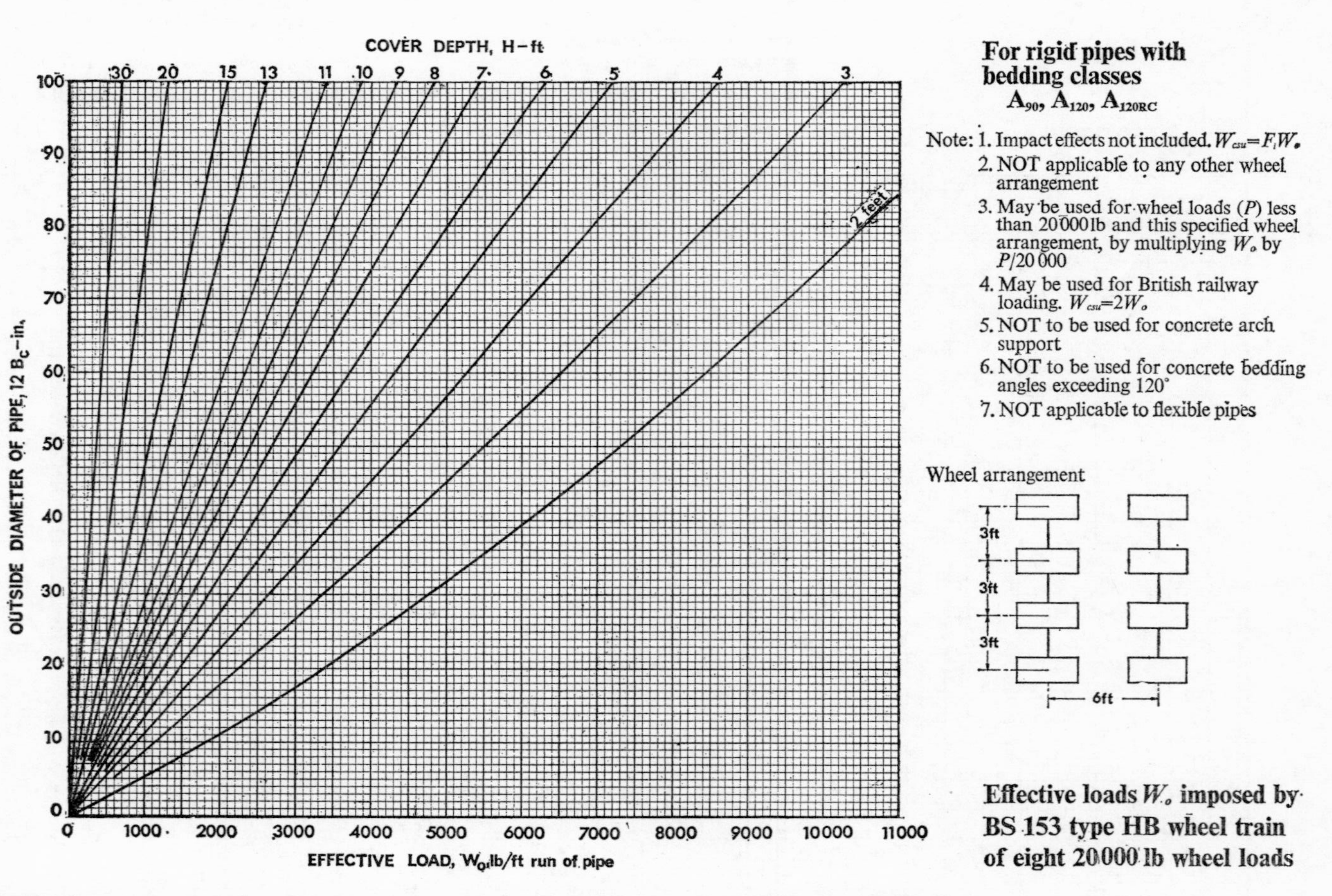

For rigid pipes with bedding classes
A_{90}, A_{120}, A_{120RC}

Note:
1. Impact effects not included. $W_{csu}=F_iW_o$
2. NOT applicable to any other wheel arrangement
3. May be used for wheel loads (P) less than 20 000 lb and this specified wheel arrangement, by multiplying W_o by $P/20\,000$
4. May be used for British railway loading. $W_{csu}=2W_o$
5. NOT to be used for concrete arch support
6. NOT to be used for concrete bedding angles exceeding 120°
7. NOT applicable to flexible pipes

Wheel arrangement

Effective loads W_o imposed by BS 153 type HB wheel train of eight 20 000 lb wheel loads

CHART C8 *(a)* *Values of W_o for Class A concrete bedding (heavy road wheel loading)*

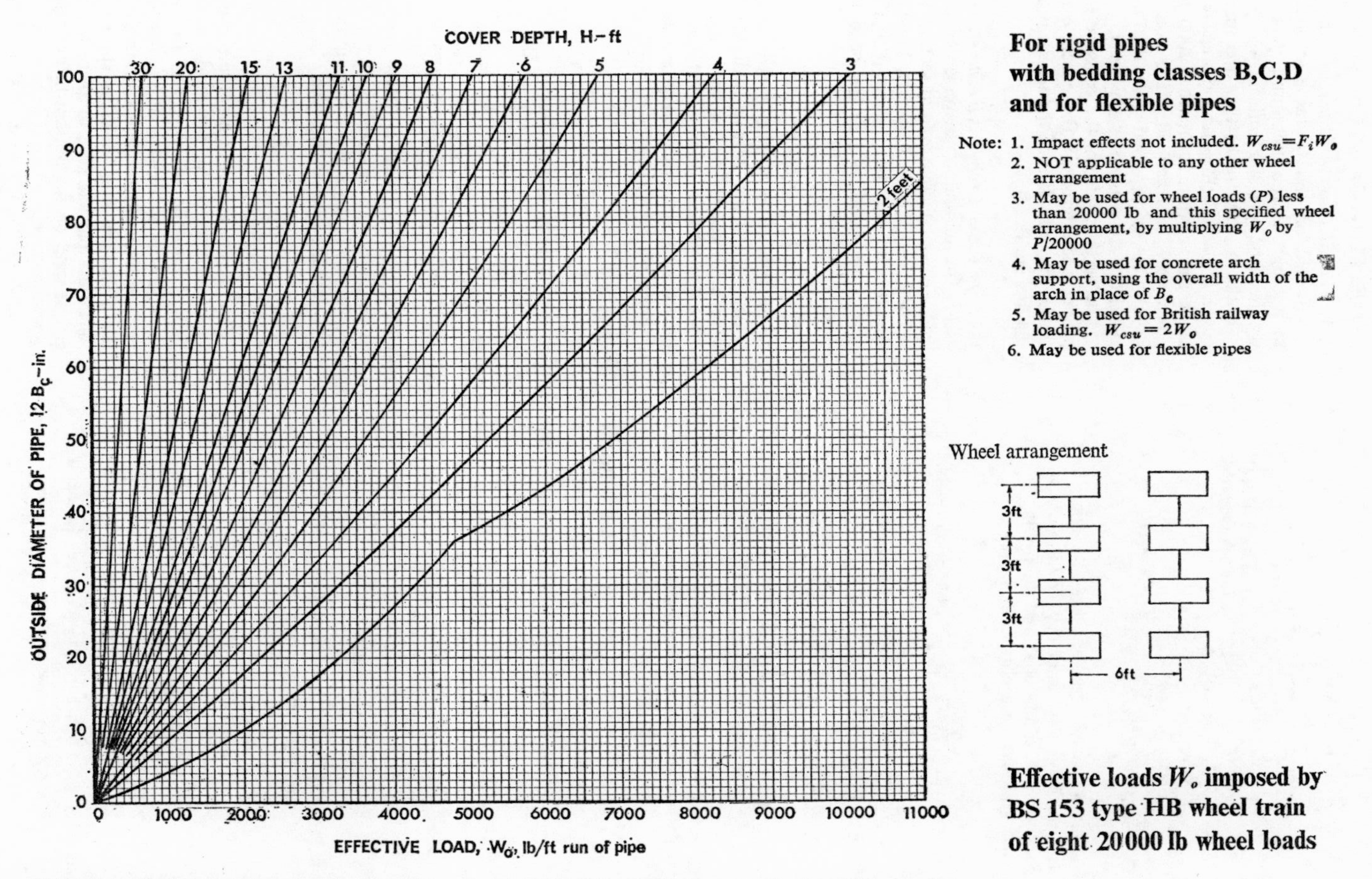

CHART C8 *(b)* *Values of* W_o *for Class B, C, D beddings (heavy road wheel loading)*

For rigid pipes with bedding classes B,C,D and for flexible pipes

Note:
1. Impact effects not included. $W_{csu}=F_i W_o$
2. NOT applicable to any other wheel arrangement
3. May be used for wheel loads (P) less than 20000 lb and this specified wheel arrangement, by multiplying W_o by $P/20000$
4. May be used for concrete arch support, using the overall width of the arch in place of B_c
5. May be used for British railway loading. $W_{csu} = 2W_o$
6. May be used for flexible pipes

Wheel arrangement

3ft

3ft

3ft

6ft

Effective loads W_o imposed by BS 153 type HB wheel train of eight 20000 lb wheel loads

CHART C9 *Trench beddings and bedding factors* F_m (Bc = outside diameter, D = inside)

CLASS A—Reinforced concrete. Bedding factor 3·4

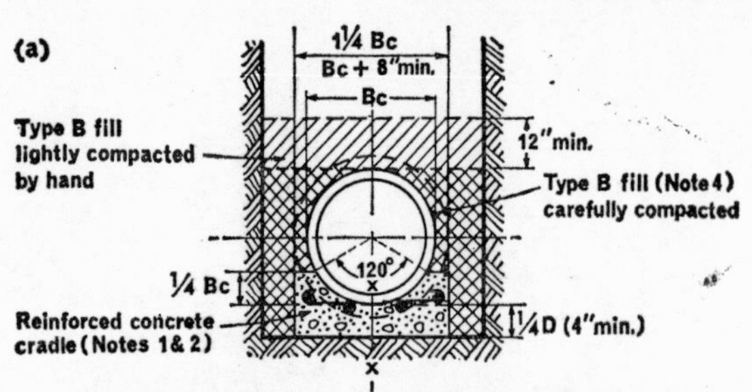

(*a*) **Reinforced concrete cradle:** Generally suitable for uniform soils and dry conditions. For rock or wet conditions, and in mining areas, use granular base course as in (*c*) (ii)

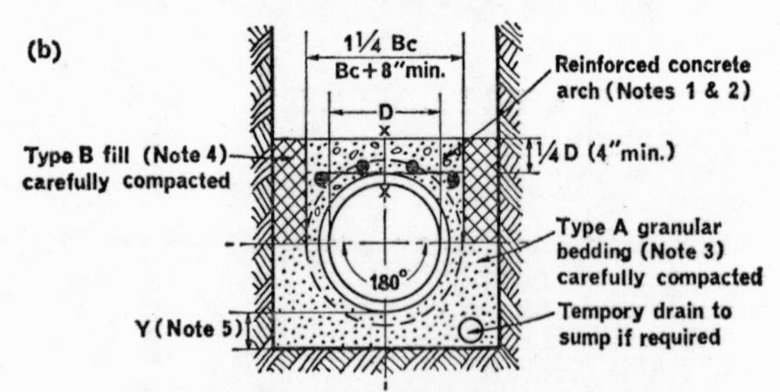

(*b*) **Reinforced concrete arch:** Generally suitable for rock or any other soil conditions, wet or dry

NOTES Charts C9 and C10

1. *For all Class A beddings:* The concrete (reinforced or plain), cradle or arch, to be monolithic with a minimum cube strength within 28 days of 3000 lb/in².

If the backfill, other than the first foot of cover, is placed before the compressive strength of the site concrete has reached 2000 lb/in², the pipes may be fractured. The concrete mix should be so designed that this danger is avoided without unnecessary delay.

The vertical sides of the concrete should be properly shuttered or the concrete extended across the full width of the trench (C9) or the shallow cut (C10).

2. *For Class A reinforced concrete cradle and arch:* The minimum transverse steel area to be 0·4% of the concrete area at section *x—x*, i.e. (0·004)(12)(0·25D)in.²/linear ft, but never less than 0·192 in.²/ft. For the arch (C9), if the steel area is increased to 1%, the bedding factor may be increased up to 4·8.

3. *Granular bedding Type A:* For rock and coarse-grained soils—ideally, broken stone or gravel, but other similar uniform material available locally may be used, e.g. crushed brick or crushed concrete. Material to pass $\frac{1}{2}$–1 in. sieve according to pipe size, and be retained on $\frac{3}{16}$ in. sieve.

CLASS A—Plain concrete. Bedding factor 2·6

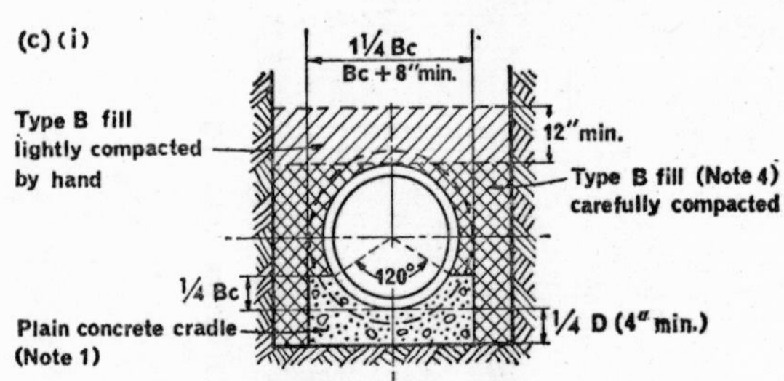

(*c*) **Plain concrete cradle:** (i) For uniform soils and dry conditions

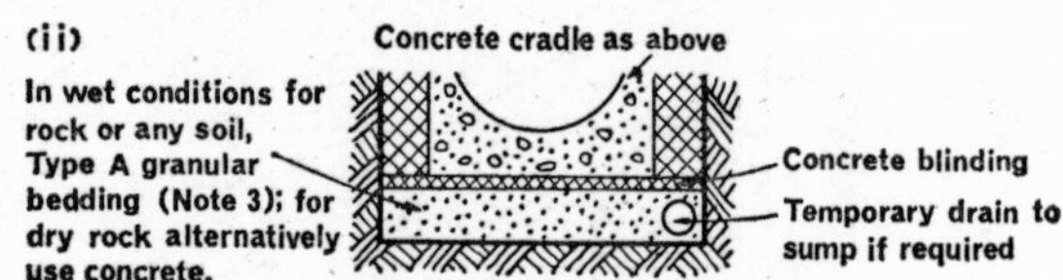

(*c*) (ii) Base course for rock in mining areas or for wet conditions in any soil. Use concrete base course only in dry stable rock, but not in mining areas even if dry

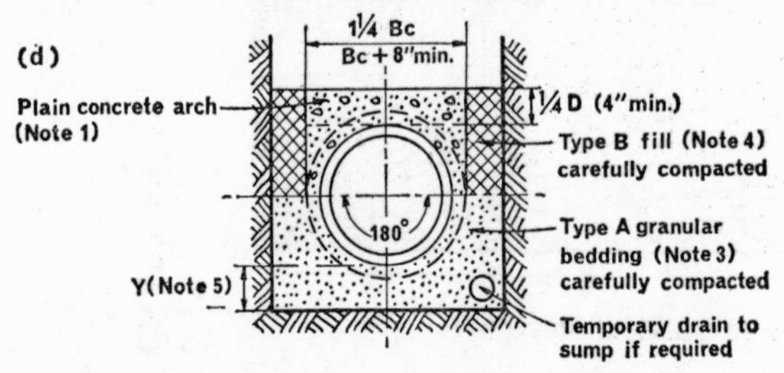

(*d*) **Plain concrete arch:** Generally suitable for rock or any other soil conditions, wet or dry

Note 3 (continued)

To prevent the intrusion of fine-grained soils such as clays, silts, fine sands or mixtures thereof into the bedding, add to the above about 1 part of free-draining coarse sand to 2 parts of stone or gravel and mix thoroughly, or use a free-draining 'all-in' mixture of similar maximum particle size, or place a blanket of coarse sand on the bottom before placing the stone or gravel.

Sands containing an excess of fine particles which would impede drainage or cause bulking during construction should not be used.

CLASS B—Bedding factor 1·9

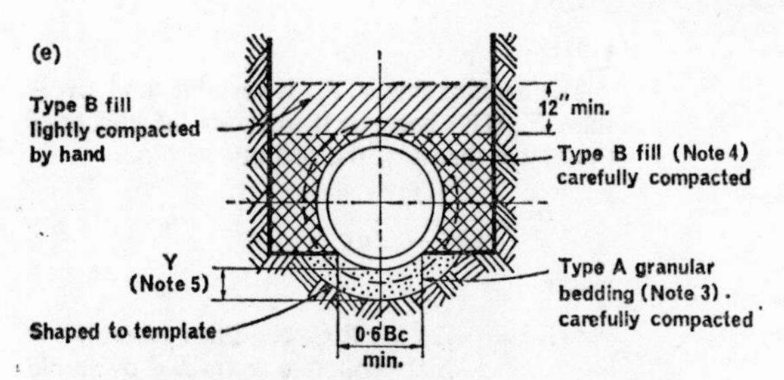

(*e*) **Hand-shaped trench bottom:** Generally suitable for uniform soils and dry conditions

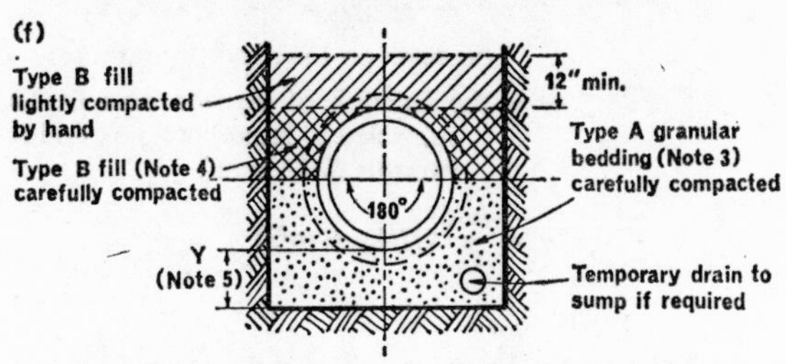

(*f*) **Machine-cut trench bottom:** Generally suitable for rock or any other soil conditions, wet or dry

Notes to Charts C9 and C10 (continued).

4. *Selected fill Type B:* Uniform readily compactible material—free from tree roots, vegetable matter, building rubbish and frozen soil, and preferably excluding clay lumps retained on 3 in. sieve, and stones retained on 1 in. sieve.

5. *Dimension Y:* In rock or mixed soils containing rock bands, boulders, large flints or stones or other irregular hard spots (C9, (*b*), (*d*), (*f*), C10, (*f*)):

$Y = \frac{1}{4}$ Bc under barrels, with a minimum of 8 in. (plus $\frac{1}{2}$ in. per foot of cover in excess of 16 ft) under both barrels and sockets.

In machine-dug uniform soils (C9 (*b*), (*d*) (*f*), C10 (*g*)):

$Y = \frac{1}{6}$ Bc, with a minimum of 4 in. under both barrels and sockets.

In hand-shaped uniform soils (C9 (*e*), C10 (*h*)):

$Y =$ 4 in. minimum under both barrels and sockets.

CLASS C—Bedding factor 1·5

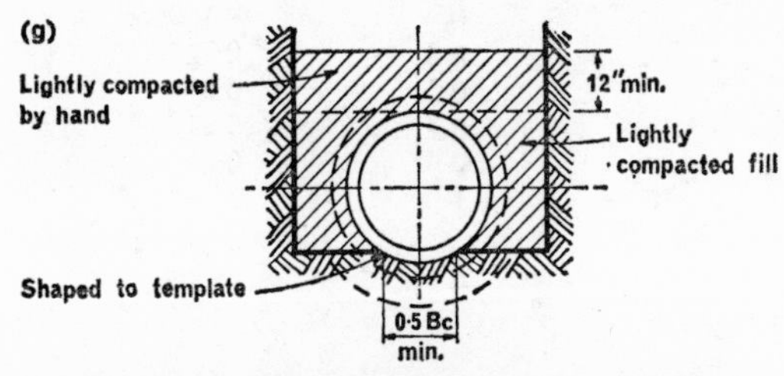

(*g*) **Hand-shaped trench bottom:** Only suitable for some cast-iron pipes in uniform fine-grained soils and dry conditions

Charts C9 and C10 are published with acknowledgments to the A.S.C.E. Manual of Engineering Practice No. 37, Figs. 74 and 76, pp. 203 and 266.

CLASS D—Bedding factor 1·1

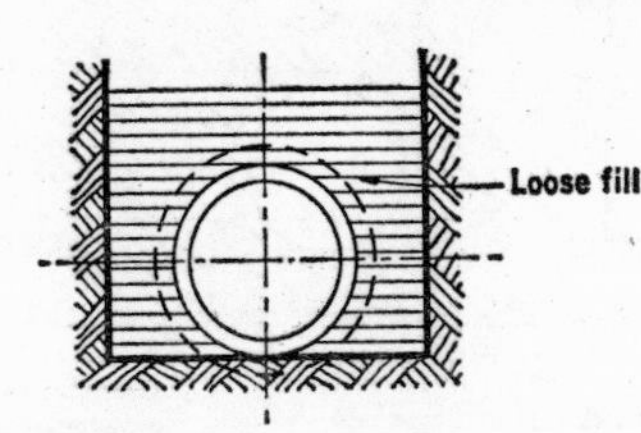

(*h*) **Flat bottom trench:** Only suitable for some cast-iron pipes in uniform fine-grained soils and dry conditions

Note

B_c = Outside Diameter

D = Inside diameter

CHART C10 *Embankment beddings and bedding factors F_p for pipes in 'very wide' trench or under embankments.* (For notes see Chart C9)

CLASS A_{120} and A_{120RC} BEDDINGS

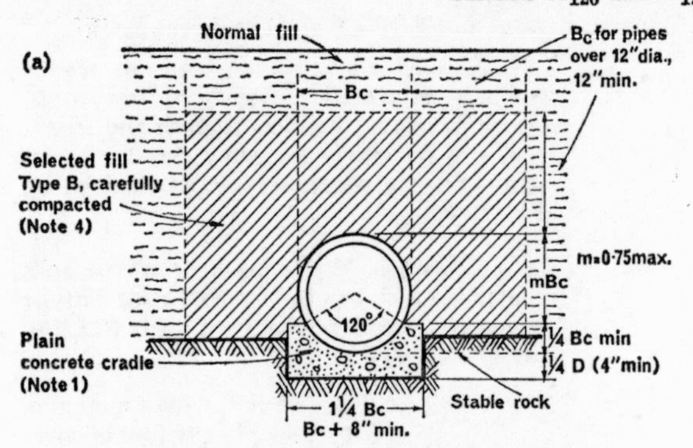

(*a*) **Class A_{120} plain concrete cradle restrained by rock:** Generally suitable only for founding on stable rock ($N' = 0{\cdot}421$). Not suitable for mining areas

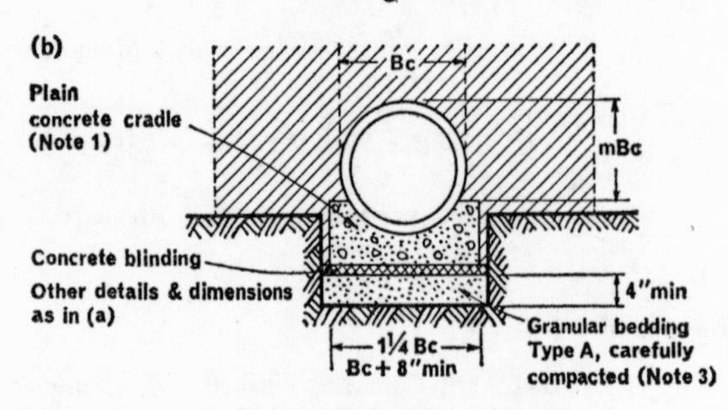

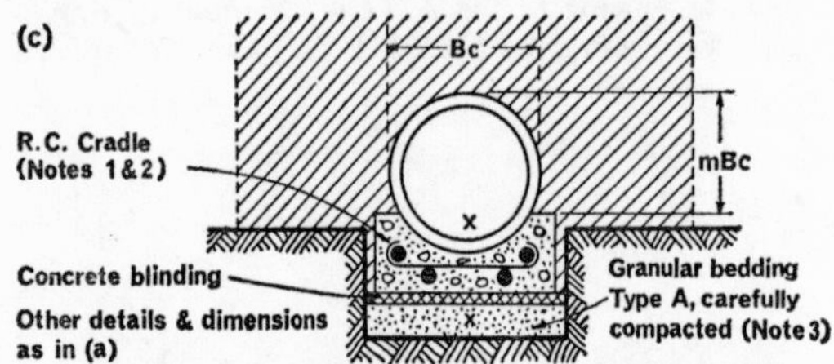

(*b and c*) **Class A_{120} plain and Class A_{120RC} reinforced concrete cradles:** Generally suitable for founding on rock in subsidence areas or on any other soils

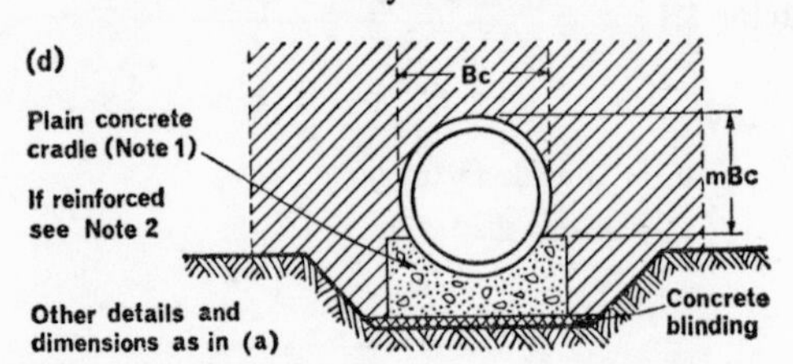

(*d*) **Class A_{120} plain or Class A_{120RC} reinforced concrete cradles:** Generally suitable for founding on uniform firm coarse-grained soils such as gravels

NOTES

The bedding factor, F_p, is variable, and larger than F_m (C9) because it includes the effects of the active thrust of the soil on the pipes.

$$F_p = \frac{1{\cdot}431}{(N' - x'q)}$$

where

$N' = 0{\cdot}421$ for R.C. beddings generally or plain concrete restrained by stable rock (see (*a*)) and

$N' = 0{\cdot}505$ for plain concrete bedding generally

x' depends on the exposure ratio m

m = the fraction of B_c (the pipe O.D.) subjected to lateral pressure and is variable (0·75 max.)

$$q = \frac{mK}{C_c}(H/B_c + m/2)$$

where

$K = 0{\cdot}33$, C_c is the appropriate coefficient for the vertical fill load, and H is the cover

x' depends on the ratio m as follows:

m	x'
0	0·150
0·3	0·743
0·5	0·856
0·7	0·811
0·9	0·678
1·0	0·638

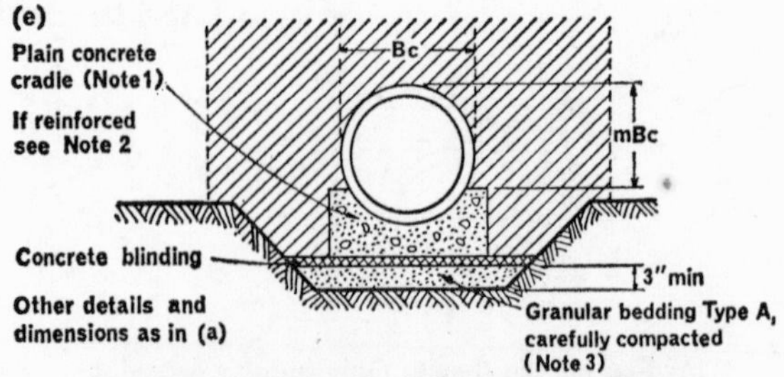

(*e*) **Class A_{120} plain or Class A_{120RC} reinforced concrete cradles:** Generally suitable for founding on fine-grained soils such as clays, silts and variable soils

CLASS B, C and D BEDDINGS

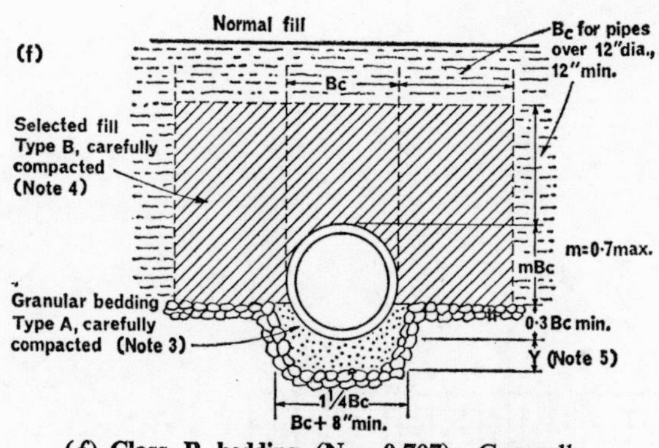

(*f*) **Class B bedding (N = 0·707):** Generally suitable for rock or mixed soils containing boulders or other large hard particles

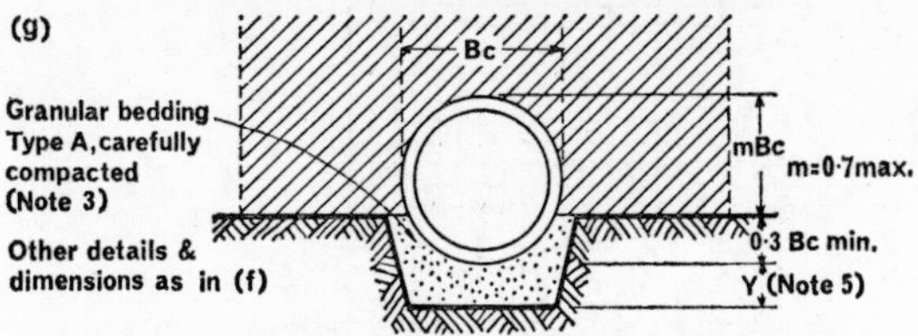

(*g*) **Class B bedding (N = 0·707):** Generally suitable for uniform soils, machine or hand dug

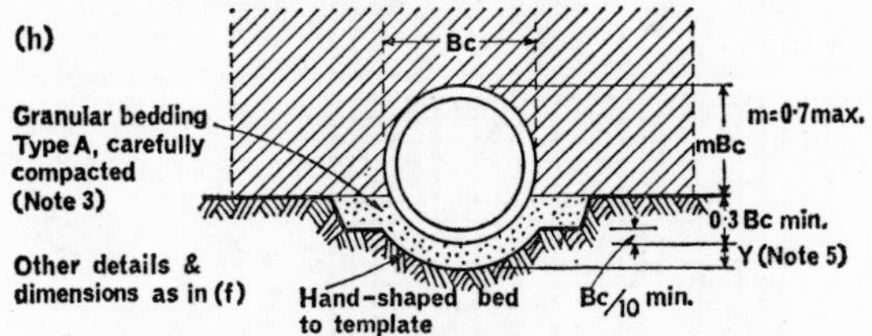

(*h*) **Class B bedding (N = 0·707):** Generally suitable for hand-shaped uniform fine-grained soils

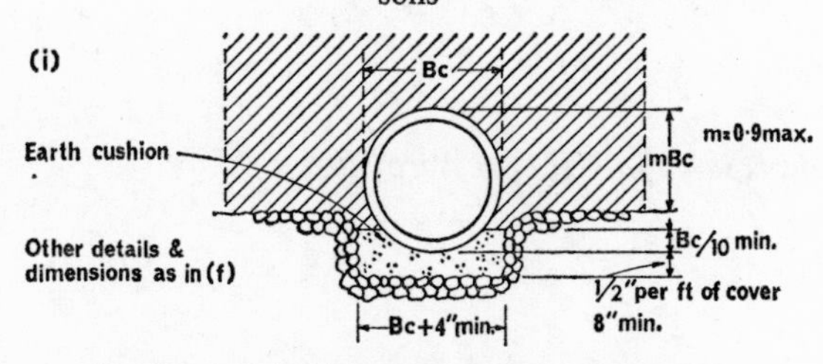

(*i*) **Class C bedding (N = 0·840):** Generally suitable for rock or mixed soils containing boulders or other large hard particles

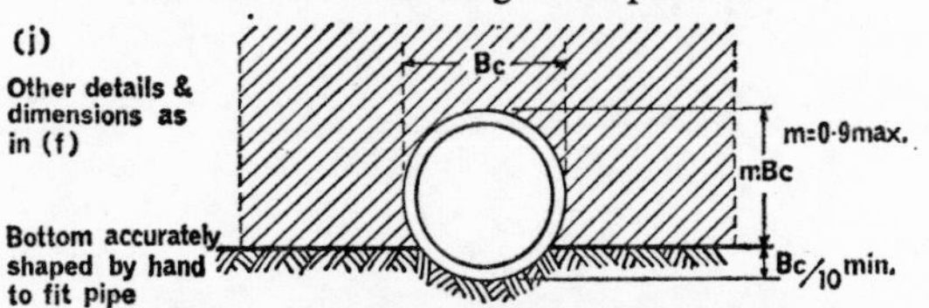

(*j*) **Class C bedding (N = 0·840):** Generally suitable for fine-grained uniform soils

$$F_p = \frac{1{\cdot}431}{(N - xq)}$$

where

$N = 0{\cdot}707$ for Class B, $0{\cdot}840$ for Class C, and $1{\cdot}310$ for Class D bedding

x depends on the exposure ratio m

m is variable (0·7 max.)

$$q = \frac{mK}{C_c}(H/B_c + m/2)$$

where

$K = 0{\cdot}33$, and C_c is the appropriate coefficient for the vertical fill load.

x depends on the ratio m as follows:

m	x
0	0
0·3	0·217
0·5	0·423
0·7	0·594
0·9	0·655
1·0	0·638

(*k*) *and* (*l*) **Class D bedding (N = 1·310)**

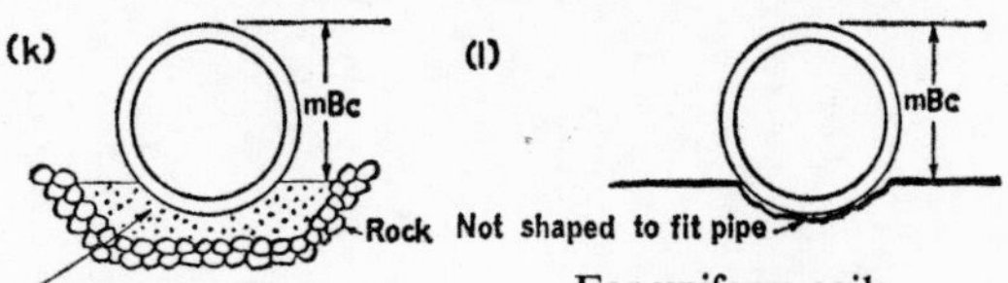

For rock or mixed soils (not recommended)

For uniform soils (not recommended)

CHART C11

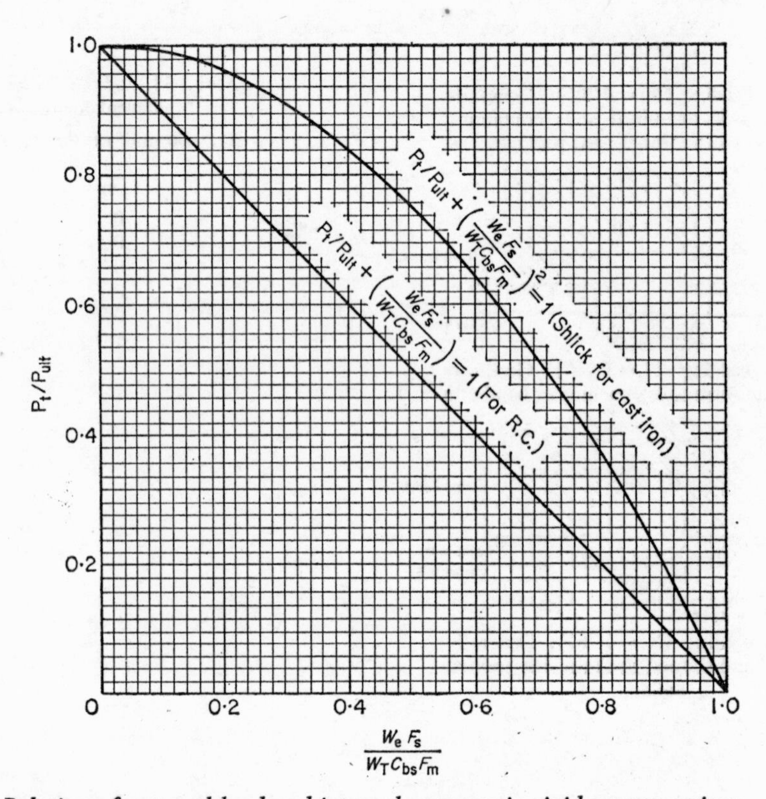

CHART C11 *Relation of external load and internal pressure in rigid pressure pipes*

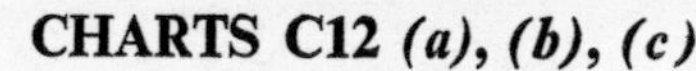

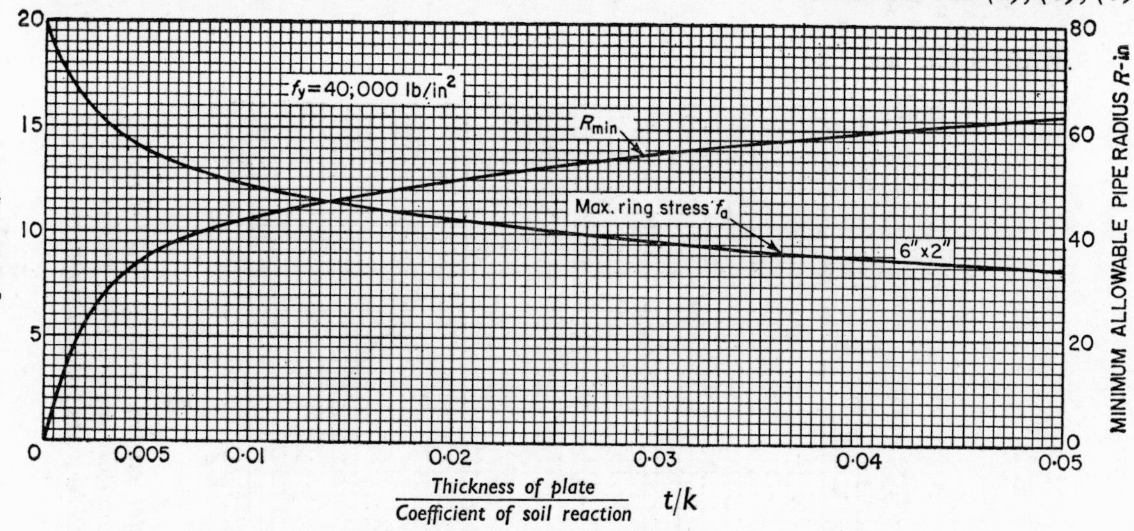

CHART C12 (*a*) *Max. allowable compressive ring stress f_a and minimum radius for* **6 in. × 2 in. corrugated steel pipe.** f_y = 40,000 lb/in.²

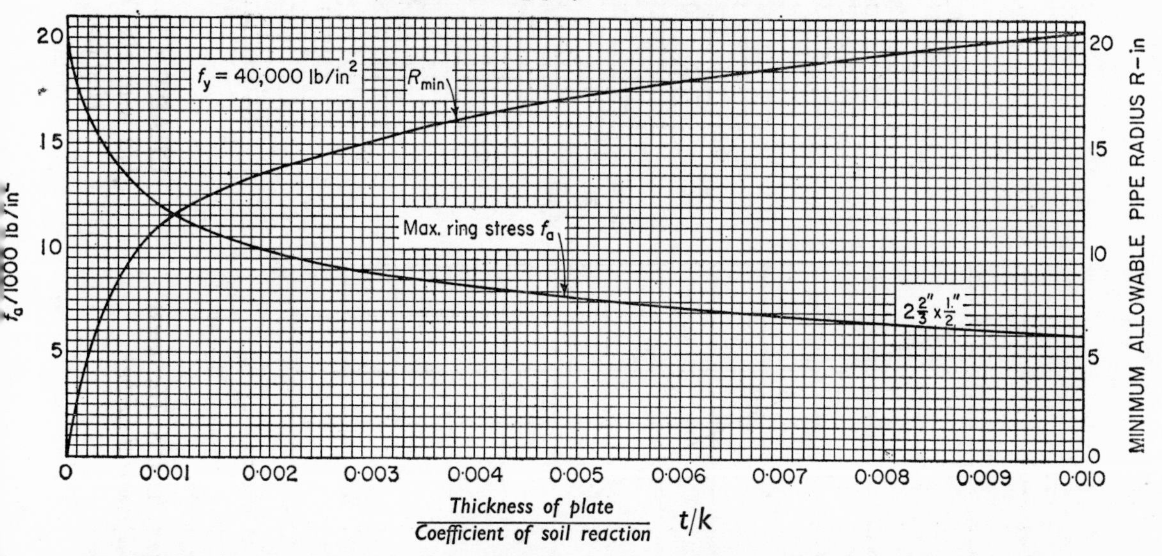

CHART C12 (*b*) *Max. allowable compressive ring stress f_a and minimum radius for* **2⅔ in. × ½ in. corrugated steel pipe.** f_y = 40,000 lb/in.²

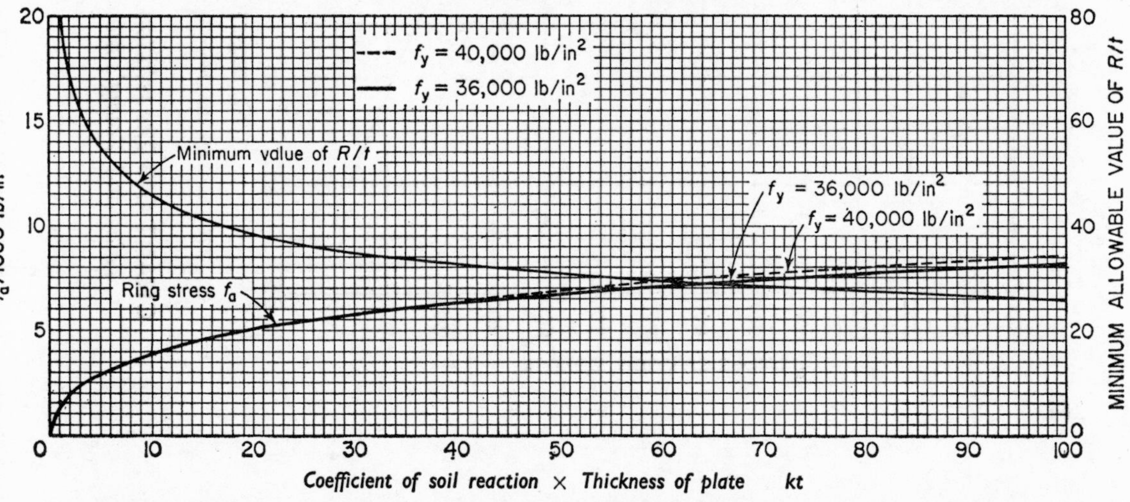

CHART C12 (*c*) *Max. allowable compressive ring stress f_a and minimum radius for* **smooth wall steel pipe.** f_y = 36,000 and 40,000 lb/in.²

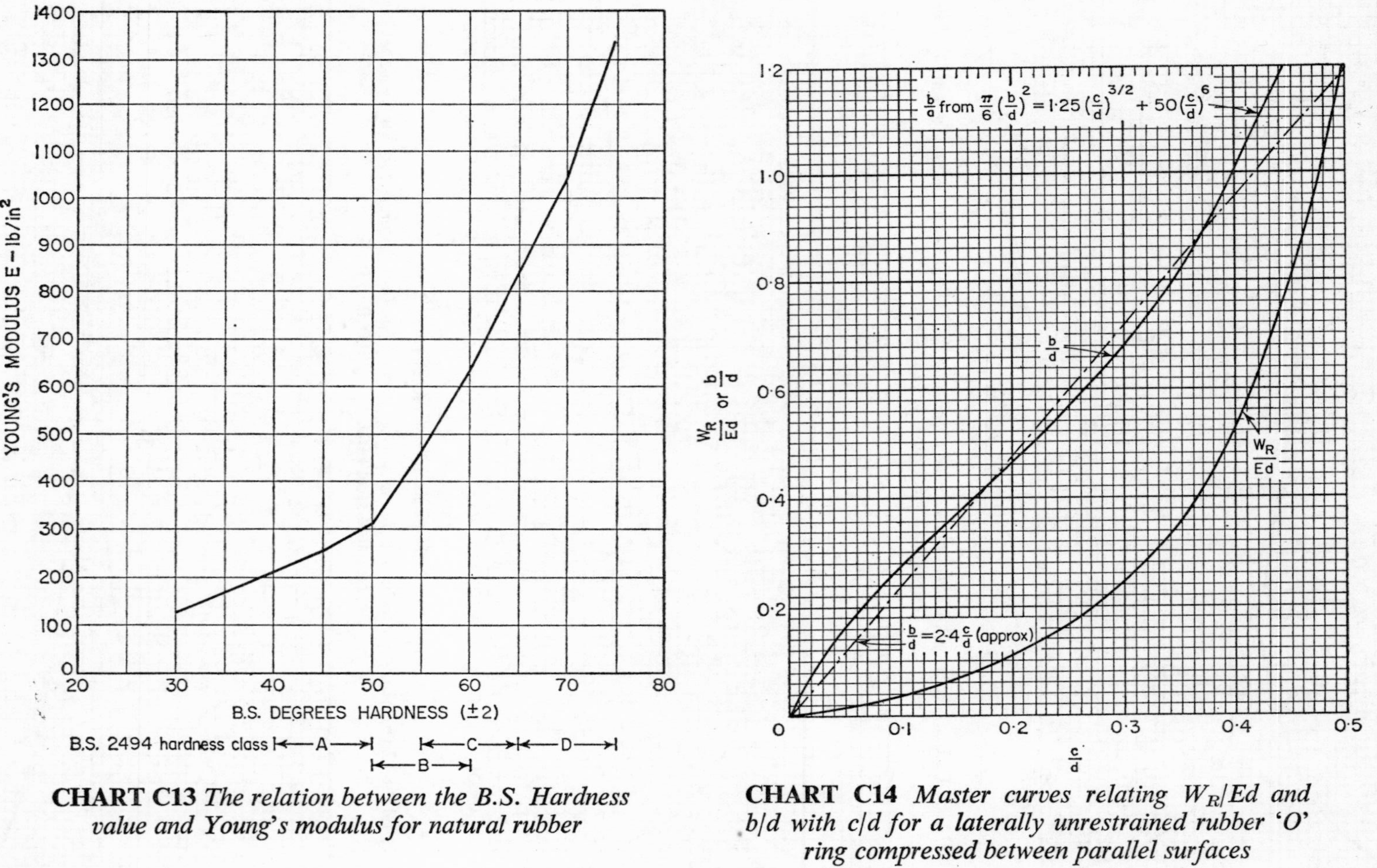

CHART C13 *The relation between the B.S. Hardness value and Young's modulus for natural rubber*

CHART C14 *Master curves relating W_R/Ed and b/d with c/d for a laterally unrestrained rubber 'O' ring compressed between parallel surfaces*

Appendix A

Worked Examples of Computations

I Examples 1–6. Rigid Pipes. Trench loading.
II Examples 7–9. Rigid Pipes. Embankment Loading.
III Examples 10–16. Flexible Pipes. Pressure and Culvert.

Appendix A
Worked Examples of Computations

I. Rigid Pipes—Trench Loading

The data assumed in the following examples such as B_c, B_d, γ_s, $K\mu$, etc. are purely illustrative. In design the actual values should always be used where possible, especially where they are greater or less than those assumed here.*

* The surcharge loads assumed in the examples are those currently used before the issue of the Second Report of the MOHLG Working Party in Oct. 1967. In actual design the values given in that report (10a) should be used (See Table D).

Example 1. *Trench Loading Conditions—Class B Bedding—36 in. R.C. Sewer*

Determine the minimum safe crushing test strength, the minimum 0·01 in. crack test load, and the minimum ultimate crushing test strength required for a 36 in. I.D. reinforced concrete sewer pipe laid on Class B bedding under a road in mixed soil, with a minimum cover of 4 ft and maximum cover of 12 ft. Assume γ_s = 120 lb/ft³, $K\mu = 0{\cdot}150$, and allow for a uniformly distributed surcharge of 500 lb/ft² as alternative to wheel loading.

External Loads

Step 1 In the absence of actual values of B_c and B_d assuming the values given in Table N, $B_c = 45\frac{3}{4}$ in. and $B_d = 76$ in. (N.B. These values would usually be excessive in practice).

Step 2 $\gamma_s = 120$ lb/ft³, $K\mu' = 0{\cdot}150$ as given

Step 3

		H_{min} (4 ft)	H_{max} (12 ft)
'Wide' trench fill load			
$H/B_c = H/45{\cdot}75$	=	1·05	3·15
Taking $r_{sd} = 0{\cdot}5$ for a Class B bed, from Chart C3 for H/B_c as above, C_c	=	1·30	4·65
$W_c' = C_c\gamma_s B_c^2 = C_c(120)\left(\frac{45{\cdot}75}{12}\right)^2$	=	2200	8100 lb/ft
'Narrow' trench fill load			
$H/B_d = H/76$		0·63	1·89
from Chart C1, for $K\mu' = 0{\cdot}15$ and H/B_d as above C_d	=	0·58	1·45
$W_c = C_d\gamma_s B_d^2 = C_d(120)\left(\frac{76}{12}\right)^2$	=	2780	7000 lb/ft

			H_{min} (4 ft)	H_{max} (12 ft)
Step 4	*Effective fill load*			
	The lower values of W_c' and W_c for H_{min} and H_{max} are		2200	7000 lb/ft
Step 5	Water load $= (0{\cdot}75)(\pi/4)(3^2)(62{\cdot}4) = W_w$	=	330	330 lb/ft
	Uniform surcharge loads for $U_s = 500$ lb/ft²			
	For effective 'wide' trench conditions at 4 ft cover,			
	$H' = 4 + 500/120 = 8{\cdot}17$ ft and H'/B_c	=	2·14	
	and from Chart C3 for $H'/B_c = 2{\cdot}14$, C_c'	=	3·15	
	$W_c'' = C'_c \gamma_s B_c^2 = (3{\cdot}15)(120)\left(\frac{45{\cdot}75}{12}\right)^2$	=	5500	lb/ft
	$W'_{us} = W_c'' - W_c'$ (from step 4) $= 5500 - 2200$	=	3300	lb/ft
	For effective 'narrow' trench conditions at 12 ft cover, from Chart C2 for $K\mu' = 0{\cdot}15$ and $H/B_d = 1{\cdot}89$ C_{us}	=		0·57
	$W_{us} = C_{us} B_d U_s = (0{\cdot}57)\left(\frac{76}{12}\right)(500)$	=		1800 lb/ft
Step 6	*Concentrated surcharge loads*			
	For 'Road' loading and Class B bedding,			
	from Chart C8 (*b*), for $B_c = 45\frac{3}{4}$ in.	W_0 =	4000	1350 lb/ft
	Multiplying by $F_i = 2{\cdot}0$	W_{csu} =	8000	2700 lb/ft
Step 7	*Total effective external design load*			
	Since W_{csu} is greater than W'_{us} at 4 ft cover,			
	$W_e = W_c' + W_w + W_{csu} = 2200 + 330 + 8000$	=	10,530	lb/ft
	Since W_{csu} is greater than W_{us} at 12 ft cover			
	$W_e = W_c + W_w + W_{csu} = 7000 + 330 + 2700$	=		10,030 lb/ft

Pipe Strength

From Step 7	W_e =	10,530	10,030 lb/ft
For R.C. pipe, assumed (Table G)	F_s =	1·0	1·2
For 36 in. pipe, assumed	C_{BS} =	0·95	0·95
For Class B bed (Chart C9)	F_m =	1·9	1·9
Minimum crushing test strength	$W_T = W_e F_s / C_{BS} F_m$ =	5800	6700 lb/ft

The required minimum 0·01 crack test load for the pipe is then 6700 lb/ft and the required minimum ultimate crushing test load is $1{\cdot}25 \times 6700 = 8400$ lb/ft.

N.B. If the computed value of the required 0·01 in. crack test load is somewhat higher than that of the nearest lower B.S. strength class, and the assumed value of B_c is greater than that of the standard pipe, the computation should be rechecked, using the lower actual value of B_c and B_d, before finally deciding on the B.S. strength class required.

Example 2. *Trench Loading Conditions—Class A_{120} Bedding—36 in. Sewer*

If the pipe in Example 1 is to have a Class A_{120} plain concrete bed, what would its minimum crushing test strength be?

External Loads

			H_{min} (4 ft)	H_{max} (12ft)
Steps 1 to 4	*Effective fill and water loads*			
	From Step 4 in Example 1,			
	Effective fill loads	$W_c' =$	2200	—
		$W_c =$	—	7000 lb/ft
	Water load	$W_w =$	330	330 lb/ft
Step 5	*Uniform surcharge load*			
	From Step 5 in Example 1,			
	Effective uniform surcharge loads	$W'_{us} =$	3300	—
		$W_{us} =$	—	1800 lb/ft
Step 6	*Concentrated surcharge load*			
	For 'Road' loading and Class A_{120} bedding,			
	from Chart C8 (*a*) for $B_c = 45\frac{3}{4}$ in.,	$W_0 =$	4000	1350 lb/ft
	Multiplying by $F_i = 2{\cdot}0$	$W_{csu} =$	8000*	2700* lb/ft
Step 7	*Total effective external design load*			
	At 4 ft cover, $W_e = 2200 + 330 + 8000$	=	10,530	
	At 12 ft cover, $W_e = 7000 + 330 + 2700$	=		10,030 lb/ft

Pipe Strength

		H_{min} (4 ft)	H_{max} (12 ft)
From Step 7	$W_e =$	10,530	10,030 lb/ft
For Class A_{120} bedding (Chart C9)	$F_m =$	2·6	2·6
With all other values as in Example 1,			
$W_T = W_e F_s / C_{BS} F_m$	=	4250	4900 lb/ft

The required minimum 0·01 in. crack load for the pipe is now 4900 lb/ft and the required minimum ultimate crushing test load is 4900 × 1·25 = 6100 lb/ft.

* N.B. The coincidence of these loads with those for Class B bedding arises because 36 in. diameter is a critical value for 36 in. wheel spacing. It would not necessarily occur with pipes of larger or smaller diameter.[11]

Example 3. *Trench Loading Conditions—Class A Concrete Arch Support—36 in. Sewer*

The pipe in Example 1 is to have a Class A concrete arch support. Determine the required minimum crushing test strength of the pipe.

External Loads

Steps 1 and 2

Using the same values as in Example 1 for B_c, B_d, γ_s and K'_u, from Chart C9, overall width of arch $B_{arch} = 1{\cdot}25\ B_c = 1{\cdot}25 \times 45{\cdot}75 = \underline{57{\cdot}2}$ in.

Crown thickness = D/4 = 36/4 = 9 in.

Overall depth of arch = $B_c/2 + 9 = 22{\cdot}87 + 9 = \underline{31{\cdot}87}$ in.

$$\text{Weight of arch} = \left[\frac{(57{\cdot}2)(31{\cdot}87)}{144} - \frac{1}{2}\cdot\frac{\pi}{4}\cdot\left(\frac{45{\cdot}75}{12}\right)^2\right](140) = 980 \text{ lb/ft}$$

assuming a concrete density of 140 lb/ft³

Step 3

		H_{min} (4 ft)	H_{max} (12 ft)
Effective conduit width B_{arch} = 57·2 in. = say 4·8 ft.			
Effective cover = $H - 0{\cdot}75$ ft	=	3·25	11·25 ft
H/B_c effective = $(H - 0{\cdot}75)/4{\cdot}8$	=	0·675	2·34
'Wide' trench fill load			
Assume r_{sd}	=	1·0	1·0
Overall depth of Pipe + arch = B_c + 9 in. = 54·75 in.			
Taking $p = 31{\cdot}87/(B_c + 9) = 0{\cdot}58$, $r_{sd}p$	= say	0·6	0·6
From Chart C3 for H/B_c as above, and $r_{sd}p = 0{\cdot}6$, C_c	=	0·7	3·5
$W_c' = C_c\gamma_s B_{arch}^2 = C_c\,(120)\,(4{\cdot}8)^2$	=	1930	9700 lb/ft
'Narrow' trench fill load			
B_d = 76 in. = 6·33 ft			
H/B_d effective = $(H - 0{\cdot}75)/6{\cdot}33$	=	0·51	1·77
From Chart C1 for $K'\mu = 0{\cdot}15$ and H/B_d as above, C_d	=	0·47	1·35
$W_c = C_d\gamma_s B_d^2 = C_d\,(120)\,(6{\cdot}33)^2$	=	2250	6450 lb/ft

---

Step 4

Effective fill load			
The lower values of W_c' and W_c for H_{min} and H_{max} are		1930 (wide)	6450 lb/ft (narrow)
Add water load, W_w	=	330	330 lb/ft
Add arch load, from Steps 1 and 2,	=	980	980 lb/ft
Total effective fill load	=	3240	7760 lb/ft

				H_{min} (4 ft)	H_{max} (12 ft)
Step 5	Effective cover			3·25	11·25 ft
	Uniform surcharge loads for $U_s = 500$ *lb/ft*^{2}				
	For effective 'wide' trench conditions at $H = 3{\cdot}25$ ft,				
	$H' = 3{\cdot}25 + U_s/\gamma_s = 3{\cdot}25 + 500/120 = 7{\cdot}42$ ft				
	From Chart C3 for $H' = 7{\cdot}42$ ft and $B_c = 57{\cdot}2$ in.				
	$H'/B_c = 1{\cdot}55$ and $\gamma_{sd}p = 0{\cdot}6$,	C_c'	=	2·15	
	$W_c'' = C'_c\gamma_s B_{arch}^2 = 2{\cdot}15\,(120)\left(\frac{57{\cdot}2}{12}\right)^2$		=	5900	lb/ft
	$W'_{us} = W''_c - W'_c$ (from Step 4) $= 5900 - 1930$		=	<u>3970</u>	lb/ft
	For effective 'narrow' trench conditions at $H = 11{\cdot}25$ ft and $B_d = 6{\cdot}33$ ft,				
	From Chart C2 for $K'\mu = 0{\cdot}15$ and $H/B_d = \frac{11{\cdot}25}{6{\cdot}33}$		=		1·77
		C_{us}	=		0·6
	$W_{us} = C_{us}B_dU_s = (0{\cdot}6)\,(6{\cdot}33)\,(500)$		=		<u>1890</u> lb/ft

Step 6	*Concentrated surcharge loads—roads*				
	Effective cover			3·25	11·25 ft
	For concrete arch with Class B bed, using Chart C8 (*b*) for Class B bedding and $B_c = 57{\cdot}2$ in.,	W_o	=	5700	1800 lb/ft
	Multiplying by $F_i = 2{\cdot}0$,	W_{csu}	=	<u>11,400</u>	<u>3600</u> lb/ft

Step 7	*Total effective external design loads*				
	Since $W_{csu} > W_{us}$ at both H_{min} and H_{max},				
	W_e = fill + water + arch + W_{csu} (Steps 4 and 6)				
	at 3·25 ft cover = 3240 + 11,400		=	14,640	lb/ft
	and at 11·25 ft cover = 7760 + 3600		=		11,360 lb/ft

Pipe Strength

				H_{min} (4 ft)	H_{max} (12 ft)
		W_e	=	14,640	11,360 lb/ft
From Step 7	For R.C. pipe, assumed (Table G)	F_s	=	1·0	1·2
	For 36 in. pipe, assume	C_{BS}	=	0·95	0·95
	For concrete arch (Chart C9)	F_m	=	2·6	2·6
	Minimum crushing test strength,				
	$W_T = W_eF_s/C_{BS}F_m$		=	<u>5900</u>	5500 lb/ft

The required minimum 0·01 in. crack test load for the pipes is then <u>5900</u> lb/ft and the required minimum ultimate crushing test load is 1·25 × 5900 – <u>7400</u> lb/ft.

Example 4. *Heading Conditions—Class B Bedding—36 in. Sewer*

The pipe in Example 1 is to be laid in heading under a road with a maximum cover of 12 ft. Determine the required minimum crushing test strength of the pipe. Assume $K'\mu = K\mu = 0{\cdot}15$; heading width = trench width; bedding Class B.

External Loads

Steps 1 and 2 Assume $B_c = 45\frac{3}{4}$ in. $= 3{\cdot}81$ ft, $B_t = 76$ in. $= 6{\cdot}33$ ft, $\gamma_s = 120$ lb/ft³, $K\mu = 0{\cdot}15$

			Roof Packing: Solid	Roof Packing: Compressible
Step 3	*Fill loads* (see Table M)	Cover H =	12	12 ft
	For *compressible* packing, $H/B_t = 12/6{\cdot}33$	=	—	1·9
	From Chart C1,			
	for $K\mu = 0{\cdot}15$ and $H/B_t = 1{\cdot}9$	C_d =	—	1·45
	$W_c = C_d\gamma_s B_t^2 = (1{\cdot}45)(120)(6{\cdot}33)^2$	=	—	7000 lb/ft
	For *solid* packing,			
	$B_d = B_c = 3{\cdot}81$ ft and $H/B_c = 12/3{\cdot}81$	=	3·15	—
	and from Chart C1			
	for $K\mu = 0{\cdot}15$ and $H/B_c = 3{\cdot}15$	C_d =	2·05	—
	$W_c = C_d\gamma_s B_c^2 = (2{\cdot}05)(120)(3{\cdot}81)^2$	=	3560	—lb/ft
Step 4	*Effective fill loads*			
	Effective fill load	W_e =	3560*	7000* lb/ft
	Water load as in Example 1		330	330 lb/ft

* N.B. These values assume no cohesion in the undisturbed soil overlying the pipes and so are probably conservative for cohesive soils.

			Solid	Compressible
Step 5	*Uniform surcharge loads for* $U_s = 500$ lb/ft²			
	For compressible packing,			
	$B_d = B_t = 6{\cdot}33$ ft, and	H/B_t =	—	1·9
	For solid packing,			
	$B_d = B_c = 3{\cdot}81$ ft,	H/B_c =	3·15	—
	From Chart C2			
	for $K\mu = 0{\cdot}15$ and $H/B_t = 1{\cdot}9$	C_{us} =	—	0·56
	and for $H/B_c = 3{\cdot}15$	C_{us} =	0·39	—
	$W_{us} = C_{us}B_tU_s = (0{\cdot}56)(6{\cdot}33)(500)$	=	—	1770 lb/ft
	$W_{us} = C_{us}B_cU_s = (0{\cdot}39)(3{\cdot}81)(500)$ (solid)	=	740	—lb/ft

			Roof Packing: Solid	Roof Packing: Compressible
Step 6	*Concentrated surcharge loads*			
	As in Example I for 12 ft cover	W_{csu} =	2700	2700 lb/ft

Step 7 *Total effective external design load*

Since W_{csu} is greater than W_{us},			
for solid packing $W_e = 3560 + 330 + 2700$	=	6590	—lb/ft
For compressible packing $W_e = 7000 + 330 + 2700$	=	—	10,030 lb/ft

Pipe Strength

From Step 7,	W_e =	6590	10,030 lb/ft
For R.C. pipe assumed (Table G)	F_s =	1·2	1·2
For 36 in. pipe assumed	C_{BS} =	0·95	0·95
For Class B bedding	F_m =	1·9	1·9
Minimum crushing test strength, $W_T = W_e F_s / C_{BS} F_m$	=	4380	6700 lb/ft

The required minimum 0·01 in. crack test loads for the pipes are then
4380 lb/ft for solid packing,
and 6700 lb/ft for compressible packing.
The required minimum ultimate crushing test loads are:
for solid packing, $1{\cdot}25 \times 4380 = 5500$ lb/ft
for compressible packing, $1{\cdot}25 \times 6700 = 8400$ lb/ft.

Example 5. *Trench Loading Conditions—Class A Arch Support, Class B Bedding—36 in. Rising Main*

The pipe in Example 3 is to be a rising main, the working pressure is 8·65 lb/in.² = (20 ft) at the lower end and zero at the upper end. The maximum test pressure at the lower end is 13 lb/in.² = (30 ft), and at the upper end 9·5 lb/in.² = (22 ft).

The minimum bursting pressure of the pipe is 90 lb/in.² Determine the required minimum safe crushing test strength of the pipe.

External Load

			H_{min} (4 ft)	H_{max} (12 ft)
	Effective cover = $H - 0{\cdot}75$ ft	=	3·25	11·25
Steps 1–4	*Effective fill loads*			
	As in Example 3	W'_c =	1930	—lb/ft
		W_c =	—	6450 lb/ft
	water load	W_w =	330	330 lb/ft
	arch load	=	980	980
	Total effective fill load	=	3240	7760 lb/ft
Step 5	*Uniform surcharge load*			
	As in Example 3, effective load	W'_{us} =	3970	—lb/ft
		W_{us} =	—	1890 lb/ft
Step 6	*Concentrated surcharge load*			
	As in Example 3	W_{csu} =	11,400	3600 lb/ft

Step 7

		H_{min} (4 ft)	H_{max} (12 ft)
Effective external design load			
(*a*) *With normal working pressure*			
Fill + water + arch + W_{csu} (Steps 4 and 6)			
at 3·25 ft cover = 3240 + 11400 =	W'_e =	14,640	—lb/ft
at 11·25 ft cover = 7760 + 3600 =	W'_e =	—	11,360 lb/ft
(*b*) (1) *With maximum test pressure*			
Fill + water + arch (Step 4) =	W'_e =	3240	7760 lb/ft
(*b*) (2) *With maximum test pressure if uniform surcharge is present*			
Fill + water + arch + W_{us} (Steps 4 and 5)			
at 3·25 ft cover = 3240 + 3970 =	W'_e =	7210	—lb/ft
at 11·25 ft cover = 7760 + 1890 =	W'_e =	—	9650 lb/ft

Internal Pressure

Step 8 Assuming that the specified pressures other than the bursting pressure are measured at the pipe crown, add 1·5 ft (=0·65 lb/in.²) to obtain mean pressures.

		H_{min}	H_{max}
(*a*) Then under normal working conditions, the effective internal pressure $p_i = p_w$	=	0·65	9·3 lb/in.²
(*b*) (1) *and* (2) Under test conditions the effective internal pressure $p_i = p_t$	=	10·15	13·65 lb/in.²

Pipe Strength for Combined Loading

		H_{min} (4 ft)	H_{max} (12 ft)
Effective cover = $H - 0{\cdot}75$ ft	=	3·25	11·25 ft
From equation (5.5), $W_T = W'_e F_{se}/(1 - p_i F_{si}/p_{ult}) C_{BS} F_m$			
For 36 in. pipe, assume	C_{BS} =	0·95	0·95
For Class A arch support with Class B bedding (Chart C9)	F_m =	2·6	2·6
Assuming first-class site work (Table G)	F_{se} =	1·6	1·6
	F_{si} =	2·0	2·0
As specified,	p_{ult} =	90	90 lb/in.²

Case (*a*)—from Step 7	W'_e =	14,640	11,360 lb/ft
—from Step 8	$p_i = p_w$ =	0·65	9·3 lb/in.²
At $H = 3{\cdot}25$ ft, $1 - p_i F_{si}/p_{ult} = 1 - \dfrac{0{\cdot}65 \times 2{\cdot}0}{90}$	=	0·98	—
At $H = 11{\cdot}25$ ft, $1 - p_i F_{si}/p_{ult} = 1 - \dfrac{9{\cdot}3 \times 2{\cdot}0}{90}$	=	—	0·79
At $H = 3{\cdot}25$ ft, $W_T = \dfrac{14{,}640 \times 1{\cdot}6}{0{\cdot}98 \times 0{\cdot}95 \times 2{\cdot}6}$	=	<u>9700</u>	— lb/ft
At $H = 11{\cdot}25$ ft, $W_T = \dfrac{11{,}360 \times 1{\cdot}6}{0{\cdot}79 \times 0{\cdot}95 \times 2{\cdot}6}$	=	—	9300 lb/ft

Case (*b*) (1)—from Step 7	$W'_e =$	3240	7760 lb/ft
—from Step 8	$p_i = p_t =$	10·15	13·65 lb/in.²
At $H = 3{\cdot}25$ ft, $1 - p_i F_{si}/p_{ult} = 1 - \dfrac{10{\cdot}15 \times 2{\cdot}0}{90}$	$=$	0·77	—
At $H = 11{\cdot}25$ ft, $1 - p_i F_{si}/p_{ult} = 1 - \dfrac{13{\cdot}65 \times 2{\cdot}0}{90}$	$=$	—	0·69
At $H = 3{\cdot}25$ ft, $W_T = \dfrac{3240 \times 1{\cdot}6}{0{\cdot}77 \times 0{\cdot}95 \times 2{\cdot}6}$	$=$	2720	— lb/ft
At $H = 11{\cdot}25$ ft, $W_T = \dfrac{7760 \times 1{\cdot}6}{0{\cdot}69 \times 0{\cdot}95 \times 2{\cdot}6}$	$=$	—	7300 lb/ft

Case (*b*) (2)—from Step 7	$W'_e =$	7210	9650 lb/ft
—from Step 8	$p_i = p_t =$	10·15	13·65 lb/in.²
$1 - p_i F_{si}/p_{ult}$ as in Case (*b*) (1)	$=$	0·77	0·69
At $H = 3{\cdot}25$ ft $W_T = \dfrac{7210 \times 1{\cdot}6}{0{\cdot}77 \times 0{\cdot}95 \times 2{\cdot}6}$	$=$	6050	— lb/ft
At $H = 11{\cdot}25$ ft, $W_T = \dfrac{9650 \times 1{\cdot}6}{0{\cdot}69 \times 0{\cdot}95 \times 2{\cdot}6}$	$=$	—	9050 lb/ft

The required value of W_T to satisfy either Case (*a*) or Case (*b*) is then the highest value, i.e. 9700 lb/ft and this is the required minimum 0·01 in. crack test load. The required minimum ultimate crushing test load, is $9700 \times 1{\cdot}25 = 12{,}100$ lb/ft.

Example 6. *Trench Loading Conditions—Class A_{120} Bedding—36 in. Rising Main*

The pipe in Example 5 is to have a concrete cradle bedding, Class A_{120}, instead of a concrete arch support. Determine the required minimum crushing test strength of the pipe.

External Loads

			H_{min} (4 ft)	H_{max} (12 ft)
Steps 1 to 4	*Effective fill load and water loads*			
	From Example 2, effective fill load	$W'_c =$	2200	— lb/ft
		$W_c =$	—	7000 lb/ft
	water load	$W_w =$	330	330 lb/ft
	Total effective fill load	$=$	2530	7330 lb/ft
Step 5	*Uniform surcharge load*			
	From Example 2	$W'_{us} =$	3300	— lb/ft
		$W_{us} =$	—	1800 lb/ft
Step 6	Concentrated surcharge load as in Example 2	$W_{csu} =$	8000	2700 lb/ft
Step 7	*Effective total external design load*			
	(*a*) *With normal working pressure*			
	Fill + water + W_{csu} (Steps 4 and 6)			
	At 4 ft cover = 2200 + 330 + 8000 =	$W'_e =$	10,530	— lb/ft
	At 12 ft cover = 7000 + 330 + 2700 =	$W'_e =$	—	10,030 lb/ft
	(*b*) (1) *With maximum test pressure*			
	Fill + water (Step 4) =	$W'_e =$	2530	7330 lb/ft

			H_{min} (4 ft)	H_{max} (12 ft)
Step 7 (continued)	*(b) (2) With maximum test pressure if uniform surcharge is present*			
	Fill + water + W_{us} (Steps 4 and 5)			
	At 4 ft cover = 2200 + 330 + 3300 =	W'_e =	5830	— lb/ft
	At 12 ft cover = 7000 + 330 + 1800 =	W'_e =	—	9130 lb/ft

Internal Pressure

Step 8	*Case (a)*, as in Example 5	$p_i = p_w$ =	0·65	9·3 lb/in.²
	Case (b), as in Example 5	$p_i = p_t$ =	10·15	13·65 lb/in.²

Pipe Strength for Combined Loading

From Equation 5.5

$W_T = W'_e F_{se}/(1 - p_i F_{si}/p_{ult}) C_{BS} F_m$

For 36 in. pipe assumed	C_{BS} =	0·95	0·95
For Class A_{120} bedding (Chart C9)	F_m =	2·6	2·6
Assuming first-class sitework, (Table G)	F_{se} =	1·6	1·6
	F_{si} =	2·0	2·0
As specified	p_{ult} =	90	90 lb/in.²
Case (a)—from Step 7	W'_e =	10,530	10,030 lb/ft
—from Step 8	$p_i = p_w$ =	0·65	9·3 lb/in.²
$1 - p_i F_{si}/p_{ult}$, as in Example 5	=	0·98	0·79
At 4 ft cover, $W_T = \dfrac{10{,}530 \times 1{\cdot}6}{0{\cdot}98 \times 0{\cdot}95 \times 2{\cdot}6}$	=	7000	— lb/ft
At 12 ft cover, $W_T = \dfrac{10{,}030 \times 1{\cdot}6}{0{\cdot}79 \times 0{\cdot}95 \times 2{\cdot}6}$	=	—	8250 lb/ft
Case (b) (1)—from Step 7	W'_e =	2530	7330 lb/ft
—from Step 8	$p_i = p_t$ =	10·15	13·65 lb/in.²
$1 - p_i F_{si}/p_{ult}$, as in Example 5	=	0·77	0·69
At 4 ft cover, $W_T = \dfrac{2530 \times 1{\cdot}6}{0{\cdot}77 \times 0{\cdot}95 \times 2{\cdot}6}$	=	2130	— lb/ft
At 12 ft cover, $W_T = \dfrac{7330 \times 1{\cdot}6}{0{\cdot}79 \times 0{\cdot}95 \times 2{\cdot}6}$	=	—	6900 lb/ft
Case (b) (2)—from Step 7	W'_e =	5830	9130
—from Step 8	$p_i = p_t$ =	10·15	13·65
$1 - p_i F_{si}/p_{ult}$, as in Example 5	=	0·77	0·69
At 4 ft cover, $W_T = \dfrac{5830 \times 1{\cdot}6}{0{\cdot}77 \times 0{\cdot}95 \times 2{\cdot}6}$	=	4900	— lb/ft
At 12 ft cover, $W_T = \dfrac{9130 \times 1{\cdot}6}{0{\cdot}69 \times 0{\cdot}95 \times 2{\cdot}6}$	=	—	8600 lb/ft

The required value of W_T to satisfy either case (*a*) or case (*b*) (2) is then the highest value, i.e. 8600 lb/ft if uniform surcharge is present during the pressure test, and this is the required minimum 0·01 in. crack test load. The required minimum ultimate crushing test load is 8600 × 1·25 = 10,750 lb/ft.

If no uniform surcharge is present, W_T = 8250 lb/ft (case (*a*)) and the required minimum ultimate crushing test load is 8250 × 1·25 = 10,300 lb/ft.

SUMMARY OF PIPE STRENGTH REQUIREMENTS—EXAMPLES 1–6
36 in. reinforced concrete pipe under a road

Example	1	2	3	4	5	6
Condition	Trench	Trench	Trench	Heading	Trench	Trench
Bedding class and factor	B (1·9)	A_{120} (2·6)	Concrete arch over, Class B under (2·6)	B (1·9)	Concrete arch over, Class B under (2·6)	A_{120} (2·6)
Function	Sewer	Sewer	Sewer	Sewer	Rising main	Rising main
Cover, Ft: min	4	4	4	—	4	4
max	12	12	12	12	12	12
Effective external load, lb/ft	10,030	10,030	14,640	6590† 10,030*	14,640	9130
Effective internal pressure, lb/in.²	0	0	0	0	0·65	13·65
Minimum 0·01 in. crack test load required, lb/ft	6700	4900	5900	4300† 6700*	9700	8600
Min F_s against cracking	1·2	1·2	1·0	1·2	1·6	1·6

* Compressible packing † Solid packing

II Rigid Pipes—Embankment Loading

Example 7. *Embankment Loading—Positive Projection or Induced Trench Conditions*

A 36 in. concrete culvert is to be laid in the bed of a stream. The valley is subsequently to be filled to a maximum height of 20 ft above the top of the pipe to form a lorry park. The bed of the stream is a sandy clay approximately 6 ft thick overlying rock. Determine the total effective external design loads for R.C. pipes if laid as a projecting conduit on (i) a Class B bed, (ii) a Class D bed or (iii) in Induced Trench conditions with Class B bed. From the cross sections the values of the projection ratio (p) have been established as 0·6, for the Class B bed, and 0·8 for the Class D bed γ_s for the fill = 125 lb/ft³. Allow for a uniform surcharge of 500 lb/ft² or B.S. 153 Type HB Road loading. Compare the computed loads with the over burden load including the uniform surcharge load.

External Loads (i) *and* (ii) *Positive Projection conditions*

		Bedding Class B	Class D
H	=	20	20 ft
B_c assuming 4 in. wall thickness = 36 + 4 + 4	=	3·67	3·67 ft
$H/B_c = 20/3{\cdot}67$	=	5·45	5·45
r_{sd} for clay overlying rock, assume from Table B	=	0·5	0·5
p as given	=	0·6	0·8
$r_{sd}\,p$	=	0·3	0·4
C_c from Chart C3 for these values of $r_{sd}p$ and $\dfrac{H}{B_c} = 5{\cdot}45$	=	7·65	7·9
$W'_c = C_c\gamma_s B_c^2 = C_c\,(125)\,(3{\cdot}67)^2$	=	12,850	13,250 lb/ft
$H' = H + 500/125$ (for $U_s = 500$ lb/ft²)	=	24	24 ft
$H'/B_c = 24/3{\cdot}67$	=	6·55	6·55
C'_c from Chart C3 for $r_{sd}\,p$ as above and $H/B_c = 6.55$	=	9·23	9·5
$C'_c - C_c$	=	1·58	1·6
$W'_{us} = (C'_c - C_c)\gamma_s B_c^2$	=	2660	2690 lb/ft
W_o from Chart C8 (*b*) for $B_c = 44$ in.; and $H = 20$ ft	=	600	600 lb/ft
F_i for Roads — assume	=	2·0	2·0
$W_{csu} = F_i\,W_o = 2W_o$	=	1200	1200 lb/ft
* Water load $W_w = (0{\cdot}75)\,\dfrac{\pi}{4}\,(3)^2\,(62{\cdot}4)$	=	330	330 lb/ft
$W_e = W'_c + W'_{us} + W_w = 12{,}850 + 2660 + 330$	=	15,840	—
$= 13{,}250 + 2690 + 330$	=	—	16,270 lb/ft

* N.B. This load is only about 2 per cent of W_c and could be ignored.

(iii) *Induced Trench Conditions*. Assume the pipes to be laid and fill placed and properly compacted to a height of $B_c = 3{\cdot}67$ ft above the top before a trench of width B_c is cut directly over the pipeline. This trench is then refilled with compressible material before the rest of the fill is placed.

H as above	$=$	20 ft	
B_c as above	$=$	3·67 ft	
$H/B_c = 20/3{\cdot}67$ as above	$=$	5·45	
$p' = 3{\cdot}67/B_c$	$=$	1·0	
r_{sd} for negative projection, from Table B assume	$=$	$-0{\cdot}3$	
C_n from Chart C4 for $p' = 1{\cdot}0$ and $r_{sd} = -0{\cdot}3$ and $H/B_c = 5{\cdot}45$	$=$	3·5	Incomplete trench condition (see Chart C4)
$W_c = C_n\gamma_s B_c^2 = (3{\cdot}5)(125)(3{\cdot}67)^2$	$=$	5850 lb/ft	
For the incomplete trench condition, $H' = H + 500/125 = 20 + 4$	$=$	24 ft	
$H'/B_c = 24/3{\cdot}67$	$=$	6·55	
C'_n from Chart C4 for $p' = 1{\cdot}0$ and $r_{sd} = -0{\cdot}3$ and $\frac{H'}{B_c} = 6{\cdot}55$	$=$	4·15	
$C'_n - C_n$	$=$	0·65	
$W_{us} = (C'_n - C_n)\gamma_s B_c^2 = (0{\cdot}65)(125)(3{\cdot}67)^2$	$=$	1080 lb/ft	
$W_{csu} = 2W_o$ as above	$=$	1200 lb/ft	
W_w as above		330 lb/ft	
$W_e = W_c + W_{csu} + W_w = 5850 + 1200 + 330$	$=$	7380 lb/ft	

Over burden load for comparison

$$W_e = H\gamma_s B_c + B_c U_s + W_w = (20)(125)(3{\cdot}67) + (3{\cdot}67)(500) + 330 = 9200 + 1835 + 330 = 11{,}365 \text{ lb/ft}$$

Comparison of loads

(i) Positive projection, Class B bed (15,840)(100)/11,365 = 139 per cent of over burden load

(ii) Positive projection, Class D bed (16,270)(100)/11,365 = 143 per cent of over burden load

(iii) Induced Trench, Class B bed (7380)(100)/11,365 = 65 per cent of over burden load.

Example 8. *Negative Projection Conditions*

A stream crossing the route of a new motor road is to be diverted into a 60 in. concrete culvert located on one side of a shallow valley in a temporarily timbered trench of the depth necessary to give an appropriate invert level. The valley is then to be filled to a maximum depth of 30 ft above the top of the pipe. Fill density $\gamma_s = 125$ lb/ft³, $K\mu = 0{\cdot}13$. The minimum mean depth of the top of the pipe below the original ground level of the trench is 10 ft. Determine the maximum load to be carried by the culvert.

Assuming a pipe wall thickness of 6 in., $B_c = 60 + 6 + 6 = 72$ in. $= 6$ ft.

Allowing for 3 in. timber sheeting and a working space of 12 in. on each side of the pipe the maximum trench width $B_d = 72 + 12 + 12 + 3 + 3 = 102$ in. $= 8{\cdot}5$ ft.

External Loads

H		=	30 ft	
$H/B_d = 30/8{\cdot}5$		=	3·53	
$p' = 10/B_d = 10/8{\cdot}5 = 1{\cdot}18$	assume	=	1·00	
r_{sd} for negative projection, from Table B	assume	=	−0·3	
C_n from Chart C4 for $p' = 1{\cdot}0$, $r_{sd} = -0{\cdot}3$, $H/B_d = 3{\cdot}53$		=	2·4	Incomplete trench condition (see Chart C4)
$W_c = C_n \gamma_s B_d^2 = (2{\cdot}4)(125)(8{\cdot}5)^2$		=	21,600 lb/ft	
For the incomplete trench condition, $H' = H + 500/125 = 30 + 4$		=	34 ft	
$H'/B_d = 34/8{\cdot}5$		=	4	
C'_n from Chart C4 for $p' = 1{\cdot}0, r_{sd} = -0{\cdot}3, H'/B_d = 4$		=	2·67	
$C'_n - C_n$		=	0·27	
$W_{us} = (C'_n - C_n)\gamma_s B_d^2 = (0{\cdot}27)(125)(8{\cdot}5)^2$		=	2450 lb/ft	
W_o from Chart C8 (*b*) of Ref. 14 for $H = 30$ ft, and $B_c = 72$ in.		=	350 lb/ft	
F_i for roads	assume	=	2·0	
$W_{csu} = F_i, W_o = 2W_o$		=	700 lb/ft	
$W_w = (0{\cdot}75)\frac{\pi}{4}(5^2)(62{\cdot}4)$		=	920 lb/ft	
Total effective external design load, $W_e = W_c + W_{us} + W_w = 21{,}600 + 2450 + 920$		=	24,970 lb/ft	

Example 9. In Example 7 at what height of fill would it be permissible for 30 ton two-wheel scrapers to pass over the pipeline assuming positive projection conditions and a Class B bed? The scraper wheels are 9 ft apart.

The pipe has been designed to carry 15,840 lb/ft (see Example 7).

Fill Loads

From Table M, $W'_c = C_c \gamma_s B_c^2$ and $\gamma_s = 125$ lb/ft³; $B_c = 3{\cdot}67$ ft; $r_{sd}p = 0{\cdot}3$ (see Example 7).

Using Chart C3

H ft	H/B_c	C_c	W'_c lb/ft
1	0·273	0·31	520
2	0·546	0·66	1110
4	1·09	1·40	2340
8	2·18	3·0	5000
16	4·36	6·1	10,200
20	5·45	7·65	12,850

Scraper Loads

The worst condition will occur when two scrapers meet with their axles over the centre of the pipe and the adjacent wheels A and B are say 3 ft apart. The surcharge load is then

$$W_{csu} = \frac{1}{\text{L}}(F_i C_{tA} + F_i C_{tB}) P F_L \text{lb/ft.}, \text{ Formulae (4.45) and (4.46),}$$

where $L = 3$ft,

beneath either of the wheels A or B where P is one wheel load = 15 tons. As the pipe is 36 in. diameter the loads will be approximately the same for the vehicles crossing the pipeline as for travelling along it. For the relatively soft soil the impact factor F_i may be assumed as 1·0.

Then for *the load at A caused by wheel at A*, using Chart C5 and Article 4.17 (*b*) (iii)

$B = 18$ in.; $L = 22$ in.

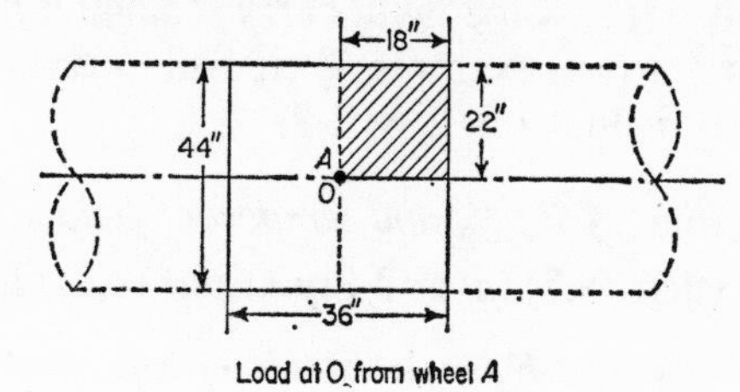

Load at O from wheel *A*

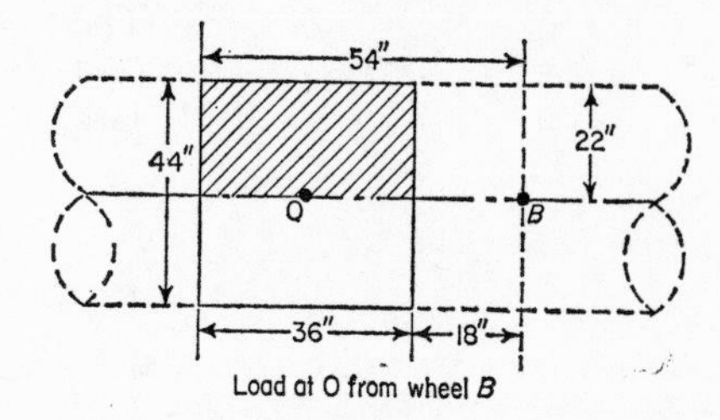

Load at O from wheel *B*

H ft	B/H	L/H	I_σ	$\Sigma I_{\sigma_A} = 4I_\sigma = C_{t_A}$
1	1·5	1·83	0·23	0·92
2	0·75	0·915	0·15	0·60
4	0·375	0·457	0·062	0·248
8	0·187	0·228	0·019	0·076
16	0·093	0·114	0·004	0·016
20	0·075	0·092	0·003	0·012

and for *the load at A caused by wheel at B*

1. $B = 54$ in.; $L = 22$ in.
2. $B = 18$ in.; $L = 22$ in.

H ft	B/H	L/H (1)	I_{σ_1}	B/H	L/H (2)	I_{σ_2}	$I_{\sigma_1} - I_{\sigma_2}$	$\Sigma I_{\sigma_B} = 2(I_{\sigma_1} - I_{\sigma_2}) = C_{t_B}$
1	4·5	1·83	0·237	1·5	1·83	0·23	0·007	0·014
2	2·25	0·915	0·193	0·75	0·915	0·15	0·043	0·086
4	1·125	0·457	0·121	0·375	0·457	0·062	0·059	0·118
8	0·562	0·228	0·047	0·187	0·228	0·019	0·028	0·056
16	0·281	0·114	0·013	0·093	0·114	0·004	0·009	0·018
20	0·225	0·092	0·009	0·075	0·092	0·003	0·006	0·012

Adding the influence values C_{tA} and C_{tB} for wheels A and B and multiplying by $P/3 = 5$ tons $= 11{,}200$ lb and by the correction factor $F_L = 1{\cdot}04$ for 36 in. pipe (see Chart C6 (*b*)), and noting that $F_i = 1{\cdot}0$.

H ft	$C_{tA} + C_{tB}$	$W_{csu} = (C_{tA} + C_{tB})(11{,}200)(10{\cdot}4)$/lb/ft	W_c' from above lb/ft	Temporary value of $W_e = W'_c + W_{csu}$ lb/ft
1	0·934	10,900	520	11,420
2	0·686	8000	1110	9110
4	0·366	4250	2340	6590
8	0·132	1540	5000	6540
16	0·034	395	10,200	10,595
20	0·024	280	12,700	12,980

Then since the pipe will carry 15,840 lb/ft when the bedding factor is F_p, (see example 7), it will carry 15,840 F_m/F_p safely at temporary values of $F_m = 1{\cdot}9$.

Value of $F_{p\ max}$ and temporary field strength of pipe.

From Article (8.2) (*a*) and Equations (8.1) and (8.2) and Chart C10

$$F_p = 1{\cdot}431/(N - xq)$$

where $N = 0{\cdot}707$ for class B bedding

$$q = mK(H/B_c + m/2)/C_c$$

and $K = 0{\cdot}33$, $m =$ (say) $0{\cdot}7$ and $x = 0{\cdot}594$ (see Chart C10).

From Example 7, at $H = 20$ ft, $H/B_c = 5{\cdot}45$ and $C_c = 7{\cdot}6$ and $H/B_c + m/2 = 5{\cdot}45 + 0{\cdot}35 = 5{\cdot}80$; and $xq = (0{\cdot}594)(0{\cdot}7)(0{\cdot}33)(5{\cdot}80)(7{\cdot}6) = 0{\cdot}104$.

Then $F_{p\ max} = 1{\cdot}431/(0{\cdot}707 - 0{\cdot}104) = 2{\cdot}37$

and $F_m = 1{\cdot}9$ (Chart C9) for Class B bedding.

The Temporary effective field strength of the pipe is then (15,840) (1·9)/2·37 = *12,700 lb/ft* and since the maximum temporary value of W_e is 12,980 lb/ft at $H = 20$ ft when F_p will be operative, the scrapers may pass over the pipeline *at any level above the initial protective fill layer say 3 ft.*

III Examples of Flexible Pipe Design

N.B. These examples are illustrative of the use of the equations and data given in the text. They have not been confirmed experimentally. The numerical data have been arbitrarily assumed and should not be taken as recommended values.

Example 10. (Condition as Article 6.2 (*a*) (2))

What minimum wall thickness excluding allowances for corrosion would be required for a 36 in. smooth walled steel pipe to be laid on the bed of a lake at a depth of 30 ft to withstand a total effective internal pressure of 200 lb/in.^{2}? For the steel, $E = 30 \times 10^6$ and the yield stress is 3 6,000 lb/in.^{2} The maximum vacuum is 12 lb/in.^{2}

Wall thickness t_e to resist external pressure

External water pressure $p_g = (31{\cdot}5 \times 62{\cdot}4)/144 \quad = 13{\cdot}6$ lb/in.^{2}
Add vacuum p_{vac} say 12 lb/in.^{2}

Max. Total external pressure (Formula 4.54) $p_{e_{max}} = 25{\cdot}6$ lb/in.^{2}

From Article 6.4 (*a*) the required minimum critical pressure is $p_c = (2)\,(25{\cdot}6) = 57{\cdot}2$ lb/in.^{2}

Using Formula (6.1), since $P_c < 581$ lb/in.^{2}

$$(2)\,(25{\cdot}6) = 50{\cdot}2 \times 10^6 \times (t_e/2R)^3.$$

Assume $R = 18{\cdot}4$ in.

whence $t_e^3 = 50{\cdot}5 \times 10^{-3}$

and $t_e = 0{\cdot}369$ in. and $R = 18{\cdot}369$

then $\dfrac{t_e}{2R} = \dfrac{0{\cdot}369}{36{\cdot}738} = 0{\cdot}01$ (acceptable see Article 6.2 (*e*).

Maximum permissible internal pressure for this thickness:

From Equation (6.18) (Article 6.5 (*a*))

$$t_{i_{min}} = \frac{2(p_{i_{max}} - p_{e_{min}})R}{f_y}$$

and since vacuum is inoperative when the pipe is under pressure, for $t_{min} = 0.369$ in.

$$0{\cdot}369 = \frac{2(p_{i_{max}} - 13{\cdot}6)\,(18{\cdot}369)}{36{,}000}$$

whence $p_{i_{max}} = 373{\cdot}6$ lb/in.^{2}

which is greater than the specified internal pressure of 200 lb/in.^{2}
∴ a value of $t = 0{\cdot}369$ in. is acceptable.

N.B. (i) The nearest B.S.534 pipe has a wall thickness of 0·375 in. and a test pressure of 750 ft = 324 lb/in.^{2}

(ii) The very small deflections and bending stress due to the submerged weight of the pipe are ignored.

Example 11. (Condition as Article 6.2 (*a*) (3))

What would be the required minimum wall thickness of a 36 in. smooth walled steel pressure pipe, laid on the trench bottom, with a cover of 5 ft, in a wide trench in fields, without controlled compaction of the backfill, the deflection not to exceed 2 per cent of diameter? The maximum internal pressure is 100 lb/in.^{2} The land is subject to seasonal flooding to a depth of 5 ft above the surface. $\gamma_s = 125$ lb/ft^3; $K_\theta = 0{\cdot}108$; $K_b = 0{\cdot}235$; vacuum = 12 lb/in.^{2}; $f_y = 36{,}000$ lb/in.^{2}; $E = 30 \times 10^6$ lb/in.^{2}

External load, when not submerged

Assume $B_c = 3{\cdot}1$ ft.
Fill load (Table M) = $H\gamma_s B_c = (5)(125)(3{\cdot}1)$ = 1940 lb/ft
Assume 500 lb/ft^2 uniform surcharge,
Surcharge load (Table M) = $U_s B_c = (500)(3{\cdot}1)$ = 1550 lb/ft
Water load $(0{\cdot}50)(\pi/4)(3)^2(62{\cdot}4)$ = 220 lb/ft
Pipe weight $(0{\cdot}50)(\pi)(3)(t/12)(480)$ and assuming $t = 0{\cdot}6$ in. = 113 lb/ft
Total load W_e = 3823 lb/ft
say 4000 lb/ft

From Equation (4·54) $p_e = p_g + p_{vac} = 0 + 12$ = 12 lb/in.^{2}
Outside radius of pipe—assume = 18·6 in.
Deflection limit = (0·02)(36) = 0·72 in.

Wall thickness

From Equation (6.5) Article 6.4 (*b*)

$$\Delta'_x = \frac{K_\theta W_e R^3}{E t_e^3 - 24 p_e K_\theta R^3}$$

i.e.

$$0{\cdot}72 = \frac{(0{\cdot}108)(4000)(18{\cdot}6)^3}{(30 \times 10^6)t_e^3 - 24(12)(0{\cdot}108)(18{\cdot}6)^3}$$

whence

$$t_e^3 = 0{\cdot}135 \text{ in.}^3$$

$$t_e = 0{\cdot}513 \text{ in.}$$

Check maximum combined compressive stress in steel due to external pressure and fill load for the unsubmerged condition.
Bending stress:
From Equation (6.6) with $K_b = 0{\cdot}235$, $K_\theta = 0{\cdot}108$, and $I = t^3/12$.

$$f_{cb} = \frac{0{\cdot}117\ W_e R E t_e}{E t_e^3 - 2{\cdot}592\ p_e R^3}$$

Inserting $R = 18{\cdot}513$ in. and the other known values,

$$f_{cb} = \frac{(0{\cdot}117)(4000)(18{\cdot}513)(30 \times 10^6)(0{\cdot}513)}{(30 \times 10^6)(0{\cdot}513)^3 - 2{\cdot}592(12)(18{\cdot}513)^3}$$
$$= 34{,}000 \text{ lb/in.}^2,$$

which is too high.
Since f_{cb} varies inversely as t^2 approximately, try

$$t_e = \sqrt{\frac{34000}{18000}}(0{\cdot}513) = 0{\cdot}70$$

Then $R = 18{\cdot}7$ in. and

$$f_{cb} = \frac{(0{\cdot}117)(4000)(18{\cdot}7)(30 \times 10^6)(0{\cdot}70)}{(30 \times 10^6)(0{\cdot}7)^3 - (2{\cdot}592)(12)18{\cdot}7)^3}$$
$$= 17{,}500 \text{ lb/in.}^2$$

and from Equation (6.7) Article 6.4 (*b*)

$$\text{Ring stress} = f_{cr} = \frac{(12)(18{\cdot}7)}{0{\cdot}70} = 320 \text{ lb/in.}^2$$

From Equation (6.8) combined stress $= f_{cb} + f_{cr} = \underline{17{,}820 \text{ lb/in.}^2}$
i.e. $< f_y/2$.
Then $t_e = 0{\cdot}70$ *is satisfactory for the unsubmerged condition.*
For the Submerged condition—

External loads

Assume $B_c = 3{\cdot}1$ ft
Fill load $W'_c = H(\gamma_s - \gamma_w)B_c$ Article 4.14 (*b*) Form. (4.34)
$= 5(125 - 62{\cdot}4)3{\cdot}1 =$ 970 lb/ft
$W_w =$ 0 ,,
As above assume W_{us} (not submerged) $=$ 1550 ,,

Ignoring pipe weight—total load W_e $= \underline{\underline{2520}}$,,

External fluid pressure (Article 4.19 (*a*))
$p_g = (5 + 5 + 1{\cdot}5)\gamma_w/144 =$ 5·0 lb/in.²
Vacuum $=$ 12·0 ,,

Total, $p_{emax} = \underline{\underline{17{\cdot}0}}$,,

Check combined compressive stress in the submerged condition.

Taking $t_e = 0{\cdot}70$ in., as in the unsubmerged condition, the compressive ring stress is (Article 6.4 (*b*)),

$$f_{cr} = \frac{(17)(18{\cdot}7)}{0{\cdot}70} = 450 \text{ lb/in.}^2$$

and since the external load has dropped from 4000 lb/ft to 2520 lb/ft the bending stress is reduced to approx. 11,000 lb/in² and a value of $t_e = 0{\cdot}70$ in. is therefore adequate for both conditions in the absence of internal pressure.

Check maximum combined tensile stress due to internal pressure and fill load.

For the unsubmerged condition.

$$p_{ie} = p_{i\,max} - 0 = 100 \text{ lb/in.}^2 \text{ (Article 6.5 (b))}$$

Assuming as above $t = 0{\cdot}7$ in.; $R = 18{\cdot}7$ in.; $W_e = 4000$ lb/ft.
From Equation (6.22) Tensile ring stress

$$f_{sr} = \frac{p_{ie}R}{t} = \frac{(100)(18{\cdot}7)}{0{\cdot}7} = 2{,}680 \text{ lb/in.}^2$$

From Equation (6.21), Tensile bending stress, given $K_\theta = 0{\cdot}108$, $K_b = 0{\cdot}235$, is

$$f_{sb} = \frac{0{\cdot}117\, W_e\, REt}{Et^3 + 2{\cdot}592\, p_{ie}R^3}$$

$$= \frac{(0{\cdot}117)(4000)(18{\cdot}7)(30 \times 10^6)(0{\cdot}7)}{(30 \times 10^6)(0{\cdot}7)^3 + (2{\cdot}592)(100)(18{\cdot}7)^3} = 15{,}300 \text{ lb/in.}^2$$

Combined stress $= f_{sr} + f_{sb} = 2680 + 15{,}300 = 17{,}980$ lb/in.² which is within the limiting stress of $fy/2 = 18{,}000$ lb/in.²

For the submerged condition.

$p_{ie} = p_{i\,max} - p_g$ since vacuum does not occur simultaneously with p_i, i.e.

$$p_{ie} = 100 - 5 = 95 \text{ lb/in.}^2$$

From Equation (6.22) Tensile ring stress

$$f_{sr} = \frac{(95)(18{\cdot}7)}{0{\cdot}7} = \underline{2540} \text{ lb/in.}^2$$

From Equation (6.21), Tensile bending stress for $W_e = 2520$ lb/ft (as above when submerged), is

$$f_{sb} = \frac{(0{\cdot}117)(2{,}520)(18{\cdot}7)(30 \times 10^6)(0{\cdot}7)}{(30 \times 10^6)(0{\cdot}7)^3 + (2{\cdot}592)(95)(18{\cdot}7)^3} = \underline{9{,}600} \text{ lb/in.}^2$$

Combined 'stress = 9600 + 2540 = 12,140 lb/in.² which is less than the limiting stress $fy/2$ = 18,000 lb/in.²

Conclusion. A wall thickness of 0·7 in. satisfies all conditions with a stress limit of 18,000 lb/in.²

Example 12. (Condition as in Article 6.2 (*a*) (4))

The pipe in Example 11 is to pass under a road and within the road boundaries the bedding and fill is to be subjected to proper compaction control. What would be the required minimum wall thickness assuming no submergence in this zone? f_y = 36,000 lb/in.² and maximum internal pressure = 100 lb/in.²

Loads and pressures

Again assuming wide trench conditions.

Fill load as before = 1940 lb/ft
Water load as before = 220 lb/ft
Pipe load as before = 113 lb/ft

Total = 2273 lb/ft

Concentrated surcharge for road loading

Since the bedding is similar to Class B, using Chart C8 (*b*) for B_c = 3·1 ft and H = 5 ft,

$$W_o = 2700 \text{ lb/ft}, \quad W_{csu} = (2)(2700) = 5400 \text{ lb/ft}$$

which is greater than the 500 lb uniform surcharge load.

Total external load = 2273 + 5400 = W_e = 7673 lb/ft.

Assuming outside radius of pipe R = 18·6 in.

Effective external pressure due to external load say of 7700 lb/ft is $p_0 = W_e/(12)(2)R$ = 7700/(12)(37·2) = 17·3 lb/in.²

Deflection limit Δ'_x = (0·02)(36) = 0·72 in.

External fluid pressure $p_e = p_{vac}$ = 12 lb/in.²

From Equation (6.13) (Article 6.4 (*c*) (i))

$$k_{min} = \frac{2{\cdot}7(p_0R + p_e\Delta'_x)}{R\Delta'_x}$$

$$= \frac{2{\cdot}7\{(17{\cdot}3)(18{\cdot}6) + (12)(0{\cdot}72)\}}{(18{\cdot}6)(0{\cdot}72)}$$

$$= 66{\cdot}6 \text{ lb/in.}^2\text{/in.}$$

From Equation (6.15), substituting t_e for A

$$t_e = \frac{(p_0 + p_e)R}{f_a} = \frac{(17\cdot3 + 12)(18\cdot6)}{f_a} = \frac{545}{f_a}$$

Try $f_a = 5000$, then $t_e = 0\cdot109$ and $kt = 7\cdot26$.

Then by successive approximations using Chart C12 (c)

For $kt=$	$f_a=$	$t_e=$	and $kt=$
7·26	3500	0·156	10·35
10·35	3900	0·140	9·3
9·3	3800	0·143	9·55
9·55	3850	0·141	9·4
9·4	3800	0·143	9·55

Then No. 9 S.W.G. with $t = 0\cdot144$ would serve but the practical limit of R/t is 100 (Article 6.2 (e)) i.e. $t_{min} = 0\cdot186$ and the nearest gauge is No.6 with $t = 0\cdot192$.

Check on the value of R. Adopting $t_e = 0\cdot192$, $kt = 12\cdot8$ and from Chart C12 (c) the minimum value of R/t is 43 i.e. $R_{min} = 8\cdot25$ in. which is less than 18 in. and $t_e = 0\cdot192$ would be satisfactory.

Check on the value of f_a. For $Kt = 12\cdot8$ from the Chart C12 (c) $f_a = 4200$ lb/in.^{2} and the actual value is from Equation (6.15) $(17\cdot3 + 12)(18\cdot6)/0\cdot192 = 2840$ lb/in.^{2} which is satisfactory.

Check tensile ring stress caused by internal pressure in the pipe when uncovered (i.e. $p_0 = 0$). Allowable stress $= f_y/2 = 36{,}000/2 = 18{,}000$ lb/in.^{2}

From Equation (6.24) (Article 6.5 (c)) the actual maximum stress is

$$f_{sr} = \frac{p_{i\,max}R}{t} = \frac{(100)(18\cdot192)}{0\cdot192} = 9500 \text{ lb/in.}^2$$

There is also a small bending tensile stress (approx. 122 psi) due to the weight of pipe and water.

The tensile stress is therefore well within the allowable stress.

Conclusion. A wall thickness of 0·192 is satisfactory for both external and internal loading and for practical stability during installation.

Example 13. (See Article (6.6))

If in Example 12 a trench is to be subsequently excavated alongside the pipe, what would be the required wall thickness assuming no submergence or surcharge loads to be operative during the danger period?

The worst condition if the pipe is kept in service would be the occurrence of vacuum pressure.

Loads and pressures (N.B. wheeled traffic prohibited over the pipeline).

Fill load, water load, and pipe weight as before $W_e = 2237$ lb/ft
External pressure = Vacuum pressure = $p_e = 12$ lb/in.^{2}
Deflection limit as before $\Delta_x = 0{\cdot}72$ in.

Wall thickness in the absence of side support from the soil. Assuming $K_\theta = 0{\cdot}108$ as for uncompacted fill and $R = 18{\cdot}6$ in.
From Equation (6.5) (Article 6.4 (*b*)),

$$\Delta'_x = \frac{K_\theta W_e R^3}{Et_e^3 - 24p_e K_\theta R^3}$$

whence

$$t_e^3 = \frac{K_\theta W_e R^3 + 24p_e K_\theta R^3 \Delta'_x}{E\Delta'_x}$$

$$= \frac{(0{\cdot}108)(2273)(18{\cdot}6)^3 + (24)(12)(0{\cdot}108)(18{\cdot}6)^3(0{\cdot}72)}{(30 \times 10^6)(0{\cdot}72)}$$

$$= 0{\cdot}08 \text{ in.}^3$$

$$t_e = \underline{0{\cdot}43} \text{ in.}$$

Check compression stress in steel. Bending stress at bottom. Assuming $K_b = 0{\cdot}235$ as for uncompacted fill and $K_\theta = 0.108$, from Equation (6.6) (Article (6.4 (*b*)),

$$f_{cb} = \frac{0{\cdot}117\, WREt_e}{Et^3 - 2{\cdot}592p_e R^3}$$

i.e.

$$f_{cb} = \frac{(0{\cdot}117)(2273)(18{\cdot}6)(30 \times 10^6)(0{\cdot}43)}{(30 \times 10^6)(0{\cdot}43)^3 - (2{\cdot}592)(12)(18{\cdot}6)^3}$$

$$= 29{,}000 \text{ lb/in.}^2$$

which is too high.

Then assume the bending stress to be approximately inversely proportional to t^2 and try

$$t = \sqrt{\frac{29000}{18000}}(0{\cdot}43) = 0{\cdot}55,\ R = 18{\cdot}55$$

and

$$f_{cb} = \frac{(0{\cdot}117)(2273)(18{\cdot}55)(30 \times 10^6)(0{\cdot}55)}{(30 \times 10^6)(0{\cdot}55)^3 - 2{\cdot}592(12)(18{\cdot}55)^3} = 17{,}000 \text{ lb/in}^2$$

From Equation (6.7)

$$\text{Ring stress } f_{cr} = \frac{(12)18{\cdot}55}{0{\cdot}55} = 400 \text{ lb/in}^2$$

Then

$$f_{c\,max} = f_{cb} + f_{cr}$$
$$= 17{,}000 + 400$$
$$= \underline{17{,}400} \text{ lb/in.}^2 \text{ which is } < f_y/2.$$

Then the required wall thickness is $t_e = 0{\cdot}55$ in. which is satisfactory for the maximum external loads. Since the wall thickness is greater than in Example 12 it is obviously adequate for the maximum internal pressure.

Example 14. (Condition as in Article 6.4 (*c*))

A corrugated steel pipe 100 in. diameter is to be installed as a culvert in the bed of a valley which is subsequently to be filled to a height of 50 ft above the top of the pipe. The bedding and fill to a height of 2 diameters and the same width on each side of the pipe is to be compacted to at least 95 per cent standard density (B.S. 1377). Deflection not to exceed 5 per cent of diameter. Compute the required value of E_s for the compacted fill and the required type and gauge of the pipe wall, assuming $E = 30 \times 10^6$ lb/in.^{2}; $f_y = 40{,}000$ lb/in.^{2}; $\gamma_s = 120$ lb/ft^3; no submergence.

Loads and pressures. Assume $B_c = 104$ in. $= 8{\cdot}66$ ft

Assume the fill load (Table M)

$$= H\gamma_s B_c = (50)(120)(8{\cdot}66) = 52{,}000 \text{ lb/ft}$$

Assume 500 lb uniform surcharge (Table M)

$$U_s B_c = 500(8{\cdot}66) \qquad = \quad 4300 \text{ lb/ft}$$

Equivalent water load (Article 3.5 (*b*))

$$(0{\cdot}5)\frac{\pi}{4}\left(\frac{100}{12}\right)^2(62{\cdot}4) \qquad = \quad 1700 \text{ lb/ft}$$

Assume equivalent pipe weight

$$(0{\cdot}50)(300) \text{ (see maker's handbook)} = \quad 150 \text{ lb/ft}$$

$$\text{Total load} = \underline{\underline{58{,}150}} \text{ say } 59{,}000 \text{ lb/ft}$$

From Article 6.4 (*c*) (i)

Effective mean external pressure at top of pipe,

$$p_0 = 59{,}000/(12)(104) = \underline{47 \cdot 5} \text{ lb/in.}^2$$

N.B. $p_{vac} = 0$ for a culvert and $p_g = 0$ for no submergence

Deflection limit

$$\Delta_x = (0 \cdot 05)(100) = 5 \text{ in.}$$

Value of k_{min}

From Equation (6.11)

$$k_{min} = 2 \cdot 7p_0/\Delta_x = (2 \cdot 7)(47 \cdot 5)/5 = \underline{25 \cdot 6} \text{ lb/in.}^2\text{/in.}$$

Value of $E_{s\,min}$

From Equation (6.14) (Article 6.4 (*c*) (i))

$$E_{s\,min} = 1 \cdot 5Rk = (1 \cdot 5)\left(\frac{104}{2}\right)(25 \cdot 6) = \underline{2000} \text{ lb/in.}^2$$

Check, from triaxial test, that E_s for the fill to be used is $\nless$ 2000 lb/in.² when it is compacted to 95 per cent proctor density, or to what higher density it must be compacted to produce 2000 lb/in.²

Wall details

From Equation (6.15) (Article 6.4 (*c*) (ii) taking $R_{\text{mean}} =$ 102/2 = 51 in., compressive ring stress is

$$f_c = p_0R/A = (47 \cdot 5)50/A \text{ lb/in.}^2$$

and from Table K using the Chart C12 (*a*) for 6 in. × 2 in. corrugated pipe, and $k_{min} = 25 \cdot 6$ lb/in.²/in., obtain:

Gauge	t in.	A in.²	$f_c = p_0R/A$ lb/in.²	t/K	Allowable stress f_a lb/in.²	Remarks
No. 7	0·1838	0·2283	10600	0·0072	13000	Too thick.
No. 8	0·1644	0·2041	11850	0·0064	13200	Acceptable.
No. 10	0·1345	0·1669	14500	0·0052	13800	Too thin.

For No. 8 gauge the compressive ring stress f_c = 11,850 lb/in.² is less than the allowable stress f_a = 13,200 lb/in.² and so is the economical choice for the computed minimum value of 'k', but see the note* below.

Check on radius

From Chart C12 (a) for $t/R = 0{\cdot}0064$, $R_{min} = 38$ in.
$\therefore R = 51$ in. is acceptable.

Check deflection

From Equation (6.11)

$$\Delta_x = 2{\cdot}7p_0/k = \frac{(2{\cdot}7)(47{\cdot}5)}{25{\cdot}6} = 5 \text{ in.}$$

---

* *Note.* If it is desired to use No. 10 g. wall thickness the value of k must be increased by increasing the compaction of the fill thus:

From Chart C12 (a) for $f_a = 14{,}500$ lb/in², $t/k = 0{\cdot}004$, and since $t = 0{\cdot}1345$, $k = \dfrac{0{\cdot}1345}{0{\cdot}004} = 33{\cdot}6$ lb/in²/in.

The corresponding value of E_s will be, from Equation (6.14).

$$E_s = 1{\cdot}5\ Rk = (1{\cdot}5)\ (51)\ (33{\cdot}6) = \underline{2600} \text{ lb/in}^2$$

The deflection would then be $(2{\cdot}7)\ (47{\cdot}5)/33{\cdot}6 = \underline{3{\cdot}8}$ in, $= 3{\cdot}8\%$ of diameter.

Example 15. In Example 14 what minimum depth of fill must be in place before it is safe to permit two wheeled, 30 ton, trailer scrapers with wheels at 9 ft centres to pass over the new pipeline? The deflection limit of 5 per cent must not be exceeded.

The basic principles to be observed are (i) that the combined soil pressure of the fill and the surcharge loading at the top of the pipe must not exceed the unit pressure, $p_0 = 47{\cdot}5$ lb/in.², for which the pipes were designed (see Example 14) and (ii) that the maximum deflection of the pipes must never exceed 5 per cent of the pipe diameter.

Fill loads and pressures. From Table M the fill load at any depth of cover, H ft is

$H\gamma_s B_c$ lb/ft and the unit pressure is then $H\gamma_s/144$ lb/in.²

and since $\gamma_s = 120$ lb/ft³

$$\text{Fill pressure} = (H)(120)/144 = \underline{0{\cdot}835H} \text{ lb/in.}^2$$

Scraper loads. $P = 15$ tons $= 33{,}600$ lb per wheel. The worst condition will occur when two vehicles meet and their adjacent wheels are say 3 ft apart and directly over the centre of the pipe. Since only the peak pressure beneath either wheel is required, consider an area 12 in. × 12 in. at the top of the pipe beneath either wheel. The pressure will be the same whether the vehicles are travelling across or along the pipe. The two outer wheels are considered to be too far away to produce any significant effect on the area considered.

From Article 4.17 (*b*) (iii), denominating the wheels *A* and *B*, the total pressure vertically beneath the centre of either will be $\Sigma I_\sigma P_A + \Sigma I_\sigma P_B$ assuming an impact factor of 1·0 and regarding the wheels as imposing point loads at the surface.

For wheel A

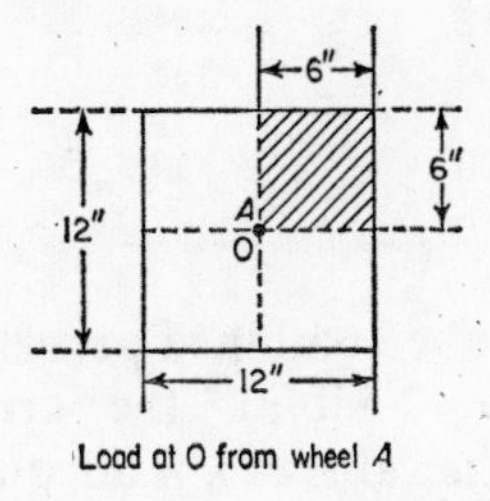

Load at O from wheel *A*

Load *A* at 0, Pressure at 0.

$B = 6$ in., $L = 6$ in. for the hatched area.

Total area affected is 4 × hatched area.

Using Chart C5 to obtain I_σ for various values of *H*,

H ft	*B/H*	*L/H*	I_σ	$\Sigma I_{\sigma_A} = 4I_\sigma$
2	0·250	0·250	0·026	0·104
4	0·125	0·125	0·007	0·028
8	0·0625	0·0625	0·002	0·008
16	0·031	0·031	0·0005	0·002
32	0·015	0·015	0·0003	0·0012
50	0·010	0·010	0·0001	0·0004

For wheel B

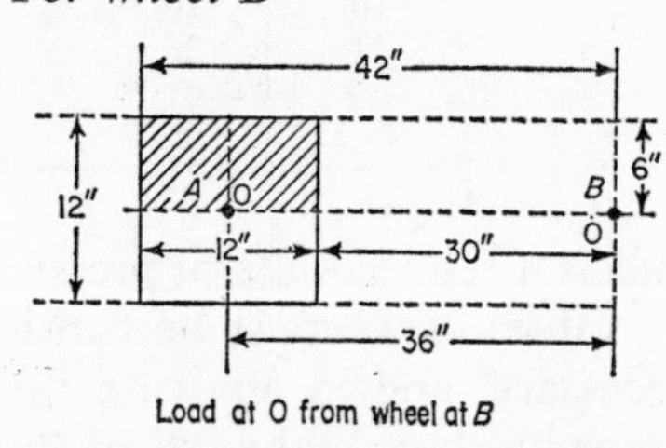

Load at O from wheel at *B*

Load at *B*, Pressure at 0

Area 1. $B = 42$ in.: $L = 6$ in.

Area 2. $B = 30$ in.: $L = 6$ in.

Total area affected is 2×(area 1 − area 2).

H	Area 1 *B/H*	*L/H*	I_{σ_1}	Area 2 *B/H*	*L/H*	I_{σ_2}	$I_{\sigma_1} - I_{\sigma_2}$	$\Sigma I_\sigma B = 2(I_{\sigma_1} - I_{\sigma_2})$
2	1·75	0·25	0·075	1·25	0·25	0·074	0·001	0·002
4	0·825	0·125	0·033	0·625	0·125	0·029	0·004	0·008
8	0·412	0·062	0·010	0·312	0·062	0·008	0·002	0·004
16	0·206	0·031	0·003	0·156	0·031	0·002	0·001	0·002
32	0·103	0·015	0·001	0·078	0·015	0·001	0	—
50	0·07	0·010	0·0003	0·05	0·001	0·001	0	—

Then adding the influence values for wheels *A* and *B*.

H ft	$\Sigma I_{A+B} = \Sigma I_{\sigma A} + \Sigma I_{\sigma B}$	$P\Sigma I_{A+B}$ lb/ft²	Total pressure from wheels lb/in.²	Fill pressure lb/in.²	Combined pressure p_0 lb/in.²
2	0·106	3560	24·8	1·67	26·47
4	0·036	1210	8·4	3·34	11·74
8	0·012	404	2·8	6·66	9·46
16	0·004	135	0·94	13·33	14·27
32	0·0012	40	0·28	26·66	26·94
50	0·0004	13·5	0·09	41·7	41·79

Then since the combined pressure is always less than the design pressure $p_o = 47{\cdot}5$ lb/in.² the scraper may pass over the pipe at any depth if the value of k is adequate to restrict the pipe deflection Δ_x to 5 in. or less.

From Equation (6.11), $k = 2{\cdot}7\, p_c/\Delta_x$ and $\Delta_x = 5$ in. as in Example 14.

Check the value of k_{min}

H ft	p_0 lb/in.²	k lb/in.²/in.
2	26·47	14·3
4	11·74	6·3
8	9·46	5·1
16	14·27	7·7
32	26·94	14·5
50	41·79	22·6

But the effect of reduced lateral pressure on the value of E_s in the fill at low values of H must be considered, resorting to triaxial tests if necessary, and in any case the scrapers should not pass until the pipe has been stabilised by the addition of fill of a depth $B_c/4 = 26$ in. (see Article 6.4 (*c*) (ii)). A careful check on the actual pipe deflection should be maintained at least in the early stages of the filling.

Example 16. A dragline excavator, weighing 48,000 lb and carried on twin tracks, each 12 ft × 2 ft and spaced at 8 ft centres, is to be used in constructing embankments of various heights over steel pipelines of various diameters. Determine the depths of fill within which the dragline may safely operate over the pipelines. $\gamma_s = 125$ lb/ft³, $\Delta_x = 5$ per cent of diameter.

The basic principles to be observed are (i) that the combined soil pressures caused by the fill and the dragline at the top of any pipeline must not exceed the pressure p_0 for which the pipes were designed and (ii) the specified maximum deflection of the pipes must not be exceeded.

It is necessary to construct a graph of the combined loading at various depths as follows: (see Fig. Example 16).

The *fill load* at any depth H ft is $H\gamma_s B_c$ and the corresponding pressure at the top of the pipe is then $H\gamma_s$ lb/ft² *irrespective of the pipe diameter*.

The transient uniform surcharge load of small extent caused by the dragline is $U_{sust} = 48{,}000/(2)(12)(2) = 1000$ lb/ft² at the surface.*

Denoting the tracks as A and B, the transient load on the pipe when its centre line is (case 1) vertically below the centre point of track A is

$$\Sigma I_{\sigma A}\, U_{sust} B_c + \Sigma I_{\sigma B}\, U_{sust} B_c \text{ lb/linear ft}$$

and (case 2) when the centre line is vertically below the c.g. of the dragline it is

$$\Sigma I_{\sigma AB}\, U_{sust} B_c \text{ lb/linear ft}$$

and the respective pressures at the top of the pipe are

$$U_{sust}(\Sigma I_{\sigma A} + \Sigma I_{\sigma B}) \text{ and } U_{sust}\, \Sigma I_{\sigma AB} \text{ lb/ft}^2$$

regardless of the pipe diameter.

Then from Chart C5 the values of I_σ are obtained as follows:

Case 1. Load beneath centre point of track A.

Load at 0

For track A

Pressure beneath centre point of track A

12′ 2′ 0 Track A 1′ 6′

$B = 2$ ft, $L = 12$ ft

H ft	$2H$	$B/2H$	$L/2H$	I_σ	$\Sigma I_{\sigma_A} = 4I_\sigma$
1	2	1	6	0·205	0·820
3	6	0·33	2	0·097	0·388
5	10	0·2	1·2	0·058	0·232
10	20	0·1	0·6	0·022	0·088
15	30	0·066	0·4	0·011	0·044
20	40	0·05	0·3	0·007	0·028
30	60	0·03	0·2	0·003	0·012
40	80	0·025	0·15	0·002	0·008
50	100	0·020	0·12	0·001	0·004

* If the dragline must work whilst standing over the pipeline and thereby induce differential loading of the tracks, the larger value of the contact pressure must be considered.

Load at 0

For track B

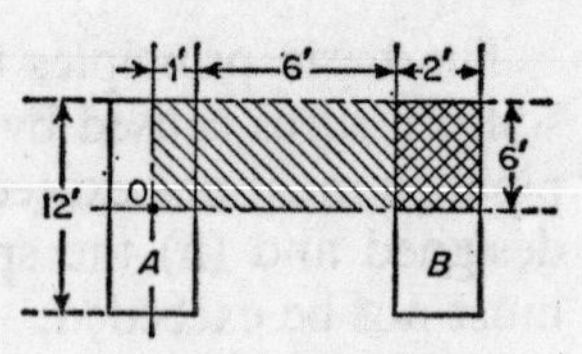

Area (1)

$B = 6$ ft, $L = 9$ ft

Area (2)

$B = 6$ ft, $L = 7$ ft

H	For Area (1) B/H	L/H	I_{σ_1}	For Area (2) B/H	L/H	I_{σ_2}	$\Sigma I_{\sigma_B} = 2(I_{\sigma_1} - I_{\sigma_2})$
1	6·0	9·0	0·249	6·0	7·0	0·249	0
3	2·0	3·0	0·238	2·0	2·33	0·235	0·006
5	1·2	1·8	0·210	1·2	1·4	0·202	0·016
10	0·6	0·9	0·132	0·6	0·7	0·117	0·030
15	0·4	0·6	0·08	0·4	0·46	0·067	0·026
20	0·3	0·45	0·053	0·3	0·35	0·044	0·018
30	0·2	0·30	0·026	0·2	0·23	0·021	0·010
40	0·15	0·225	0·015	0·15	0·17	0·012	0·006
50	0·12	0·18	0·009	0·12	0·14	0·008	0·002

Summation of pressures $= U_{sust}(\Sigma I_{\sigma A} + \Sigma I_{\sigma B})$ lb/ft², where $U_{sust} = 1000$ lb/ft².

H	Pressures: Track A lb/ft²	Track B lb/ft²	Tracks A and B lb/ft²
1	820	0	820
3	388	6	394
5	232	16	248
10	88	30	118
15	44	26	70
20	28	18	46
30	12	10	22
40	8	6	14
50	4	2	6

Case 2. Load beneath c.g. of dragline

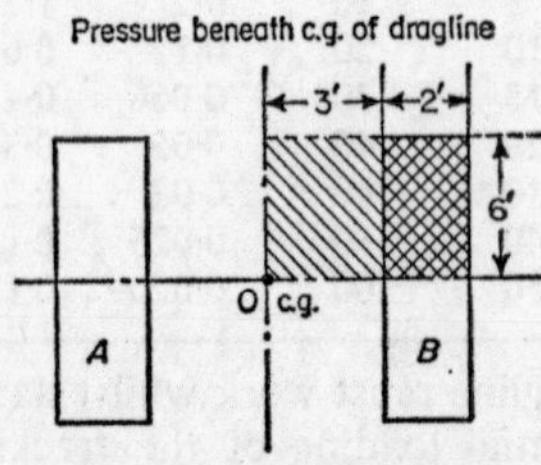

Area (1) $B = 5$ ft, $L = 6$ ft

Area (2) $B = 3$ ft, $L = 6$ ft

As before $U_{sust} = 1000$ lb/ft²

H	For Area (1) B/H	L/H	I_{σ_1}	For Area (2) B/H	L/H	I_{σ_2}	$\Sigma I_{\sigma_{AB}} = 4(I_{\sigma_1} - I_{\sigma_2})$	Pressure lb/ft²
1	5	6	0·249	3	6	0·246	0·004	4
3	1·66	2	0·22	1	2	0·200	0·100	100
5	1·00	1·2	0·185	0·6	1·2	0·143	0·164	164
10	0·50	0·6	0·095	0·3	0·6	0·063	0·128	128
15	0·33	0·4	0·054	0·2	0·4	0·033	0·084	84
20	0·25	0·3	0·034	0·15	0·3	0·020	0·056	56
30	0·166	0·2	0·015	0·10	0·2	0·009	0·024	24
40	0·125	0·15	0·009	0·075	0·15	0·005	0·016	16
50	0·10	0·12	0·006	0·06	0·12	0·004	0·008	8

The total pressure on the pipes is then the sum of the pressures caused by the fill load and the greater of the surcharge loads in Case 1 and Case 2:

Cover H ft	Fill pressure $H\gamma_s$ lb/ft²	Max. Track Pressure lb/ft²		Combined Pressure lb/ft²	Remarks
1	125	820	Case 1	945	centre point of Track A or B over C.L. of pipeline
3	375	394		769	
5	625	248		873	
10	1250	128	Case 2	1378	c.g. of dragline over C.L. of pipeline
15	1875	84		1959	
20	2500	56		2556	
30	3750	24		3774	
40	5000	16		5016	
50	6250	8		6258	

These values are plotted on the adjoining graph which can now be used to determine the depths of fill at which the dragline may safely operate over any size of pipeline, as follows:

Case A. If the pipes are designed for a cover of 6 ft or less, the combined pressure is always greater than p_0 and the dragline cannot pass over the pipeline safely at any depth, unless the pipe is adequately strutted or tied.

Case B. If the pipes are designed for a cover of 8 ft, the dragline can pass over the pipeline at any value of H between 3 ft and 6 ft 6 in. (subject to adequate strutting or tying of the pipe on the development of an acceptable value of 'k' (see below).

Case C. If the pipes are designed for a cover of 20 ft or more, the dragline can pass over the pipeline at any depth greater than 3 ft (subject to the development of an acceptable value of 'k').

Deflection and 'k' value. Since B_c is the overall diameter of the pipe in feet, the limiting deflection of the pipe is $0{\cdot}05\ B_c(12) = 0{\cdot}6\ B_c$ in. and from Equation (6.11) $0{\cdot}60\ B_c = 2{\cdot}7p_0/k$.

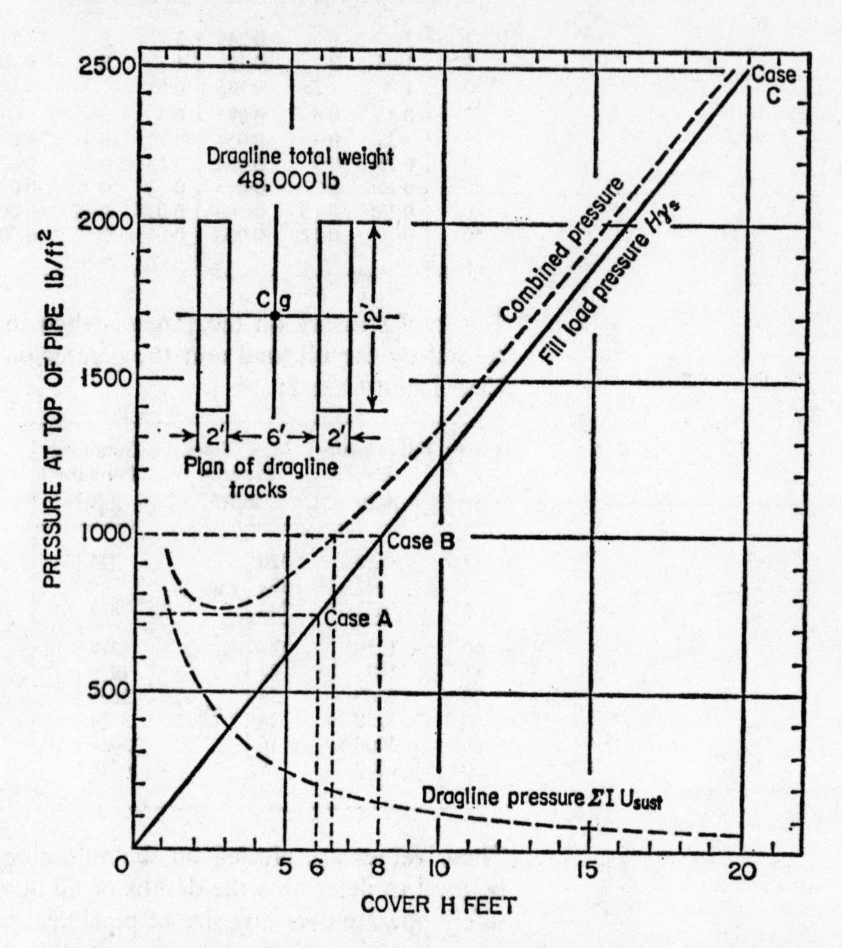

For Case B. The design value of $p_0 = (8)(125)/144 = 6{\cdot}95$ lb/in.² (i.e. 1000 lb/ft²), and $k = (2{\cdot}7)(6{\cdot}95)/0{\cdot}60\ B_c$ lb/in.²/in. $= 31{\cdot}3/B_c$ and from the graph, since the combined pressure is less than 1000 lb/ft² at values of H less than 6·5 ft the maximum required value of k is $31{\cdot}3/B_c$ lb/in.²/in.

At $H = 3$ ft the combined pressure is $769/144 = 5{\cdot}3$ lb/in.² which is the approximate minimum and the required value of k is: $k = (2{\cdot}7)(5{\cdot}3)/0{\cdot}60\ B_c = 23{\cdot}8/B_c$.

For Case C. The design value of $p_0 = (20)(125)/144 = 17{\cdot}3$ lb/in.² or more and $k \ngtr (2{\cdot}7)(17{\cdot}3)/0{\cdot}60\ B_c$ lb/in.²/in. $\ngtr 78/B_c$.

At $H = 3$ ft the required value of k_{min} will be the same as for Case B, i.e. $23{\cdot}8/B_c$.

For large strutted or tied pipes. Equation (6.11) is not applicable whilst the struts or ties are in place.

For Case A the struts and walings must be strong enough to carry the combined load and they must not be removed until the dragline is no longer operating.

For Cases B and C the struts or ties may be removed when k reaches or exceeds the value required by the combined load.

Appendix B

The Selection of Pipes and Joints for Sewers and Drains in Areas Subject to Coal Mining Subsidence

(Extracts from B.R.S. Note No. B. 164, (1956))

Summary

The forces and deformations to which a pipeline is potentially subject when laid in an area where mining subsidence may occur, and the general manner in which pipes made of various available materials respond are outlined. An order of preference for pipe material is established depending on the relative flexibility of the material and available sizes of pipe. The available types of flexible joints are also briefly discussed. The special requirements for the strengthening of plain concrete pipes are considered in the light of the development of the much stronger reinforced or prestressed concrete pipe and suggestions regarding suitable methods of bedding and laying and connecting to manholes or other structures are made. Finally corrosion and temperature effects are briefly considered.

This note has been prepared at the request of the Institution of Civil Engineering Mining Subsidence Sub-Committee on Public Utilities.

The Selection of Pipes and Joints for Sewers and Drains in Areas Subject to Coal Mining Subsidence

(*1*) *Potential Forces on, and Deformation of the Pipeline*

The special conditions imposed on an underground pipeline in areas subject to coal mining subsidence are:

(i) *due to the passage of a transverse wave:*

(*a*) a slow and more or less uniform longitudinal bending and stretching of the line in the vicinity of the wave. This action travels along the line as the wave advances;

(*b*) a slow and more or less uniform shortening of the pipe behind the wave as the ground reaches its new level.

(ii) *due to the passage of a parallel wave:*

(*a*) Possible disturbance to the uniformity of the bedding with consequent local longitudinal bending and local concentrations of crushing load at the 'beam' reaction points.

(*b*) Severe permanent bending and stretching at the ends of the wave.

(*c*) Possible permanent lateral displacement with stretching of the line.

(iii) *due to the passage of a wave inclined to the pipe axis < 90°:* A combination of the effects described in (i) and (ii) above.

(iv) *in the vicinity of a 'pillar' or at a stopped or an abandoned face; or at the edges of subsidence 'pockets'.*

(*a*) In a line at right angles to the face, severe permanent longitudinal bending and stretching, with possible concentration of movement at particular joints.

(*b*) In a line parallel and near to the face on the subsiding side, permanent and possibly uneven lateral and vertical displacement with possible disturbance to the uniformity of the bedding and with more severe distortion in both planes at the ends of the pillar or face. Concentration of movement at particular joints may occur in either or both planes.

(*c*) In a line inclined to the face at angle < 90° a combination of the effects noted in (iv) (*a*) and (*b*) above.

(v) *due to geological or induced faults involving abrupt changes of level at the surface.*

(*a*) In a pipeline at right angles to the fault line: very severe permanent longitudinal bending and stretching accompanied by local disturbance to the bedding and high shearing forces on the pipe in the vicinity of the fault plane.

(*b*) In a pipeline parallel to the fault line and on the downthrow side. Effects similar but possibly more severe and variable than those described in (ii) (*a*), (*b*) and (*c*) above.

(*c*) In a pipeline inclined to the line of fault at an angle < 90°. A combination of the effects described in (v) (*a*) and (v) (*b*) above and possibly more severe than either.

Under any of the above conditions, the magnitude of the effects on the pipeline will depend upon the magnitude of the total subsidence, the rate at which it occurs and the number of subsidence cycles. Their adversity will be aggravated by uneven subsidence, by changes of direction of the pipe axis either horizontally or vertically, by the presence of branches, junctions, manholes, valve chambers and by excessive depth of cover.

Summarising, the pipeline may be called upon to suffer:

(*a*) Severe longitudinal (axial) bending tending to rupture it transversely.

(*b*) Possibly severe crushing overloads tending to rupture it longitudinally or to cause buckling or even complete collapse.

(*c*) Considerable axial lengthening and shortening tending to cause drawn joints, (especially at bends, branches, manholes or valves) or axial buckling or crushing, upheaval or broken sockets; these effects being aggravated by local resistance to sliding due to projections beyond the pipe barrel, or other anchorages.

(*d*) Severe transverse shear forces.

(2) Pipe Behaviour and Desiderata

(i) The manner in which a pipeline responds to the various movements and forces to which it is subjected will depend on:

(*a*) The strength, thickness and flexibility of the material used for the pipe wall.

(*b*) Whether the pipe itself is flexible or depends for its flexibility on flexible joints.

(*c*) When flexible joints are used their spacing and their ability to permit adequate angular and axial displacements of the pipe without damage to the pipe or leakage.

(*d*) The degree of axial anchorage imposed on the line and the positions of potential anchorages.

(ii) The ideal pipeline would evidently be smooth, jointless, flexible, elastically extensible, thick enough to have high resistance to crushing and shear, chemically stable and inert to corrosive agents in groundwater or sewage. Accordingly the ideal material used for the pipe wall should be tough, ductile, weldable and have high tensile strength with a high yield point and low Young's modulus.

(iii) The ideal material has not yet been produced though some plastics would appear to be approaching it for small bore pipes.

Most available metals, whilst having desirable elastic characteristics, have undesirable chemical characteristics and are costly. Conversely the brittle materials such as glazed earthenware (clayware), which are admirable chemically and relatively cheap are quite unsuitable elastically.

The choice of material is therefore resolved into a compromise between elastic characteristics, resistance to corrosion and cost, relative to the desirable life of the pipeline and local conditions. Ultimately, however, the choice must be governed by economics, the size of the pipe and sound engineering judgement based on knowledge and experience of local conditions.

(iv) To meet the special conditions of loading imposed on a pipeline by ground subsidence the order of preference of the pipe material is, disregarding corrosion resistance, (I) 'Flexible', (II) 'Rigid'. For large pipes the choice is limited to mild steel and ductile iron in Class I; to R.C. and prestressed concrete, plain concrete or cast iron in Class II. Of these, only steel can be welded and it is handicapped by the necessity of protection against corrosion in any soil. Such protection should be flexible as, if brittle, it may be cracked or displaced when the pipe is deformed.

For pipes made of materials having little or no axial extension suitable flexible joints at small intervals, are essential. The smaller the distance between joints the greater the axial extensibility and flexibility of the line and the lower the crushing overload which may develop due to irregularities in bedding. This provision may materially increase the cost of the line, as may the necessity for special sockets or collars, longer than those specified in the various British Standards, to cope with the expected large axial movements of individual pipes, failing which joints may 'pull' locally with consequent leakage and consequential damage.

(3) Joints

(a) For Non Pressure Pipes

Of the flexible joints currently available the simplest and possibly the least expensive is the single rolling or sliding rubber ring type in which the rubber ring is squeezed between the exterior surface of the spigot and the interior surface of the socket. This joint has proved satisfactory for concrete pipe sewers over many years and it is now being used by continental pipe makers for pressure pipes. It has the further merit of requiring no metal parts.

Hot poured and cold caulked bituminous or rubber-bitumen compositions have also been used but we have little evidence of their behaviour in service. The necessity for the joint to open and close, possibly several times in the course of the life of the pipeline places severe limitations on these compositions.

(*b*) *For Pressure Pipes*

There are numerous mechanical flexible joints on the market all depending more or less on rubber sealing rings in various forms. Choice will again depend on the degree of angular movement provided and the ability of the joint to open and close without leakage, and cost.

(*4*) *The Strengthening of Concrete Sewer Pipes—Special Requirements*

For sewer pipes the Ministry of Housing and Local Government currently requires that relatively large concrete pipes be strengthened by the addition of *in-situ* concrete in the form of bedding and haunching or surround. These requirements assume B.S. plain concrete pipes with a proof crushing load of only 1350 lb per linear foot.* Glassfibre or steel reinforced and prestressed pipes however are capable of withstanding very much greater loads than the corresponding plain pipe and indeed may be designed to withstand any practicable loading without recourse to *in-situ* concrete.

If *in-situ* concrete is used it must evidently be interrupted at every pipe joint if the full potential axial flexibility of the line is to be achieved. This means additional site work, careful inspection and higher cost. Furthermore the strengthening effect of the *in-situ* concrete, which depends on its thickness and quality, cannot be estimated with any confidence, and, in the event of a fracture or the necessity for realignment, the repair work is complicated by its presence and it may prevent the recovery of the pipe.

It would appear preferable therefore to use stronger pipes without *in-situ* concrete, even in the 'Rigid' Class. Alternatively where the subsidence conditions are likely to be mild, 'Rigid' pipes entirely embedded in a continuous reinforced concrete casing with suitable hinges and expansion-contraction joints at junctions and at intervals in the main line might be considered. It would be necessary to design the casing as a continuous beam capable of sustaining either positive or negative bending moments at any point in the length, the longitudinal steel area being such as fully to exploit the cross section of the composite structure for vertical loads, and to provide lateral bending resistance by placing both top and bottom steel as near the corners as is practicable (*see Fig. 11.5*).

* N.B. The Ministry later (by Form K29) alternatively required that the loads and pipe field strength be computed and equated.

(5) Pipe Bedding and Laying

In view of the probable disturbances to the pipe bed, which may be expected, and the detrimental effects of irregular bedding on the pipe loads, particular attention to uniformity of the bed material over an adequate depth vertically is desirable in order to prevent or reduce the formation of hard spots. Where the trench bottom is of uneven resistance (e.g. alternating shale and sandstone) or where the strata in which the pipe is laid are likely to deform unevenly as the subsidence occurs it will be advisable to excavate well below formation level and repack with soft uniform material such as sand, fine gravel, ashes (preferably free from sulphur), fluedust, sandy clay or broken shale and surround the pipe with the same material. The fill above and below the pipe might well extend over depths equal to the pipe diameter, but should not be less than 12 in.

In laying jointed pipes it is necessary that the spigot ends are so placed that they may move equally in or out of the socket without fouling the socket on the one hand or breaking the joint on the other.

The depth of cover should be kept as small as possible in the interests of maintenance. Where practicable pipes may with advantage be laid above ground.

(6) Connections to Manholes and Other Structures

Whatever the material of the pipe, flexible connections are necessary where the pipe enters or leaves a manhole or other structure. The last joint should preferably be in the manhole wall and the last pipe should be short and the joints at either end of it should allow for ample axial movement in either direction.

(7) Chemical Attack and Temperature Effects

Unprotected Portland Cement concrete in any form is susceptible to external attack by acid groundwater (e.g. drainage from mines or from peaty areas) and by sulphates if present in the soil. It is also susceptible to internal attack at the exit from local depressions or rising mains where, owing to inadequate velocity, sludge can become septic and encourage bacterial action on the sulphur compounds in the sewage. This action results in the formation of sulphuric acid at exits in the presence of air and the pipe is attacked above the normal water line. Aluminous cement concrete is largely free from these forms of attack but is adversely affected by temperatures greater than about 80°F in the presence of moisture.

Some polyester resins, such as might be used in glass fibre reinforced concrete pipes, are susceptible to attack by alkaline groundwater and possibly by alkaline effluents in cracked pipes.

The highly stressed wire in prestressed concrete is peculiarly susceptible to stress corrosion in conditions where groundwater containing chlorides, sulphates or acids can saturate the concrete in which it is embedded. Reinforcing bars are subject to some corrosion under similar conditions but the effect is likely to be much milder. Steel and spun iron need protective coatings in any soil.

Pitch fibre is unaffected by most corrosive agents but is preferably not used where the effluent temperature is likely to be sustained above about 150°F (e.g. laundry wastes).

The rubber in joint rings, if manufactured to an appropriate specification, appears to be adequately resistant to most naturally occurring corrosive agencies. Natural rubber should not be used however if the effluent contains considerable quantities of oils or fats. Synthetic rubber rings have been successfully used on oil pipe lines.

If pipes are laid above ground thermal movements will occur and distortions due to differential temperature and moisture content are probable, especially with concrete pipes. These movements may be additional to subsidence movements in all senses. If carried by saddles, continuous welded pipes must not be restrained by the saddles, and provision for rectifying any relative movement of the saddles as the result of subsidence is advisable. Jointed pipes are probably better laid with a full length bearing on the ground, the bed being prepared as suggested above for pipes in trenches to minimise the effects of irregularity.

Protective coatings should either be plastic or have thermal coefficients as nearly as possible the same as those of the pipe material otherwise they may crack or spall. Natural rubber joint rings should not be exposed to sun and air above ground.

Where the pipes have shallow cover or where the effluent in underground pipes is subject to sustained temperature variations, similar movements of the pipe are to be expected.